PHARMACEUTICAL THERMAL ANALYSIS
Techniques and Applications

ELLIS HORWOOD BOOKS IN BIOLOGICAL SCIENCES

General Editor: Dr Alan Wiseman, Department of Biochemistry, University of Surrey, Guildford

SERIES IN PHARMACEUTICAL TECHNOLOGY

Editor: Professor M. H. RUBINSTEIN, School of Health Sciences, Liverpool Polytechnic

* In preparation

PHARMACEUTICAL THERMAL ANALYSIS
Techniques and Applications

JAMES. L. FORD
B.Sc., M.R.Pharm. S., Ph.D.
Principal Lecturer in Drug Formulations and Design
Liverpool Polytechnic

and

PETER. TIMMINS
B.Pharm. M.R.Pharm. S.Ph.D.
Manager, Pharmaceutical Development, International Development Laboratory
E.R. Squibb and Sons, Merseyside

ELLIS HORWOOD LIMITED
Publishers · Chichester

Halsted Press: a division of
JOHN WILEY & SONS
New York · Chichester · Brisbane · Toronto

First published in 1989 by
ELLIS HORWOOD LIMITED
Market Cross House, Cooper Street,
Chichester, West Sussex, PO19 1EB, England
The publisher's colophon is reproduced from James Gillison's drawing of the ancient Market Cross, Chichester.

Distributors:

Australia and New Zealand:
JACARANDA WILEY LIMITED
GPO Box 859, Brisbane, Queensland 4001, Australia

Canada:
JOHN WILEY & SONS CANADA LIMITED
22 Worcester Road, Rexdale, Ontario, Canada

Europe and Africa:
JOHN WILEY & SONS LIMITED
Baffins Lane, Chichester, West Sussex, England

North and South America and the rest of the world:
Halsted Press: a division of
JOHN WILEY & SONS
605 Third Avenue, New York, NY 10158, USA

South-East Asia
JOHN WILEY & SONS (SEA) PTE LIMITED
37 Jalan Pemimpin # 05–04
Block B, Union Industrial Building, Singapore 2057

Indian Subcontinent
WILEY EASTERN LIMITED
4835/24 Ansari Road
Daryaganj, New Delhi 110002, India

© **1989 J. L. Ford and P. Timmins/Ellis Horwood Limited**

British Library Cataloguing in Publication Data
Ford, J. L.
Pharmaceutical thermal analysis
1. Drugs. chemical analysis
I. Title II. Timmins, P.
615'.19015

Library of Congress Card No. 88–37189

ISBN 0–7458–0346–6 (Ellis Horwood Limited)
ISBN 0–470–21219–5 (Halsted Press)

Typeset in Times by Ellis Horwood Limited
Printed in Great Britain by The Camelot Press, Southampton

Table of contents

To our wives and children
Linda and Katharine Ford
Alison, Ashley and Benjamin Timmins

Here is the endeavour of our absences

Preface

Thermal analysis has long been the Cinderella of the instrumental techniques available to the pharmaceutical researcher. It is one of the most widely used, yet most often misunderstood, methodologies called on to resolve problems created during the use of other techniques. The results thrown up often create more problems than they solve and thermal analysis often requires other confirmatory analytical techniques. Yet it is one of the most common techniques used in the preformulation and development of medicines.

The precise terminology of the method is often ignored by pharmaceutical researchers and consequently obsolete terms such as thermograms are still too frequently used. Whilst it is not the aim of this book to convert to the new nomenclature, attempts have been made to bring the terminology extracted from the literature more in line with that proposed by the Nomenclature Committee of the International Confederation of Thermal Analysis (ICTA). We apologize for any unnoted errors in this respect, advise the interested reader to peruse that Committee's recommendations published regularly in the *Journal of Thermal Analysis* and *Thermochemica Acta*, and remind everyone that pharmaceutical researchers tend to hold their own common usage of expression and consequently the old heresy often still reigneth.

This book therefore neither concentrates on thermal analytical instrument design and development nor is a discourse on the available instrumentation. The interested reader is asked and recommended to explore W. W. Wendlandt's publications and especially those cited in Chapter 1. Instead the authors' intention is that the book should fully review the uses made of thermal analysis in the specific arena of pharmaceutical research. Each of the authors uses thermal analysis as a regular analytical technique and being from different backgrounds (one being very much rooted in industry while the other is a dedicated academic) we believe that the full pharmaceutical exploitation of thermal analysis is for the first time presented in one text book. Previously the pharmaceutical applications have been limited to at best a chapter and usually a few paragraphs in books overviewing the techniques, or dispersed throughout chemical, biological, physical and pharmaceutical research journals.

Preface

We were surprised at the amount of pharmaceutically relevant literature when we began collating information for this book. We therefore hope this book will be a valuable timesaver to the novice encountering thermal analysis for the first time and also refresh seasoned researchers in familiar fields or introduce them to new ones.

Wirral, Merseyside   James Ford
February 1989   Peter Timmins

1

Instrumentation for thermal analysis

[A variety of instrumentation is available for carrying out thermal analysis in its various modes.]This chapter introduces some of the basic types of instrumentation available and describes their principles of operation. The widely applicable modes, i.e. differential thermal analysis (DTA), differential scanning calorimetry (DSC), thermogravimetric analysis (TG), hot-stage microscopy (HSM), thermomechanical analysis (TMA) and dynamic mechanical analysis (DMA) are covered in addition to some less common techniques such as evolved gas analysis (EVA), thermosonimetry (TS) and combination techniques, e.g. thermogravimetry–mass spectroscopy. It is not the aim of this chapter to describe fully the operational modes of these instruments and the interested reader should consult either the relevant commerical manufacturer's literature or the excellent text of Wendlandt (1986).

1.1 DIFFERENTIAL THERMAL ANALYSIS (DTA)

DTA is often considered inferior to the related technique DSC, as it can only readily provide qualitative data. However, useful information, e.g. phase transitions, can be obtained and DTA is still very much used by pharmaceutical scientists.

A typical DTA instrument comprises a furnace having sample and reference cells, a differential temperature controller with its associated amplifier and output system (e.g. chart recorder) and control equipment for the furnace temperature programme and furnace atmosphere (Fig. 1.1). All commercial instruments measure the differential temperature of the sample against temperature during a temperature programme (e.g. linear termperature rise with time).

The Stanton Redcroft DTA systems centre around a solid alumina block having platinum/rhodium or chromel/alumel temperature-detecting thermocouples (Fig. 1.2). The thermocouples are joined so that the differential temperature between reference and sample, in addition to the actual sample temperature, can be monitored. The chromel–alumel type system with its temperature range of − 150 to 500°C (Stanton Redcroft model 671B) is very suited to evaluation of pharmaceutical materials because of its range and sensitivity (Dunn, undated).

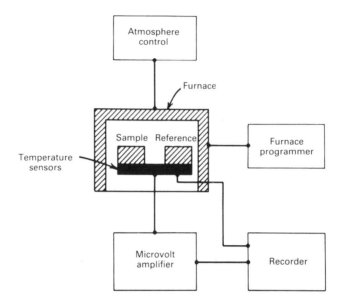

Fig. 1.1 — Schematic diagram of a typical DTA apparatus (reproduced with permission from Wendlandt, 1986).

The DuPont 1600 DTA cell is a module that plugs into the DuPont 910 cell base (also utilized with a DSC module). Within the cell the sample and reference cups rest on top of two independent thermocouple pedestals and are surrounded by the programmable furnace (DuPont, undated). The instrument appears to be recommended by DuPont for inorganic materials rather than pharmaceuticals, primarily because its operational range is ambient to 1600°C.

1.2   DIFFERENTIAL SCANNING CALORIMETRY (DSC)

Two types of instrumentation are commercially available: the heat-flux DSC, typified by the DuPont systems, and power-compensation DSC, typified by the Perkin-Elmer systems. Setaram and Mettler systems are also of the heat-flux type. The construction of these types of system will be described with reference to the commercial systems manufactured by DuPont (heat-flux) and Perkin-Elmer (power compensation).

1.2.1   Heat-flux systems

Fig. 1.3 shows a diagram of the DuPont 910 DSC cell in cross-section (DuPont, undated). A constantan disc provides the primary means of transferring heat to sample and reference positions whilst also functioning as one element of the temperature-measuring thermo-electric junctions. During a scan the sample and reference are contained in pans which are positioned on raised platforms on the constantan disc. Heat is transferred through the disc and through the sample pan to

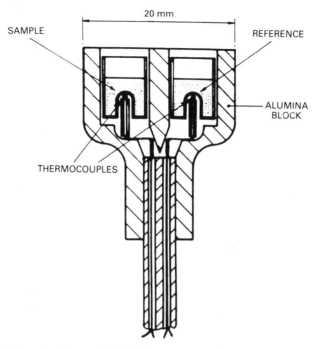

Fig. 1.2 — Schematic diagram of the basic components of the Stanton Redcroft DTA 673 system (reproduced with permission from Dunn, undated).

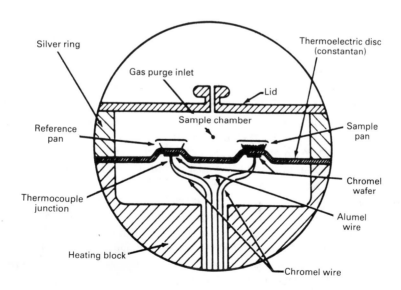

Fig. 1.3 — Cross-section of DuPont 910 DSC cell (courtesy DuPont).

the contained sample and reference.] The differential heat flow is monitored by chromel–constantan area thermocouples formed by the junction of the constantan disc and the chromel wafer. (There is a chromel wafer covering the underside of the raised platforms beneath both the sample and reference pans.) Sample temperature is monitored directly via chromel–alumel thermocouples formed from chromel and alumel wires connected to the underside of the chromel wafers. Software linearization of the cell calibration is utilized to maintain calorimetric sensitivity.

The cell has a volume of 2 ml and can be utilized with various non-corrosive inert atmospheres as well as oxidizing and reducing atmospheres. Available sample pans (hermetic, open or sealed) allow sample volumes of 0.1 ml which can be up to 100 mg depending on sample density.

Enthalpy values can be determined with a calorimetric precision of ±1% and a temperature precision of ±0.1°C. The 9900 Thermal Analyser (controller system with IBM compatible computer) can provide for heating rates of 0.01 to 200°C/ minute in 0.01°C increments when utilized with the 910 cell.

DuPont have available a similarly constructed 912 dual DSC cell having two sample positions, allowing the simultaneous running of the two samples against a single reference. In running a large number of similar samples, this can give improved sample throughput.

Related pressure cells are available for both single and dual DSC systems to allow for scanning under elevated pressures by employing a gas-tight version of the cell. The use of controlled pressure is often of great use in thermal analysis. This is especially so in the application of reactions which are pressure-sensitive such as accelerated oxidation of lubricants and polymers, evaluation of catalysts and the resolution of overlapping transitions (Gill, 1984). Reactions which produce their own partial pressure, such as dehydration mechanisms, could benefit from controlled-pressure facilities since the integrity of a sample pan and lid, even a volatile-sample pan and lid, cannot be guaranteed. The development of uncontrolled pressure may further modify the nature of the dehydration. In conjunction with the 9900 Computer and Thermal Analyzer or the 9000 Thermal Analyser and module interface, expansion by adding on other thermal analysis modules is possible, with multitasking of up to four modules possible through the computer-assisted version. Data output can be to the computer screen, a dot matrix printer or an X–Y plotter.

Multitasking can include data from experiments in progress, plotting real-time or stored data, analysing data from previous experiments, or setting-up the instrument for the next experiment. With multimodule software, the 9900 system can control up to four analysis modules simultaneously whilst still providing multitasking capabilities through the appropriate software.

Reference should be made to the heat-flux Mettler system (DSC20, DSC30 and TA3000) (Mettler, 1986). This utilizes a unique five-junction thermocouple, vapour-deposited on a ceramic die to detect differential temperature. The DSC-20 operates in the range ambient to 600°C and the DSC-30 in the range − 170 to 600°C, easily encompassing the region of interest for most pharmaceuticals. Temperature reproducibility of ± 0.2°C is claimed with a precision of ± 0.1°C. Calorimetric reproducability is claimed at ± 0.5% with an accuracy of ± 2%. Computer control and data reduction is available for the Mettler systems (Mettler, 1986).

1.2.2   Power compensation systems

Perkin-Elmer engineers designed and patented the first power-compensated differential scanning calorimeter, the DSC-1, which was introduced in the early 1960s. As new technologies became available refinements were introduced early in new instruments such as the DSC-1B, DSC-2 and DSC-4 calorimeters. The newest design, the DSC-7, still utilizes the Perkin-Elmer power-compensation system and incorporates recent advances in microcomputer and micro-electric technologies including the PE 3700 computer (Perkin-Elmer, 1986).

The basic principle of the construction of the DSC cell (Fig. 1.4) relies on sample

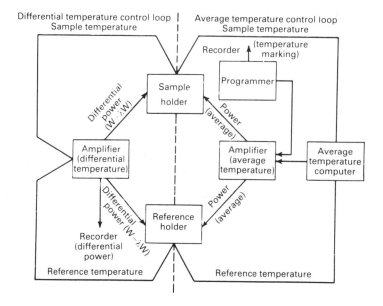

Fig. 1.4 — Schematic diagram of a Perkin-Elmer DSC instrument (reproduced with permission from Wendlandt, 1986).

and reference having separate heaters. Sample and reference are maintained at nominally the same temperature via the system operated through platinum resistance thermometers and resulting in different amounts of heat being supplied to each specimen as appropriate. The difference in power output to the heaters is monitored.

Calorimetric accuracy and precision of ±1% and 0.1% respectively are claimed (Perkin-Elmer, 1986) with temperature accuracy and precision both of 0.1°C being indicated. Heating rates of 0.1 to 200°C/minute are possible in 0.1°C increments with a temperature range of −170° to 725°C. Inert or active gas atmospheres at ambient pressure can be employed. Through the PE 3700 data station, unattended operation or simultaneous operation and data analysis is possible. Other analytical

modules can be added on through a TAC7 intelligent microprocessor controller. The
data system produces hard-copy output on a multiple-pen printer plotter.

1.3 THERMOGRAVIMETRIC ANALYSIS (TG)

TG utilizes a thermobalance, which allows for ongoing monitoring of sample weight
as a function of temperature. This may involve a controlled heating or cooling
programme or a maintained fixed temperature. Instrumentation is typically a
balance with some sort of output to record the weight, e.g. strip chart recorder or a
data acquisition system (Wendlandt, 1986). A furnace surrounds the sample holder
and ancillary controls to modulate such as furnace temperature and operational
atmosphere are available (Fig. 1.5). The actual nature of the equipment can vary

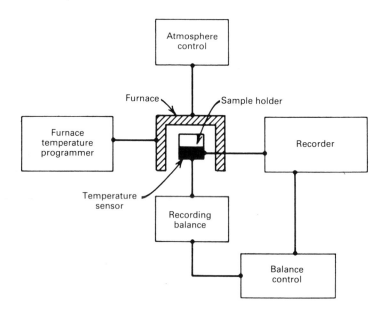

Fig. 1.5 — Schematic diagram of a modern thermobalance (reproduced with permission from
Wendlandt, 1986).

with application, e.g. maximum operating temperature and sample size, but for
pharmaceutical studies temperatures of up to 350°C and sample sizes of 5–20 mg are
generally adequate.

The balance utilized for TG should be a good analytical instrument with the
required accuracy, precision and sensitivity, good mechanical (and electronic)
stability, a rapid response to weight change, and resistance to vibration. The balance
should be rugged, yet simple in construction to minimize costs and aid maintenance.

Wendlandt & Gallagher (1981) defined three classes of thermobalance; namely deflection-type instruments, null-type instruments and resonance-frequency-change instruments. The last are very specialized and rarely used in pharmaceutical studies. The principle of operation is that mass deposited on or lost from the surface of a highly polished crystal results in a shift of an oscillatory frequency. They will not be considered further.

Deflection-type instruments depend on the measurements of the balance-beam deflection about its fulcrum due to mass changes. Electromechanical, photographic or displacement-transducer measurement of the deflection can be employed. Few commercial systems utilize this approach.

The most widely available instruments utilize the null-balance principle. Movement of the balance beam from its null position is sensed and a restoring force applied to restore the beam to the null position. The change in restoring force, proportional to the change in the sample mass, is monitored. Further elaboration of contruction of thermobalances is best done through consideration of the commoner commercial instruments.

Perkin-Elmer (1986) have introduced a modular computer-controlled analyser in the TGA7, replacing the original TGS-1 and well-established TGS-2 instruments. The configuration is with the sample hanging down from the balance inside the furnace. The computer software allows for derivative recording, automatic temperature and mass calibration and zeroing. The furnace has a low thermal mass to allow rapid, linear heating and cooling rates (0.1 to 200°C/minute in 0.1°C increments). Forced air cooling in conjunction with low thermal mass allows for rapid cooling down at the end of the experiment. Cooling from 1000°C down to 50°C in less than 15 minutes is possible. The manufacturer claims the microbalance to be the most sensitive in the industry (Perkin-Elmer, 1986), capable of detecting weight changes as small as $0.1\,\mu g$ with 0.1% accuracy. This demands that the balance be resistant to ageing, temperature and vibration effects. The balance is thermally isolated from the furnace to minimize interference.

All types of samples, powders, liquids, films and fibres can be analysed with static or dynamic atmospheres provided by most gasses. Ambient or reduced pressure operation is possible. Typical of most modern systems, this computer-controlled thermogravimetric analyser provides for simple, precise and convenient operations.

The DuPont 951 thermogravimetric analyser uses an alternative balance/furnace configuration. The use of horizontal balance avoids chimney effects that may affect results and also collection of condensate from the sample on balance hang-down mechanisms. The location of the temperature-sensing thermocouple in close proximity to the sample, as opposed to close to the sample holder as in most other commercial instruments, provides for accurate readings of the sample temperature, although combustion of the sample can result in spurious results (DuPont, undated). The horizontal balance facilitates furnace pre-heating of the apparatus before insertion of the sample, and removal of the furnace before it is completely cooled, which allows replacement with a spare furnace, enabling fast turnaround and thus high productivity.

The furnace temperature range is ambient to 1200°C with the balance having a sensitivity of $2\,\mu g$. Varied gases can be utilized (a quartz furnace tube is an option for use with corrosive atmospheres) and the axial purging inherent in the horizontal balance construction allows for very high purge rates up to 1000 ml/minute.

Stanton Redcroft have TG 750 and TG 770 instruments available (with operating ranges of ambient to 1000°C and ambient to 1500°C respectively). These compact instruments will operate with 1–10 mg samples held in a platinum crucible suspended from one arm of the electronic microbalance. The whole balance/sample assembly is held in a purgible glass bottle that forms a gas-tight seal into the furnace, surrounding the sample, during operation (Dunn, undated). The construction reduces error including buoyancy effects to a minimum and water-cooling of the furnace enables rapid cooling rates, e.g. from 1000°C to 100°C in less than four minutes. The Stanton Redcroft TG 780 is a simultaneous DTA–TG instrument operating over the ambient to 1500°C range (Dunn, undated). By use of this instrument direct correlation of TG and DTA results is possible. A diagram of the TG 780 head is given in Fig. 1.6. The

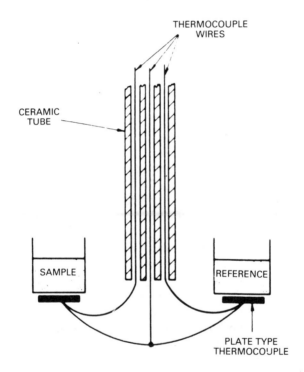

Fig. 1.6 — Schematic diagram of the Stanton Redcroft TG 780 head (reproduced with permission from Dunn, undated).

instrument is based on an electronic microbalance, with two pans for sample and reference, suspended from one arm of the balance. DTA thermocouples and the sample temperature-measurement thermocouples are welded directly to the sample platforms (Dunn, undated).

Setaram have available the TAG 24 G 1 simultaneous symmetrical TG and DTA thermoanalyser and a TG-DSC 111 combining TG with DSC (Setaram, undated). A

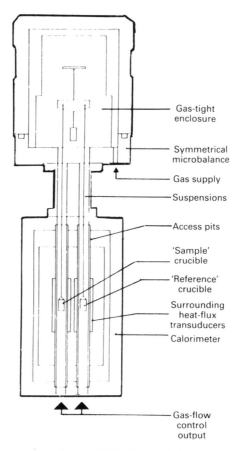

Fig. 1.7 — Schematic diagram of the Setaram TG-DSC111 unit (courtesy Setaram).

diagrammatic representation of the head of the TG-DSC 111 is given in Fig. 1.7. The TG instrument can take samples of up to 500 mg and has a detection threshold of 1 μg. The DSC has a sensitivity of 10 μW. The instrument operates over the range $-123°$ to 827°C at scanning rates of up to 30°C/minute.

Mettler have the TG50 Themogravimetric Analyzer available for use in conjunction with the TC10A Thermal Analysis Processor. The latter provides for control of the TG module as well as data collection and evaluation (Mettler, undated). The TG50 is based on a Mettler M3 microbalance and offers ambient to 1000°C operating range with heating/cooling rates of up to 100°C/minute. The air-cooled furnace can be brought down from 1000°C to 100°C in 12 minutes at the end of a run. Balance reproducibility is ± 1 μg.

Many other thermogravimetric analysers are commercially available from Netzsch, Cahn, Theta, Harrop, Columbia, Shimadzu, Rigaku, ULVAC-Riko, Linseis, Spectrum Products and Hungarian Optical Works. These instruments, more rarely found in pharmaceutical laboratories in the UK and USA than those already described, are detailed by Wendlandt and Gallagher (1981) and Wendlandt (1986).

1.4   HOT-STAGE MICROSCOPY (HSM)

HSM, also known occasionally as thermoanalytical microscopy, is a valuable supportive tool when used in conjunction with other techniques. It can be utilized to ascertain the nature of events leading to endotherms or exotherms on DSC traces or weight changes in TG.

Decomposition with gas evolution, and especially loss of water of crystallization from hydrates, can be observed by mounting crystals in silicone oil and observing during a temperature programme. Gas or vapour bubbles can be observed emanating from crystals at temperatures correlating with decomposition or desolvation endotherms (exotherms) in DSC or DTA.

Stanton Redcroft and Mettler both have furnaces available for mounting on the stage of a suitable microscope. Still or video cameras can be used to monitor events on the microscope slide during a temperature programme. It should be strongly emphasized that in the characterization of pharmaceutical materials HSM is an invaluable tool in the support of other thermal techniques.

1.5   THERMOMECHANICAL ANALYSIS (TMA)

TMA can measure dimensional changes in a sample (such as expansion or contraction) as a function of temperature, under a programmed temperature change, or as a function of time when the sample is subjected to a constant temperature (isothermal hold). More properly, TMA measures deformation of a sample under a constant load whereas volumetric-change determination (expansion or contraction) is a thermodilatometric measurement. Both techniques can be carried out in the same instrument, usually by employing different probes. Large, flat-faced probes are used to examine the extension or expansion of a material under heat, whilst narrower, flat-faced or concave-shaped probes are used to examine the penetration changes into a material. Pointed probes are used to examine flexure changes. Thus events such as expansion, deformation, flexure, softening and stretching of the sample, usually polymeric in nature, may be determined. The stress applied may be the tension, compression, flexion or torsion modes depending on the form of the specimen and the design of the holder. Penetration is used for thin films and coatings on substances whereas fibres and films are normally clipped so that a tensile load may be applied. The pharmaceutical relevance of these techniques will increase because of continued interest in polymeric drug-delivery systems and the potential of polymers in medical prostheses.

Typical commerical instruments have the sample held within the furnace, the sample being placed on a quartz sample holder and contacted by the appropriate quartz probes. The probe connects to a linear variable differential transducer (LVDT). Changes in the sample will result in movement of the probe and production of a signal by the LVDT. The signal can be outputted to a chart recorder or a data acquisition system (Fig. 1.8).

An indication of the probe types used in commercial equipment is shown in Fig. 1.9. The DuPont 943 TMA can handle plugs, sheets, pellets, films, fibres or powders over the range $-180°$ to $800°C$. A sensitivity of $0.5\ \mu m$ is claimed. The Perkin-Elmer

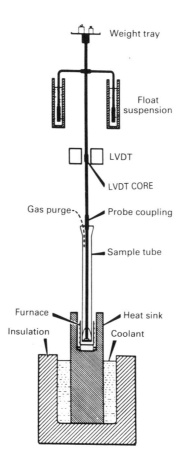

Fig. 1.8 — Perkin-Elmer TMS-2 TMA apparatus (from Wendtlandt, 1986).

TMA7 can operate with smaller samples over the range −170 to 1000°C and has a sensitivity of 0.4 μm/cm. Both DuPont and Perkin-Elmer instruments can operate with various gas purges and have equivalent linearity, sample-loading and probe types (Perkin-Elmer, 1986; DuPont, undated).

Other commercial instruments are available as the Mettler TMA 40, again with similarities to the DuPont and Perkin-Elmer instruments already described. The three systems so far described are used with computer-aided thermal analysis controllers for programming, data acquisition, data storage and data analysis. Other TMA systems are available from Stanton Redcroft, Rigaku, Netzsch, Harrop, Theta, Sinku-Riko (ULVAC), Orton, Linseis and Hungarian Optical Company (Wendlandt and Gallagher, 1981).

Cassel (1977) detailed the various operational modes of TMA. These included expansion, penetration, extension, stress–strain testing, flexure analysis, compressive compliance testing, and dilatometry.

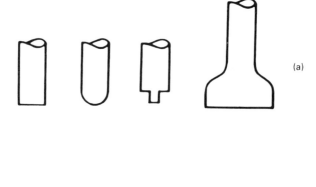

(a)

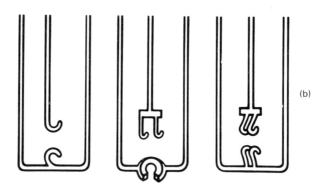

(b)

Fig. 1.9 — Typical DuPont TMA probes showing probes for (a) compression and expansion modes, and (b) tension mode (courtesy DuPont).

1.5.1 Dynamic mechanical analysis (DMA)

A special subset of TMA is provided by DMA or dynamic thermomechanometry, equipment for which is available from DuPont or Polymer Laboratories Ltd and in the Tetrahedron Universal Relaxation Spectrometer. Of these only the DuPont module is widely available and is therefore considered here.

The DMA 983 from DuPont can measure the mechanical response of a material as it is deformed under an oscillatory load. Thus the viscoelastic behaviour of a material can be characterized (DuPont, undated). The sample under evaluation is clamped between two parallel arms which are so fixed as to allow movement only in the horizontal plane. The arms are driven by an electromagnetic motor at an oscillation rate selected by the operator. An LVDT measures both displacement and frequency. In operation the whole sample and mechanical system is enclosed in a radiant heater and dewar to provide precise temperature control (Fig. 1.10). Four modes of operation are possible: fixed frequency, stress relaxation, resonant frequency and creep. The fixed-frequency operation can provide for reliable determination of frequency dependence and end-use performance prediction whereas the resonant frequency mode allows for detection of subtle transitions.

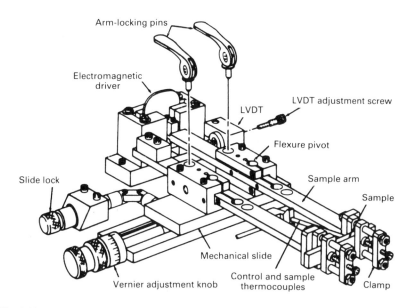

Fig. 1.10 — The mechanical system components of the DuPont 983 DMA system (courtesy DuPont).

Stress relaxation will measure the ultra-low-frequency molecular relaxation of polymers and composites whilst creep analysis will measure flow at constant stress. The latter allows determination of load-bearing stability of materials. TMA is highly applicable to the characterization of polymeric materials allowing for assessment of expansion coefficients, glass transition temperature, and, with DMA, detection of subtle transitions in polymers not observed by routine DSC or TMA.

1.5.2   Torsional braid analysis (TBA)

One final subset of TMA is TBA which utilizes the torsional mode and torsional strain as the applied stimulus. The measured responses are the decay of oscillation frequency as the modulus parameter and the rate of decay as the parameter for damping. The method was originally designed for the study of polymers and is based on the intermittent application of a torsional strain. A fibreglass braid supports the sample and the method is therefore used for samples which are too weak to support their own mass. The development of TBA was described by Gillham (1974). It was derived from a freely oscillating torsional pendulum which possessed advantages such as inherent simplicity, the use of a low frequency ($\simeq 1\,\text{Hz}$) which permits easy comparison with static methods such as DSC and DTA, and the high resolution of transitions. Disadvantages included a long-time scale due to the low thermal conductivity of organic and polymeric specimens and the requirement for relatively large samples. Additionally the temperature range used is limited because of the inability of materials to support their own weight near critical load-limiting transitions such as glass or melting transitions. TBA uses small samples, approximately

10 mg, supported on an inert multifilamented substrate, the braid. The specimens are prepared by dissolving the polymer in an organic solvent and impregnating the braid, which consists of 3000–4000 filaments. Gillham (1974) considered that this arrangement allowed the picking up of relatively large amounts of solution and minimized flow due to gravity. Braids are used in an attempt to balance any twists in the component yarns before evaporating the solvent off. This is accomplished above the boiling point of the solvent and into the fluid state of the polymer. The rigidity and loss moduli are used to determine the properties of the material under test. Because the specimen is small compared with the torsional pendulum, far more rapid heating rates may be employed. Because an inert support is used, the method may be used to discriminate transitions in the liquid, rubbery and solid states.

1.6 LESS COMMON THERMAL ANALYSIS TECHNIQUES

EVA is used where a knowledge of the type and amount of gases produced during a heating programme can be determined.

Stanton-Redcroft have developed applications for interfacing their DTA and TG equipment to a mass spectrometer (Anonymous, 1984) and DuPont and Nicolet have developed an interface between their respective TG and fourier-transform infra-red spectrometers. The combination of a Bio Rad FTIR spectrometer with a Stanton Redcroft Thermobalance is also reported (Anonymous, 1987). Interfacing between a DuPont 951 thermogravimetric analyser and a DuPont mass spectrometer is reported by Chiu and Beattie (1980) and earlier by Zitomer (1968).

These types of combination techniques enable EVA but are neither simple nor inexpensive techniques for the routine pharmaceutical laboratory. However they can be utilized to confirm the nature of crystallization in a solvate or identifying the nature of gases produced during thermal decomposition.

Thermosonimetry measures the noise produced during changes in (normally) crystalline materials subjected to a heating or cooling programme. Such noise can be produced by propagation of cracks in crystals, which can come from thermal stress or eruption of gases or vapours (e.g. loss of solvent of crystallization). Crystallization from a melt can also produce such noise signals. The technique counts acoustic events on a ring-down counting system. Several parameters may be measured including emitted noise, rate of emission as a function of temperature, and frequency content. The emitted sounds arise from thermal stresses which are imposed on the substance by the temperature programme and the induced strains may be released by processes such as chemical decomposition, melting, and solid-state transformation (Clark, 1981). However, mechanical strain-release processes involve motions and creation of structural imperfections; for example, microcracks, dislocations and grain boundaries occur which are considerably less energetic than chemical and physical processes. These, although undetected by DSC or DTA methodologies, were detected by thermosonimetry (Clark, 1981). In practice the sounds are emitted as mechanical vibrations during and prior to thermal events occurring within the substance. These consist of a rapid cascade of decaying signals each of so-called 'ring-down' shape. Sound quantity is measured by ring-down counting in which each component of each signal is digitised and is registered as a single count. The total count or the count rate (counts per second, cps) is displayed as the ordinate versus temperature as the abscissa. Clark (1981) showed that the magnitude of a count of a ring-down signal is a combination of the initial amplitude, the rate of amplitude decay and the frequency. Thus the thermosonimetric count is a composite measure which may have convenience but has no theoretical significance. Thus a thermosonimetry technique based on ring-down counting cannot be used quantitatively. However, measurements of amplitudes, amplitude distributions and frequency distributions may be used to quantify the nature of the material under test. Thus frequency distributions of thermal events may be used to characterize the material under examination.

Thermosonimetry has particularly had some very limited application to pharmaceutical substances with the work of Clark & Fairbrother (1981). A thermosonimetry apparatus is schematically shown in Fig. 1.11 (Clark, 1978).

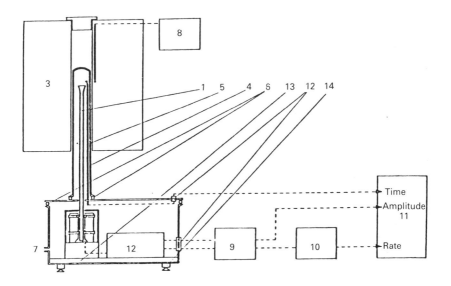

Fig. 1.11 — General layout of a typical thermosonimeter. Key: 1, silica waveguide; 2, preamplifier; 3, furnace; 4, silica sheath; 5, thermocouple (Pt/Pt,Rh); 6, silicone rubber vacuum seal rings; 7, vacuum/gas port; 8, temperature programmer; 9, main amplifier; 10, rate meter; 11, recorder; 12, vacuum lead-throughs; 13, acoustic foam; 14, vibration mounts (reproduced with permission from Clark, 1978).

Thermoacoustimetry, thermoelectrometry, thermophotometry and emanation thermal analysis are minor, very specialized techniques described by Wendlandt and Gallagher (1981) and Wendlandt (1986) that would appear not to be particularly applicable to the pharmaceutical laboratory and so will not be covered further in this book.

The types of data obtained from the different types of apparatus described are covered in the subsequent chapters.

REFERENCES

Anonymous (1984) *Lab. Pract.*, **33**(5), 23.

Anonymous (1987) *Lab. Pract.*, **36**(8), 21.

Cassel, B. (1977) *Polymer Testing by TMA*, Perkin-Elmer TAS 20, Perkin-Elmer, Norwalk, Connecticut.

Chiu, J. & Beattie, A. J. (1980) *Thermochim. Acta*, **40**, 251–259.

Clark, G. M. (1978) *Thermochim. Acta*, **27**, 19–25.

Clark, G. M. (1981) *Proceedings of the 2nd European Symposium on Thermal Analysis* ed. Dollimore, D. Heyden, London, 85–88.

Clark, G. M. & Fairbrother, J. E. (1981) *Proceedings of the 2nd European Symposium on Thermal Analaysis'* ed. Dollimore, D. Heyden, London, 255–258.

Dunn, J. G. (undated) Stanton Redcroft Technical Information Sheet No. 101. *An introduction to thermal methods of analysis*, Stanton Redcroft, London.

DuPont Brochure E-73498 (undated) *DuPont 9000 and 9900 Thermal Analysis Systems.*

Gill, P. S. (1984) *Am. Lab.*, **16**(1) 39–49.

Gillham, J. K. (1974) *Amer. Inst. Chem. Eng. J.*, **20**, 1066–1079.

Mettler (1986) Mettler Technical Literature; *Mettler TA 3000 Thermal Analysis System ME-724074.*

Perkin-Elmer (1986) Perkin-Elmer Technical Literature; *DELTA Series Thermal Analysis System VP118610*, Perkin-Elmer Corp, Norwalk, Connecticut.

Setaram (undated) *Technical Information on TG-DSC111.*

Wendlandt, W. W. (1986) *Thermal analysis*, 3rd. Edition. (Chemical Analysis Series, 19, Monographs on Analytical Chemistry and Applications ed. Elving, P. J. & Windfordover, J. D., J. Wiley & Sons, London.

Wendlandt, W. W. & Gallagher, P. K. (1981) Instrumentation. In *Thermal characterization of polymeric materials*, ed. Tsu, E. A., Academic Press, New York, 1–90.

Zitomer, F. (1968) *Anal. Chem.*, **40**, 1091–1095.

2

Information derived from thermal analytical data

It is the aim of this chapter to outline the general applicability of thermo-analytical techniques. Transitions that may be encountered are discussed and introduced as phenomena rather than by the specific technique employed. Pharmaceutical uses will follow in the later chapters. The overall aim is to guide the reader in the choice of examining specific problems. The choice between DSC and DTA is less clear than in the past because many of the early DTA instruments were incapable of giving results that could be converted to calorimetric values. However the decrease in sample size from once more than 100 mg to less than 2 mg with modern instrumentation allows easy conversion of DTA data to calorimetric values.

No matter whether DSC or DTA is used, the obtained peak, whether endotherm or exotherm, has several characteristic temperatures which may be used to describe it (Fig. 2.1). The onset temperature is that temperature when the transition just leaves the baseline. The peak temperature is the temperature represented by the apex of the peak. Where the peak transition returns to the base line may be referred to as the recovery temperature (Brancone & Ferrari, 1966). Additionally the extrapolated onset represents that temperature corresponding to the intersection of the pre-transition baseline with the extrapolated leading edge of the endotherm or exotherm of that transition. Kambe *et al.* (1972) used similar values for the onset point, the extrapolated onset point, and, in the case of sharp transitions, the peak point. However for very blunt peaks the latter temperature was derived from the intersection of the extrapolated lines corresponding to the up and down curves of the transition. At the start of a scan the input of the sample and reference pans will not be identical due to differences in sample size, heat capacity or thermal conductivity. Consequently there will be a small drift, either endothermic or exothermic, at the start of any run.

Table 2.1 summarizes the general uses of thermal analytical techniques which are expanded upon in this chapter.

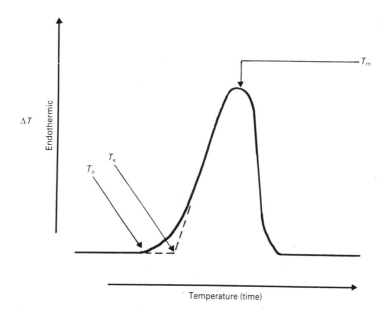

Fig. 2.1 — Typical endotherm showing the positions of onset temperature (T_o), extrapolated
onset temperature (T_e) and peak temperature (T_m).

Table 2.1 — Summary of pharmaceutically important information derived from
thermal analysis

Reaction	DSC	DTA	HSM	TG	TMA	DMA
Melting points	+	+	+	−	+*	+*
Desolvation — bound	+	+	+	+	−	−
— adsorbed	+	+	+	+	−	−
Glass transition	+	+	−	−	+	+
Heats of transition	+	+?	−	−	−	−
Purity determination	+	+?	+?	−	−	−
Compatibility	+	+	+?	+?	+*	+*
Decomposition kinetics	+	+	−	+	−	−
Polymorphic transitions	+	+	+	−	−	−

Key: + applicable, − inapplicable, +? potential application, * polymers only.

2.1 SINGLE COMPONENT SYSTEMS

The characterization of materials usually involves properties such as melting points, boiling points and transition temperatures of processes such as dehydration (desolvation) and decomposition and their related heats of transition.

2.1.1 Transition temperatures

The greatest accuracy will be achieved with a small sample size, proper sample encapsulation, slow scanning speeds and correct instrumental calibration. This is considered in detail in Chapter 3. Transition temperatures are usually quoted at the vertex of the exotherm or endotherm. The problems of lag corrections are also discussed in Chapter 3. Care should be exercised when comparing results obtained by different methods of analysis. Optical methods may not measure those phenomena which correspond to the maximum rate of change of the transition. For example in melting processes it is usual to take the temperature when the last trace of solid melts as the melting point of a system using HSM. In contrast the peak temperature of a DSC scan corresponds to the maximum melting rate.

2.1.1.1 *Melting points*

A melting transition would be instantaneous for a theoretical 100% pure material (Barall, 1973), i.e. truly isothermal. Usually a range is detected. Most organic materials do not melt isothermally and this creates difficulties in describing a constant event which may be ascribed to transition temperatures. Thus, for melting, that temperature when the first detectable liquid phase is detected could be used. The last trace of solid anisotropy coincides to ±0.05°C with the endothermal minimum using DTA–optical microscope studies for small samples (Barrall, 1973). It is essential to maintain good thermal contact between the temperature-sensing device and the sample. In older DTA models using a thermocouple in direct contact with the sample, this could be accomplished either by prior melting of a few crystals of the sample under test directly onto the thermocouple prior to analysis or by using an inert diluting agent. When the sample is encapsulated the sample pans are of high thermal conductivity. Whatever method is used it is essential that the thermal lag is known and reproducible.

2.1.1.2 *Boiling points*

Some provision should be made to equilibrate the solid and its vapour. Ideally the instrumental atmospheric pressure must be controlled. Few commercial instruments are designed to facilitate the accurate determination of boiling points unless separately modified (Barrall, 1973). Heats of vaporization are considerably greater than those of fusion. Therefore smaller sample sizes are used with reduced instrument sensitivity. The vertex of a boiling endotherm has little significance, being related to instrumental design, rate of vapour diffusion and sample weight. The boiling point is usually taken as the extrapolated onset temperature. Its value needs correction for lag times and instrument readout. Additional readings should be taken from a manometer incorporated into the instrument to allow barometric correction. Several scans should be completed at several pressures. Typical data treatment includes plotting *ln* pressure against reciprocal temperature (Barrall, 1973) which should give a straight line from approximately 5 to 760 torr. The slope of this line will be $-E/R$ where R is the gas constant and E is the energy of vaporization. Such plots average out experimental error.

2.1.1.3 Solvates
Thermal analysis has proved useful when materials incorporate solvent into their crystal lattice to form solvates (or water to form hydrates). On heating, the solvent is usually released at a discrete temperature or, if desolvation is a multistage process, at a series of discrete temperatures. The event will be registered as endotherms on DTA or DSC from which heats of desolvation can be determined. Events can be confirmed by HSM. The sample can be suspended in an inert vehicle, e.g. liquid paraffin, and solvent release will be detected by gas formation. The removal of solvent will often be accompanied by a breakdown in crystal structure or by migration of the desolvation front into the crystal. TG is most useful since not only does a loss in weight reflect solvent loss at the defined temperature but the stoichiometric weight of the solvent may be calculated.

2.1.1.4 Liquid crystals
Many pharmaceutical materials, especially semi-solids, display orders of state intermediate between the solid and liquid states. These are liquid crystals or mesophases which are thermodynamically stable states and may represent a condition of incomplete melting. A simple classification of two main classes (Brennan & Gray, 1974) suffices. Lyotropic liquid crystals occur when certain compounds are treated with solvent and, rather than dissolving to an isotropic solution, form liquid crystals. Ampiphilic surface-active agents are a classic example. Thermotropic liquid crystals occur when certain solids are heated to a temperature above which their crystal lattice becomes unstable. This class is subdivided into those with enantiotropic mesophases when the liquid crystals reform on either heating or cooling and those with monotropic mesophases which are only formed when the isotropic liquid is supercooled. Three liquid crystal structures may be differentiated. Smectic mesophases are liquid crystals of the highest order and consist of a two-dimensional order of parallel matrices of molecules arranged in layers (Brennan & Gray, 1974). Nematic mesophases are a thread-like state more mobile than the smectic mesophase. Although the molecules are parallelly orientated the layered arrangement does not form. Cholesteric mesophases occur in optically active materials and are regarded as a twisted nematic layer which forms helical structures (Brennan & Gray, 1974).

When mesophase-forming solids are heated, more than one mesophase may exist. Thus liquid crystals may display both smectic and nematic mesophases or smectic and cholesteric phases and are termed polymesomorphic. When more than one mesophase exists the smectic mesophase always exists at the lower temperature. These phase changes occur at well-defined temperatures. Additionally two forms of both the cholesteric and smectic phases may exist.

A combination of DSC or DTA with HSM (with polarizing facilities) has proved useful in interpreting changes involving liquid crystals. Optical methods suffer from distortions imposed by the experimental conditions. Purity determinations may also be made on the transitions, as described in Chapter 5.

Examples are cholesteryl palmitate and *p*-azoxyanisole. The latter displays an endotherm at 118°C, due to fusion and conversion to the nematic liquid crystalline state, and a small endotherm at 136°C due to conversion of the nematic mesophase to isotropic liquid (Brennan & Gray, 1974). On cooling, the isotropic to nematic mesophase transition occurred at approximately the same temperature (Brennan & Gray, 1974) but the nematic state could be supercooled by 30°C before crystallization occurred, indicating that liquid crystal transitions are easily nucleated.

Cholesteryl palmitate displays well-recognized transitions at 77.5°C solid→ smectic), at 77.0°C (smectic→cholesteric) and at 81.7°C (cholesteric→isotropic liquid). On heating, only the initial and final transitions were detected, but on cooling all three were displayed although the recrystallization transition was depressed by many degrees of supercooling (Brennan & Gray, 1974). The smectic →cholesteric transition is a monotropic transition as it is only normally observed on cooling due to the close proximity of its transition to that of the solidus melting endotherm.

2.1.1.5 Isomerization
Brancone & Ferrari (1966) used DTA (DSC is equally applicable) to study isomerization in ethambutol dihydrochloride. Both the D and L forms displayed endotherms at 75 and 200°C. However the D,L configuration had endotherms at 170 and 180°C whilst the *meso*-configuration showed endotherms at 40 and 200°C.

2.1.1.6 Glass transition
Many solids when heated above their melting point (T_m) and rapidly cooled through it do not immediately crystallize but form a supercooled liquid. Depending on the cooling rate recrystallization may be retarded by only a few degrees but there is a possibility that this supercooled state will itself pass through a heat-capacity change, known as glass transition, thereby further inhibiting recrystallization. Cooling curves should therefore not be used to determine melting (freezing) points. There has been considerable interest in glasses becasue their high energy state contributes to fast dissolution rates.

The properties of glasses are markedly affected by their thermal history and the temperatures involved in their preparation. Factors such as preparation temperature, cooling rates and subsequent annealing conditions contribute to variations in the observed glass transition temperature (T_g). Because organic materials tend to soften at the T_g and form a liquid with poor ductile properties either DTA or DSC is used to characterize their T_g. Polymers often undergo enthalpies of relaxation which display as endothermic peaks. Many drugs, although of low molecular weight, show similar trends. Their glass transitions are apparent either as endothermic drifts in the heating curve or even as small endothermic peaks. The theory of glass transitions and the glassy state of simple inorganic and organic materials and polymers is very complicated and beyond the scope of this book. To gain further information the interested reader is invited to examine the publications of Haida et al. (1977), Suga & Seki (1974) and Chen et al. (1980). Values of the T_g for various drugs (Ford, 1987) include chloramphenicol (28°C), glutethimide (0°C), griseofulvin (89°C), indomethacin (41°C), and paracetamol (24°C).

Serajuddin et al. (1986) characterized the glassy state of indapamide. DSC of a melt that had been quench-cooled through its T_g showed only an endothermic change at 98°C equivalent to the T_g. An endotherm developed at its T_g as the cooling was more slowly controlled and its area increased as the rate decreased from 10 to 1°C/min. The influences of annealing were examined in quench-cooled samples following storage at 85°C (Fig. 2.2). The area of the endotherm that developed

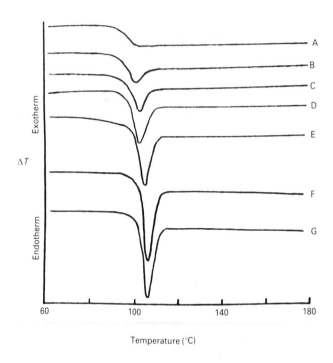

Temperature (°C)

Fig. 2.2 — DSC scans of indapamide glass showing the effect of the duration of annealing at 85°C on the size of the endotherm at the glass transition temperature. Key: annealing time A,0; B,2; C,4; D,21; E,45; F,168; and G,336 hours (reproduced with permission from Serajuddin *et al.*, 1986).

increased storage to reach a maximum after 168 hours. The explanation (Serajuddin *et al.*, 1986) was that large-scale molecular motion is frozen when a glass is quench-cooled and the system therefore has an excess of enthalpy relative to the equilibrium glassy state. Equilibrium is reached during annealing by enthalpy relaxation, i.e. the strain induced by rapid cooling is released. The maximum relaxation of enthalpy due to annealing at any temperature may be calculated with reference to the area under endotherms such as shown in Fig. 2.2.

Fukuoka *et al.* (1986) reported that T_g of glassy indomethacin, calculating it as the intersect of the slope of the endothermic change with the projected baseline. Glasses were prepared by heating above the T_m of indomethacin and then cooling to 270 K. Subsequent heating gave an increase in heat capacity of samples around the T_g. Different heating rates produced different structural relaxation within the glasses. The recovery, as evidenced by a decrease in the area of the endotherm associated with the T_g, became more apparent at lower heating rates. DSC was completed at 20 or 40 K/min to reduce these influences. The T_g increased as the heating rate increased and a linear relationship was apparent when the logarithm of the heating rate was plotted as a function of $1/T_g$. The calculated activation energy of the glass transition was 212.5 kJ/mol. The kinetic principles behind this calculation are given in Chapter 4. The value of the T_g was influenced by the cooling rate, e.g. the temperature was 320 K when quenching was used and increased to 324 K when a cooling rate of 0.62 K/min was employed. Thus a relaxation process took place during cooling. Annealing at temperatures below the T_g gave similar results as described above for indapamide since an increase in endotherm area occurred on storage. Curiously the T_g concomitantly increased. By following the influence of temperature on the endotherm area, Fukuoka et al. (1986) showed that the rate of area increase, related to the rate of relaxation, was fastest at 303 K but was barely evident at 283 and 313 K. Therefore annealing temperatures may effect the rate of relaxation and the T_g values of organic glasses.

2.1.2 Specific heats

One of the most important uses of DSC is in the determination of the heat capacity of a sample. The method is described (Perkin-Elmer 1966), detailed by Gray (1968) and much of the subsequent work has been based on this technique. To fully understand the method requires a knowledge of the principles behind the operation of power-compensation DSC (see Chapter 1). Most values derived from thermal analysis are derived by power-compensation DSC (Laye, 1980) although for DTA the steady-state signal ($T_s - T_r$, temperature differential between sample(s) and reference(r)) established when both the sample and reference are subjected to the same linear temperature increase is proportional to the difference in heat capacity between the sample and reference.

A typical scan for power-compensation DSC is given in Fig. 2.3. The ordinate scale is proportional to dH/dt (e.g. cal/s) and the related thermodynamic read-out (dH/dt, e.g. cal/°C) when adjusted to correspond to 1 g of material (at constant pressure) becomes the specific heat of the material, C_p. This is a function which slowly varies with temperature. When a constant scanning rate ($\equiv T'$) is employed, the ordinate displacement due to the sample heat capacity is dH/dt, which is equivalent to $T'dH/dT$ and this equals $T'c_p$ for a sample size of 1 g. This provides the fundamentals for calculating c_p. The DSC ordinate displacement is therefore proportional to $T'c_p m$ where m is the mass of the sample and this relationship may be used to compute optimal sample weight, scanning speed, and instrumental range setting (Perkin-Elmer, 1966). Scanning rates of up to 20°/min may be used. Experimentally the deflection due to the heat capacity of the sample and its pan and lid are measured and therefore a no-sample run must be made in empty sample pans. Ideally this should obviously be reproducible irrespective of the sample pans. However since the weights of these pans vary, albeit even slightly, it is usual to determine the reference baseline in the same sample pan and lid intended to be used for the sample. Additionally the thermal losses between sample pan and references must be the same from run to run and be independent of the sample holders. Consequently it is usual to use protective covers, positioned in exactly the same manner, for each of the runs. Since radiation losses occur from the outside of the

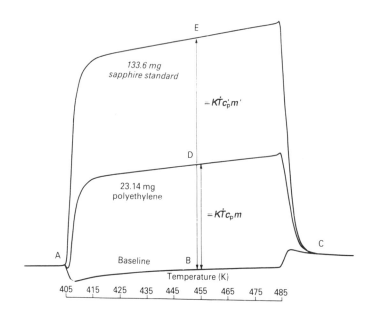

Fig. 2.3 — Typical specific heat determination, molten linear polyethylene 405–485 K, scanning at 20°/min, range 8 (reproduced with permission from Perkin-Elmer, 1966).

sample pans, a covered holder should display the same losses irrespective of contents. The no-sample runs may be checked with the sample runs for reproducibility by isothermal scanning and after the temperature range of interest. Sample sizes should be in the range 1 to 5 mg (Gray & Brenner, 1965). The usual experimental methodology required to derive specific heats is as follows:

(1) Run empty sample pans and pan covers in reference and sample holders isothermally at a temperature below that of the range of interest to establish a flat baseline at about 10% of full scale deflection. Scan at the desired rate to a temperature above that of interest. The baseline will be displaced slightly due to the differences in capacity of the sample pans. Continue to run the trace isothermally. Again there will be a slight displacement of the baseline. The overall curve is given by ABC (Fig. 2.3). Should a large baseline displacement occur extra pan covers may be added to the reference before repeating the run for a new baseline.

(2) Repeat (1) for a known weight of the sample using the same sample pans and covers. The isothermal positions before and after the scanning should be the same as without the sample. If they are displaced by the same amount, corrections may be made without error. The holder covers may need replacing if there is major displacement. The first and final portions of the curve ADC (Fig. 2.3) are periods of adjustment until equilibrium is reached and cannot be used to determine c_p. These were estimated to take 80 seconds for the Perkin-Elmer

DSC 2 and correspond to periods of 25° and 6°C at scanning speeds at 20° and 5°C/min respectively (Perkin-Elmer, 1966).
(3) Procedure (2) is repeated with a known weight of standard material using the same sample pans and covers. Sapphire is an ideal standard in the range 0 to 1200 K because its heat capacity over this range is well known. Curve AEC (Fig. 2.3) represents a typical calibration.

At any point where equilibrium has been established the ordinate deflection may be represented by y which is equal to $KT'c_p m$, where K is a calibration constant. Thus the relationships exist for the sample (s) and the reference (r) at any temperature of interest in equations (2.1) and (2.2)

$$y_r = KT'c_{pr}m_r \tag{2.1}$$

$$Y_s = KT'c_{ps}m_s \tag{2.2}$$

c_{ps} can be determined from equation (2.3)

$$c_{ps} = y_s m_r / y_r m_s \, (c_{pr}) \tag{2.3}$$

The estimated error of the determined value of c_p is ±2%. The error can be reduced if the scanning range is reduced. Where c_p values are required over a wide range these are best determined by several experiments of shorter scanning duration. An alternative, although less accurate, method of determining c_p is to measure the area of the curve ABCD and, using the fusion of a known standard such as indium, calculate the apparent calorific value of the area and divide it by the temperature interval. This only gives an average value of c_p in the range of interest.
   The effective range is 100–1000 K (Laye, 1980) and the technique is not absolute, requiring comparison with a standard(s). The fundamental requirements of the process include (Laye, 1980):

(1) A reproducible baseline. Older DSC instruments display a curved baseline but this is not important if it is reproducible since the relative magnitude of the signals is used to measure heat capacity.
(2) Temperature assignment. This is important and causes problems in older apparatus where calibration with several standards over a wide temperature range is recommended. The temperature lag may vary with the temperature. Sample geometry and its interface with the pan should not contribute to this lag and a failure to return to the isothermal baseline after scanning indicates possible geometry problems. Thermal lag will not markedly influence the heat capacity unless obtained in the vicinity of the glass transition, where the heat capacity may be strongly temperature-dependent. Poor packing may increase the thermal lag due to air pockets in the sample but pressing the sample may induce atypical changes in the sample morphology.
(3) Slow scanning speeds. When employed these tend to reduce the thermal lag.
(4) Weighing of the sample pan before and after scanning. Decomposition and sublimation invalidate the results.

Laye (1980) estimated that in early power-compensation DSC the degree of uncertainty was 1–3%. This can be reduced by care to 1% and has decreased further with increased sophistication of modern instruments. Gaur *et al.* (1978) described the use of a computer-interfaced DSC to determine heat capacities and suggested that sapphire was the reference to be used from 200–1000 K but not below because its heat capacity is highly dependent on temperature. Benzoic acid was more useful at below 200 K but above this temperature could not be used due to its high vapour pressure. Takahashi (1985) emphasized the importance of using as many reference materials as possible due to the temperature dependence of the proportionality constants, especially at low temperatures.

Gaur *et al.* (1978) pointed out that although larger samples give better accuracy too large a sample will give noise due to poor packing and may cause excessive temperature gradients within the sample. Ordinate displacements may be increased using higher sensitivities but this in turn may give rise to higher noise levels. There is therefore an optimum amount of sample that can be used. Alternatively faster heating rates will give higher ordinate displacements, giving better precision in the heat capacity values, but rates faster than 10°C/min may produce temperature gradients within the sample.

2.1.3 Heats of transition

Heats of transition may be determined by either DSC or quantitative DTA. The prerequisites for accuracy are accurate weighing and accurate estimation of the area under the curve. The former is dependent on the quality of analytical balance used but the latter, at one time dependent on the accuracy of the operator in determining the area under the curve by planimetry or weighing, has been automated by the use of microprocessors sited in the thermoanalytical apparatus. Planimetry may lead to an error as high as 5% (Barrall & Johnson, 1970). Care has to be taken to ensure that the correct area is detected by the computer.

The method requires calibration with materials of known heats of fusion. The transition area of a known weight of such a material, often indium fusion, is compared with the transition area of a known weight of the sample under test. The heat of transition is computed from equation (2.4).

$$\Delta H = \frac{\Delta H_r . W_r . A . C_r . R.}{A_r . C . W . R_r} \qquad (2.4)$$

where ΔH is the heat of transition of the sample, ΔH_r is the heat of transition of the reference, W, A C and R are the weight of sample, the area of sample transition, the chart speed and the range setting used for the sample respectively, and W_r, A_r, , and R_r are the corresponding values for the reference. This expression may be considerably simplified when the chart speed and range settings are the same for sample and reference. A calibration constant K ($\equiv \Delta H_r . W_r / A_r . R_r$) may be derived summating all the calibration settings. In many older instruments K is temperature-dependent and it is advisable to use a calibrant whose transition temperature is near that of the sample. Certain experimental constraints should be employed to obtain accurate determination of transition energies. (1) The scan speed should be identical for reference and sample. (2) The transition energies of the sample and reference should be similar. (3) The transition energy of the sample should be approximately known since it controls sample size. Very highly energetic transitions may be incompletely recorded by the instrument although this may be avoided by using small samples. (4) If sublimation or gaseous products are evolved these should be entrapped within the sample pan to prevent weight loss leading to incorrect post-transition baseline. (5) Samples subject to decomposition should be run in a nitrogen atmosphere to minimize decomposition.

The baseline construction used to determine the area of transition is up to the operator and two baselines may be employed (Barrall & Johnson, 1970). The *scanning baseline* is constructed between the first departure from the scanning baseline prior to the transition to the last sensible departure post-transition when the endotherm or exotherm is therefore completed and reaches the scanning baseline. This requires a sensible regular approach from the operator to produce reliable results when microprocessors are not used. Sharp transitions can readily be determined by this method. However when the sample becomes increasingly impure due to decomposition some correction has to be made to the determined area. Although the peak area is usually determined by drawing a straight line between the points deviating from the base line in the leading and trailing regions there is probably no reasonable rationale for doing this (Takahashi, 1985). The *isothermal baseline* is constructed by drawing the best straight line between isothermal points obtained before and after the transition. The construction of this baseline is similar to that described in section 2.1.2 in that the sample is run isothermally to obtain the baseline prior to scanning through the transition, and the isothermal baseline is also obtained subsequent to the scanning. If this method is used, the normal specific heat or enthalpy change of the sample is integrated into the enthalpy change of the sample from the transition heat. Generally enthalpies are determined in sealed sample pans to avoid the problems caused by sublimation, thermal emissivity and change in sample shape.

Most pharmaceuticals are very pure materials and consequently the areas under a curve are clearly defined by an interpolated baseline that joins up the instrumental baselines before and after transition. Whilst this is acceptable for the determination of the melting point of pure materials, the greater the impurity content the greater the discrepancies that exist, due in part to the failure of instruments to detect early melting (see Chapter 5). These problems have usually been ignored in the pharmaceutical literature, except when purity evaluation has been accomplished.

An alternative approach was proposed by Takahashi (1985) who converted the melting curve of indium to the curve of specific heat versus temperature. By integrating the specific heat capacity thus measured, enthalpy increase as a function of temperature was plotted. By taking linear enthalpy increases in the solid state and liquid state and extrapolating to the melting region, the heat of fusion was estimated from the difference in the extrapolated lines of the enthalpy at the extrapolated onset melting temperature.

In addition to providing transition enthalpies DSC and quantitative DTA may be used to determine the entropy of fusion, since this merely represents the enthalpy of fusion divided by the melting temperature at which it occurs. For instance, Zordan *et al.* (1972) derived the enthalpies of fusion for urea and thiourea as 3.22 and 3.3 kcal/

mole giving the derived entropies of fusion as 8.0 and 7.28 cal/mole/degree respectively.

Staub & Schnyder (1974) determined the heat of of vapourization by quantitative DTA. Their method involved weighing into a sample cup which had a pinhole in its cover, scanning across the boiling point and measuring the area of the boiling endotherm. Modifications to the method include holding the temperature isothermally at the boiling point or placing a small metallic sphere onto the hole in the lid to reduce evaporation loss (Staub & Schynder, 1974). Two major sources of error are unknown sample loss between weighing and measurement and a constantly changing baseline implying weight loss during heating and consequently changes in the heat capacity. The estimated errors in the method are in the order of 10%. Staub & Schnyder (1974) used the pinholed lid technique in combination with dynamic scanning and isothermal holding to determine heat of volatilization. Scanning was accomplished from an isothermal hold at ambient up to the temperature at which the heat of vaporization was required at which point isothermal holding was used to maintain this temperature. The area under the curve represents the heat necessary to raise the temperature to the boiling point and the heat required for partial evaporation of the sample. The area under the isothermal hold corresponds only to heat required to volatalize the remaining part of the sample. Obviously this is an unknown weight. However by repeating the run on a sample of different weight in the same container and otherwise treating the sample identically, even including the size of the pinhole which controls the evaporation rate, only the area under the curve corresponding to the isothermal hold will be significantly altered provided the time lapses between weighing and running the experiment are the same. By maintaining these 'unknowns' in the same manner the difference in the two areas under the curve will correspond to the volatilization of the weight of sample equivalent to the difference between the original weighings and consequently the heat of volatilization can be calculated for this incremental difference. The typical accuracy in the measurements was claimed to be ±2% for *n*-decane, *N*-butanol and water (Staub & Schnyder, 1974).

2.1.4 Sublimation
Materials which sublime during heating should be examined in volatile sample pans. The example of camphor (Barrall & Johnson, 1970) emphasizes the importance of proper encapsulation.

2.2 POLYMERS
Polymers may be characterized by several thermo-analytical techniques. DSC and DTA give information on melting, recrystallization and glass transitions. Evolved gas analysis or detection add information such as thermal stability. However mechanically based methods provide further information on, e.g. liquid–liquid transition temperatures (Tll) which are not readily detectable by other methods. These methods are thermomechanical analysis (TMA), dynamic thermomechanometry (DMA) and its complementary method, torsional braid analysis (TBA). A T_g is found in all amorphous polymers and in amorphous regions of partially crystalline polymers. The T_g of the latter is independent of the degree of crystallization but the

magnitude of the transition decreases with an increase in crystallinity with the result that the transition becomes difficult to detect in highly crystalline polymers. A polymer at temperatures above its T_g is limp and flexible but a polymer below its T_g is brittle and characterized by stiffness, hardness and often high optical clarity and transparency. The brittleness is due to a limitation in molecular motion which is restricted to very short chain segments or side-chains. The glass transition represents a change in the polymer from a brittle to a less brittle, rubbery state and is regarded as a second-order transition because it reflects changes in secondary thermodynamic properties such as expansion coefficients and heat capacity. The changes occur usually over a temperature range and consequently both the method used to derive the T_g and the portion of the thermal curve used to describe it are important and must be defined. Many authorities consider the transition is a reversible process equivalent to liquefaction and heating the polymer through its T_g will convert the polymer from a glassy state to a liquid state.

The T_g is controlled by three molecular factors (Brennan, 1973). Polymers with a stiff backbone and rigidly held, bulky side-groups possess a high T_g. These generally restrict polymer movement. Thus polyethylene with no bulky side-groups has a low T_g whilst polystyrene with large bulky side-groups has a high T_g. Polymers whose chains display attractive forces will have a high T_g. Polymers with dangling side-groups which loosen polymer structure exhibit a high T_g. The T_g of homologous polymers of different molecular weights increase with increase in molecular weight. Generally a straight-line relationship exists between T_g and the reciprocal of molecular weight.

Polymers fall into two categories (Brennan, 1973). In type I polymers two or more phases exist in the melt. In graft copolymers one polymer chain is attached to the backbone of a different type of chemical chain. Block copolymers consist of alternating, long sequences of two different polymers. On thermal analysis each of these groups shows properties similar to those of physical mixes, e.g. both the T_gs are apparent. In type II the copolymers exist as random alternating groups along the polymer chain and are typified by no phase separation on heating. Only one T_g is apparent and lies between the T_gs of the individual polymers. Although the relationship between T_g and composition is not always linear, only one particular blend will give a particular value and therefore the T_g may be used to define a specific blend.

Certain materials, termed plasticizers, may be added to polymers to lower the T_g and reduce brittleness. If the plasticizer reduces the T_g from a temperature above ambient to one below, the material will change from a brittle glass to a soft pliable material. Any plasticizer will possess a very much lower T_g than the polymer. Equation (2.5) is commonly used to describe the effect of a plasticizer:

$$1/T_g = W^1/T_g^1 + W^2/T_g^2 \qquad\qquad (2.5)$$

where T_g is the glass transition of the polymer plasticizer blend, W^1 and W^2 are the weight fractions and T_g^1 and T_g^2 are the glass transition temperatures of the polymer and plasticizer respectively. The equation predicts a straight-line relationship between the T_g and the weight fraction of either the polymer or the plasticizer. The use of equations similar to equation (2.5) is more fully dealt with in Chapters 7 and 9.

A knowledge of the T_g is used to assess the degree of cure and polymerization, since if the latter is attempted at temperatures below the T_g of the polymer, the reaction rate will be very low or zero. Since the T_g is marked by changes in the thermal capacity, the expansion coefficient and rigidity, it follows that techniques such as DSC or DTA, dilatometry or TMA or DMA will quantify the T_g of a particular polymeric system.

2.2.1 DSC and DTA of polymers
Fig. 2.4 shows a typical scan for a polymer. The T_g is evident as a small endothermic

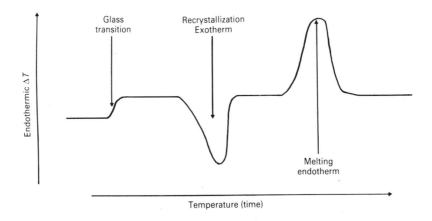

Fig. 2.4 — Typical DSC or DTA scan of a glassy polymer showing glass, recrystallization and melting transitions.

rise and is represented by the midpoint of the rise measured from the extension of the pre- and post-transition baselines, i.e. when the transition assumes half the value of this change. The increase in heat capacity is due to the onset of molecular movement, increasing the degrees of freedom of the polymer and thereby altering its heat capacity. Brennan (1973) quoted T_gs for silicone rubber, glycerine, polystyrene and poly(methyl methacrylate) as 148, 196, 370 and 392 K. Many polymers exhibit a small endothermic peak at their T_g. This effect is due either to a bulk stress relaxation in the system (when the peak will be observed only in the first heating through the T_g) or to a free volume effect at the molecular level (Brennan, 1973). In the latter case the post-transition baseline is extrapolated through the peak to determine the T_g. The change in the scan on a second or subsequent scans is of vital importance and demands that when T_g values are quoted the polymer under examination must have documented thermal history, even to the extent of storage. Ellerstein (1966) used DSC to show that an increase in peak temperature occurred as the ageing process continued. Rak *et al.* (1985) actually used DSC endothermic peak temperatures to define the T_g of poly(d, l-lactic acid)s.

As already mentioned the determination of the T_g of a crystalline polymer is very difficult due to the relatively small number of molecules that undergo the transition. Rapid quench-cooling the polymer through its T_g has been adopted in an attempt to avoid crystallization. When this is accomplished the DSC scan appears similar to that in Fig. 2.4. The polymer subsequently crystallizes at a temperature above T_g and melts at its melting point. The method was used by Ford (1987) to examine the T_g of polyethylene glycol.

DSC may also be used to estimate the cure of thermoset polymers. Gill (1984) showed that the T_g of an epoxy thermoset increased in temperature with an increase in cross-linking and was therefore used to estimate the cure. A cross-linking exotherm defined the kinetics and predicted cure rate. The type of calculations required were similar to the kinetic principles outlined in the Chapter 4.

The DSC and DTA methods of the American Standards Institution are among those most used to determine the T_g, the temperatures of fusion and recrystallization and their related enthalpies for polymers. Only DSC and quantitative DTA provide the enthalpies. Both the relevant standards (ASTM D3417-83, ASTM D3418-82) consider that DTA and DSC are methods which are applicable to polymers in granular form (preferably smaller than 60 mesh avoiding grinding if possible) or use any fabricated form into which the material may be cut. Reactions with air may also give invalid results necessitating the use of a nitrogen atmosphere. The standards recognize that grinding of samples, or other techniques required to effect particle size reduction, often introduces thermal effects due to friction or orientation into the polymers, changing their thermal history. The true heats of fusion may only be determined in conjunction with structural investigation and consequently specialized crystallization techniques are required (ASTM D3417-83). Controlled heating and cooling of the materials at a controlled rate in a controlled atmosphere are prescribed. DSC is used to study first-order transitions (i.e. fusion, crystallization) and the influence of annealing. Increases in particle size may result in variation in the derived enthalpies. Changes in the heating or cooling rates and final heating or cooling temperatures have a profound effect on the measured heats of transition. In practice the sample should be small and representative of the bulk material since large samples may cause error in the heat measurement. Standardization is essential to remove the effects of previous thermal history from the sample. ASTM D3147-83 recommends that samples should preliminarily be heated from ambient to 30°C above their melting point at 10°C/min to remove previous thermal history effects, held at the temperature for 10 min, cooled at 10°C/min to at least 50°C below the recrystallization temperature, recording the cooling curve to derive the heat of crystallization, then reheated at 10°C/min as soon as possible to provide the heating curve and allow computation of the heats of fusion. Care should be taken to ensure that such prior heat treatment does not wipe out the transitions of interest. In such cases the preliminary cycle should be omitted. Repeat runs on the same samples are made to ensure that thermal history does not alter the transition temperatures. ASTM D3417-83 recommends power compensation or heat flux DSC be used to determine enthalpies. In addition ASTM D3418-82 advocates DTA (or quantitative DTA) in the determination of the transition temperatures of polymers. Thermal analysis also gives valuable information on the oxidation, decomposition and curing rate of polymeric systems.

Four characteristic temperatures may be used to define melting or crystallization transitions of polymers (Fig. 2.5). T_f is the extrapolated onset temperature, T_m is

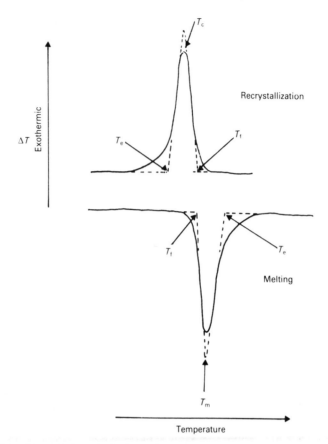

Fig. 2.5 — Transitions of a hypothetical material showing extrapolated onset temperature (T_f), melting peak temperature (T_m), crystallization peak temperature (T_c) and extrapolated end temperature (T_e) during heating/cooling.

temperature at the melting peak, T_c is the crystallization peak temperature and T_e is the extrapolated end temperature. The technique recommended for the determination of T_g is similar but recommends recording a preliminary cycle at 20°C/min to a temperature high enough to erase the previous thermal history, quench-cooling to 50°C below the T_g and holding the sample for 10 min prior to repeat heating at 20°C/min. Fig. 2.6 shows that four temperature indices may be measured. These are T_f, T_g, T_e and T_{mid} which are the extrapolated onset temperature, the glass transition temperature, the extrapolated end temperature and the midpoint temperature. Rather than a peak corresponding to the transition, a new baseline will be established. The standard considered that for most applications T_f is more meaningful and may be designated as the T_g rather than T_{mid}.

2.2.2 Thermomechanical analysis (TMA)
Thermomechanical analysis involves the use of a constant non-oscillatory load. The resultant deformation of the sample is examined. The method evaluates changes

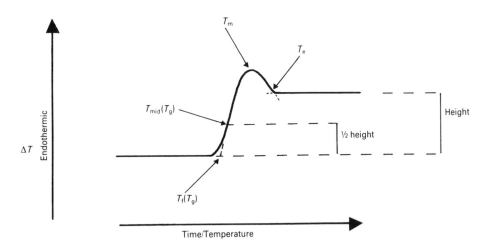

Fig. 2.6 — Transitions of a glass-forming material showing extrapolated onset temperature (T_f), midpoint temperature (T_{mid}), glass transition temperature (T_g) and extrapolated end temperature (T_e).

such as expansion and contraction. The applied stress produces changes in shape or size as a result of either energy dissipation caused by a relative movement of molecules (viscous response) or energy storage which is released on the removal of the stress (elastic response). The total TMA response is a combination of the expansion behaviour and the viscoelastic effect (Daniels, 1981). The viscoelastic response is influenced by a variety of factors including molecular weight and its distribution, branching, polymer–filler interactions, and the presence of plasticizer. Since the viscous response is time-dependent and the eleastic response is independent of time, the viscoelastic response can be resolved into its two components by the use of a time probe and oscillatory stress, i.e. DMA. The principle of the method is that it examines changes in the properties of materials under mild shear provided by the addition of low-weight masses to a probe in direct contact with the material under test. The type of probe varies considerably.

There are only a few examples of TMA being used effectively for pharmaceutical polymers. Brennan (1977) described data for polyethylene films finding that differences in film orientation produced different values of the T_g. A film orientated in the transverse direction to stress expanded slightly until 133°C when shrinkage occurred due to the film orientation imposed upon the sample. At 138°C the sample expanded due to melting. In the randomly orientated sample the film expanded gradually up to 134°C when gross sample melting occurred producing rapid expansion. Baker (undated) used both expansion and penetration modes to determine the T_g of polyethylene terephthalate. Values of 70 and 72°C respectively were obtained. A fixed penetration may be used to measure elasticity and recovery. Maurer (1978) used the penetration mode to examine creep recovery in small samples which were subjected to a load of 100 g until a fixed penetration of 0.064 cm was achieved at 23°C. The load was then removed and creep recovery monitored. The percentage recovery at 4 mins was used as an arbitrary measure of elasticity. Maurer (1978) used the technique to distinguish the molecular weight of polymers. Other examples are discussed in Chapters 8 and 9.

2.2.3 Dynamic thermomechanometry (DMA)

Dynamic mechanical properties relate to the linear viscoelastic response of a polymer to an externally applied excitation which is usually oscillatory at a prescribed frequency. By varying the temperature and frequency a spectrum of the dynamic response can be obtained (Koo, 1974). Since the technique assumes that the material will behave in a viscoelastic manner, the experimental techniques applied must involve only very small perturbations within which the visco-elastic response is assumed to hold. Small strains are therefore imposed on the sample and an elastic or complex modulus or damping factor is usually the determined response. The effective upper temperature limit occurs when the sample becomes too soft to respond linearly. The mathematical concepts and scope of the technique are discussed by Koo (1974). Thus the modulus curve of an amorphous polymer will exhibit a sharp drop in excess of two decades at the T_g, which corresponds to an intense damping maximum. Such large drops are not apparent in crystalline polymers due to the retention of crystalline domains which are unaffected by the transition. The damping maximum is associated with internal molecular motion and may be assigned to various transitions. Techniques such as X-ray diffraction and IR (infrared) spectroscopy are required to differentiate the participating molecular units. Nonetheless DMA may be used to examine for cross-linking (damping peak shifts to a high temperature and broadening of the peak implies a wide distribution of segment lengths between cross-links), crystallinity, plasticization and the composition of copolymers or blends. Thus in block and graft copolymers the identity of each component is maintained. DMA has become popular as a method of thermal analysis due to the advent of instrument automation permitting rapid, non-subjective user interpretation (Gill, 1984) and data analysis capability providing automatic calibration and precise quantitative measurements. DMA measures the properties of polymers when an oscillatory load is applied.

Daniels (1981) reviewed the principles behind commercial and research instruments used for DMA. Three constructional principles were outlined. (1) Continual application of the load and measuring the modified oscillatory response. Intermittent application of the load may be used and the decay of oscillation monitored via viscous damping. (2) The specimen is driven in forced oscillation or allowed to assume its natural frequency. (3) The mode of the applied stress is flexure, torsion/compression or torsion. Based on these simple principles a plethora of instrumentation has been developed. It is preferable to use an instrument which gives a continuous or rapidly sampled measure of modulus/ damping as the specimen temperature is varied (Daniels, 1981) Two major categories are recognized based on either natural vibration or forced vibration as the applied stimulus (Daniels, 1981). Instruments in the latter group were Rheovibron, dynamic mechanical spectrometer, and the dynamic mechanical thermal analyser (Stanton–Redcroft). Those in the first category were vibrating reed, complex modulus apparatus, dynamic mechanical analyser (DuPont) and torsional braid analyser. Only the last two techniques have at present pharmaceutical applications and the latter is considered in section 2.3.1.1.

In the DuPont DMA a flexural stress of fixed amplitude is applied to the specimen, the modulus is related to the resonant frequency of the system and the damping to its power absorption. The method uses sinusoidal stress as the applied stimulus, natural frequency as the measured response, resonant frequency as the parameter for modulus and power absorption as the parameter for damping. The DMTA of Stanton–Redcroft used the flexural mode with sinusoidal strain as the applied stimulus, actual strain as the measured response, strain amplitude as the parameter for modulus and phase difference as the damping parameter.

The equations describing DMA (Lear & Gill, undated) are extremely complicated. The pharmaceutical interest lies not in their derivation but in how the results may be used to evaluate polymers. Nonetheless they warrant some consideration to understand the rationale of the data generated by the method. Thus if an oscillatory stress is applied to a viscoelastic material the resulting strain in the material will be oscillatory with the same frequency but out of phase with the stress in a manner which is dependent on the relative elastic and viscous responses (Daniels, 1981). For a purely elastic material the phase difference is zero. The elastic and viscous components are related by the equation (2.6):

$$E = E' + iE'' \qquad (2.6)$$

where E is the appropriate bulk modulus for the type of applied stress and E' and E'' are the elastic and viscous components respectively. When δ is the phase difference, $E' = E\cos\delta$ and $E'' = E\sin\delta$. Thus the relationship equation (2.7) is developed:

$$\tan\delta = E''/E' \qquad (2.7)$$

The technique separates the viscoelasticity of a material into the two components of modulus E: a real part (E') and an imaginary part which is the damping or viscous component (E''). The viscoelastic properties are important because they may be correlated with end-use parameters and its measurement is a very sensitive indicator of the internal structure. DMA is particularly useful in measuring the T_g and secondary transition temperatures, polymer blend compatibility, the detection of plasticizer in polymers and a prediction of the physical ageing of amorphous polymers.

The term viscoelastic covers the properties of a material intermediate between a solid (elastic) and liquid (viscous). These materials will display viscoelastic behaviour such as creep and stress relaxation. The elastic properties can be best represented by a spring in which all energy during deformation is stored and can be regained by releasing the stress. The modulus of such a system is described by a plot of stress versus strain. The viscous form is best represented by a dashpot which resists stretching at a force proportional to the strain rate. The slope of the stress versus strain rate is proportional to viscosity. The deformation energy is released as heat and cannot be recovered. Impact-resistant materials have a large viscous component to dampen out harmful vibrations and elastomers have a low viscous component since continuous flexing will result in heat build-up and loss of structural integrity.

Tan δ reflects the balance between the viscous and elastic components. Low tan δ values indicate elastic materials whereas high values indicate liquid-like materials which do not easily vibrate. Comparing tan δ values over a temperature range allows monitoring of the change of the viscoelastic properties of materials. In DMA the polymer is subjected to a sinusoidal stress and the deformation is analysed with respect to the energy lost or stored per cycle. The energy stored per cycle is defined as the dynamic storage modulus (G' or E' for shear or Young's modulus) and the dynamic loss modulus (G'' or E'') defines the heat dissipated in a viscous manner. The ratio of the loss modulus to the storage modulus is the loss tangent (tan δ) and is a unitless parameter defining the ratio of energy dissipated to energy stored per cycle.

These energy-transforming processes are classified into three types (DuPont E42397). Type I motions (usually α-transition) involve movement of long segments of the molecules, forcing chain movement. They are usually equivalent to the T_g. Type II motions involve shorter chain segments and occur below the T_g. Impact resistance is usually associated with transitions of this type. Type III motions involve short chain motions but not usually of the main polymer chain.

DMA is particularly useful in examining transitions below the T_g. Generally the stiffness of polymers (flexural modulus E') will decrease with increasing temperature, with a major decrease occurring at the T_g. This will be however represented as a peak on the tan δ curve. The apparent T_g will move to higher temperatures as the frequency increases. DMA allows the determination of T_g of even crystalline temperatures when the transition energy would be very low due to the small proportion of amorphous material within the sample.

In principle Wetton (1981) considered that a sample could be scanned over either a wide frequency range or a wide temperature range to study its relaxational behaviour. He considered that it was normal practice to scan temperature at a nominal frequency and that limited frequency scans could be made around the transition of interest. Normal convention for describing loss events is to assign them as α, β, γ and δ in order of decreasing temperature and to subscript them as either a or c depending on whether the event occurred in the amorphous or crystalline segments of the polymer. Fig. 2.7 gives a typical plot for polystyrene indicating an α_a peak due to a glass to rubber transition equivalent to the onset of freedom of rotation about main chain bonds, a β_a peak due to phenyl group liberation around it symmetry axis and a γ_a peak due possibly to chemical defects. The δ_a peak was not explained. Polymers such as poly(methyl methacrylate) exhibit a major transition in main-chain mobility at their T_g which corresponds normally to a 1000 decrease in modulus and a damping peak of magnitude about 2 in a temperature range above the static T_g. Its exact position depends on the frequency. At a frequency of 0.01 Hz the loss peak will correspond to the T_g but it will shift upwards by some 5–7°C per decade increase in frequency (Wetton, 1981). The melting points of crystalline polymers register as a multistage process in DMA. Thus following both E' and tan δ for polyethylene oxide of high molecular weight through its melting point (Wetton, 1981) indicated melting at 72°C but also a pre-melting peak modified by crystallization conditions and probably related to less crystalline regions within the polymer which may be due to either thinner or more defected lamellae. The existence of more than one transition of copolymers and blends may be attributed to the miscibility/compatability of the

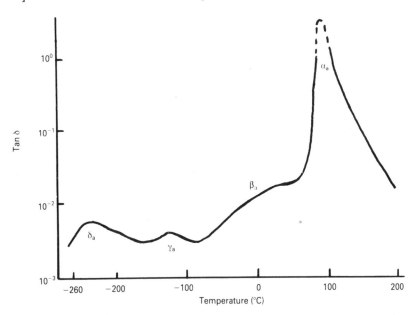

Fig. 2.7 — Dynamic mechanical data for polystyrene in the temperature plane at approximately 1 Hz. The main T_g relaxation is α_a (reproduced with permisison from Wetton, 1981)).

polymers. Compatible polymers will display only one T_g throughout their blends at temperatures intermediate to the T_g of the homogenous components. Partial miscibility and phase-boundary mixing will be readily detected by the method.

2.2.3.1 Torsional braid analysis (TBA)

The concept of TBA was introduced in Chapter 1. Although complicated theoretical relationships hold for the torsional pendulum, these have often been simplified for the TBA to parameters such as relative rigidity ($1/P^2$), and the mechanical damping index ($1/n$) where n is the number of oscillations counted for a decay of a wave between two arbitrary but measured boundary amplitudes. The precise mathematical relationships were described by Gillham (1974). The resultant damped oscillations are followed examining the logarithmic decrement (P) and the period (s) for each damped wave. Other parameters similar to those previously described in the text including the loss modulus may be derived. Use of the relative rigidity gives an estimate of the value of G'. Generally relative rigidity or logarithmic decrement are followed as a function of either temperature or time. Thus storage and loss functions may be presented as a function of temperature or time to produce thermomechanical spectra.

The loss modulus spectra for poly(methyl methacrylate)s (Gillham *et al*. 1974) are characterized by a series of peaks which may be attributed to the onset of motion of internal structural elements of the material. Thus a sharp peak was displayed at the T_g but also a β-transition in the glassy state. This latter is due to motion of the ester side-groups with increasing temperature. The intensity of the transition varied according to whether the polymer was in the syndiotactic, atactic or isotactic state. The syndiotactic polymer has its T_g some 65°C higher than the isotactic material. Physical properties may be related to such transitions since the polymer may be brittle below the β-peak, tougher above it and solid only until the T_g. Such changes are of utmost importance in implant materials and especially those intended for prosthetics. TBA possesses the advantage of being used for measurements through the load-limiting transitions, i.e. the T_g for amorphous polymers and the melting point of crystalline polymers, and is more sensitive than conventional DTA or TG (Gillham *et al.* 1974). This was shown for cellulose triacetate by Gillham (1972). The T_g at 190°C was accompanied by a drastic decrease in rigidity, a maximum in damping and a DTA endothermic shift. Crystallization or chain-stiffening accounted for a decrease in rigidity at above 200°C and was followed by a DTA exotherm maximum. Melting was noted by a decrease in rigidity, a maximum in damping and a DTA endotherm at 290°C. Subsequent cross-linking and chain-stiffening was apparent as a decrease in damping, an increase in rigidity, a DTA exotherm and weight-loss on TG analysis. TBA may also be used in the study of thermosetting processes (Gillham, 1974).

TBA, because it is capable of working at temperatures above the T_g may show transitions which are not detectable by other mechanical or dynamic testing methods and are barely discernible by conventional DTA and DSC. These are so called liquid–liquid transitions (T_{11}) and may be due to chain entanglement in the liquid state of part or the whole of the molecular chain. Such $T_{11} > T_g$ transitions have been reported for polystyrene (Gillham & Boyer, 1977; Stadnicki & Gillham, 1976) but not for any pharmaceutical examples and their relevance is at present unclear. Roller (1981) also considered that TBA is superior to DSC or DTA. Values of T_g were considered not only to be a function of the instrument or technique used in their measurement but were often operator-dependent. Thus when DSC is used to determine the T_g, extrapolation of the straight-line portions of the endothermic shift has been reported at both the upper and lower end of the region. Alternatively the midpoint of the shift can be used. Each approximation will give a different value for the transition and Roller (1981) considered that many T_gs were reported without specifying which methodology was used.

Several criteria used to define T_g values from TBA curves were described by Roller (1981). T_g was indicated as a maximum in log decrement (Δ). T_s was indicated as the onset of the T_g region as indicated by the intersection of the lower temperature straight-line portions of the log relative rigidity versus temperature curve. T_{s1} is the first deviation of the log relative rigidity versus temperature plot from a straight line. T_1 is the end of the T_g region as indicated by the intersection of the higher temperature straight-line portions of the log rigidity versus temperature curve. The width of the T_g region may be defined as $T_1 - T_s$.

2.3   MULTICOMPONENT SYSTEMS

The construction of phase diagrams from either DTA or DSC data is relatively simple but knowledge of the fundamentals behind their construction is essential before correct interpretations can be made in purity determinations, in predictions of stability or compatibility, in drug–polymer interactions and in the characterization of solid dispersions.

2.3.1 Phase diagram determination

The classification of the types of interactions that occur in phase diagrams used in this chapter is based on that used by Chiou & Riegelman (1971) and Grant & Abougela (1982). Thus the diagrams may represent simple eutectic mixtures without solid solution formation, solid solutions, glassy solutions or dispersions, complexes and amorphous precipitates displaying little crystalline structure. Glassy states are discussed in section 2.1.1.6.

There are many variations in the methods used to analyse phase interactions. The simplest approach is to prepare a physical blend of the two ingredients by gentle trituration. Although the work put into the system is relatively low care should be taken to ensure that no polymorphic transitions have been induced. Obviously more intimately blended systems may be prepared either by prior fusion and solidification or by dissolving the ingredients in a solvent in which they are mutually miscible and then removing the solvent by evaporation. The probability of producing polymorphic forms and solvates different to the starting materials cannot be underestimated and the crystal state of the resultant mass may be extremely sensitive to processes such as trituration. There is no general rule as to which method is the most acceptable and it is up to the investigator to decide the most appropriate method for the sample under study. Complex formation should be detectable by each method but may be either absent on the examination of physical mixes or manifested as a recrystrallization exotherm at temperatures intermediate between the eutectic melt horizontal and the endotherm corresponding to fusion of the excess ingredient.

✱ Phase diagrams are usually determined from heating-curve data. However not all materials solidify immediately but may form a supercooled or glassy liquid whose metastability varies considerably such that on reheating recrystallization exotherms, a glass transition or a steady baseline may be apparent. Such phenomena create greater problems when attempts are made to determine phase equilibria from cooling curves. Supercooling by only a few degrees is very common and may lead to poorly reproducible scans and phase diagrams that are impossible to decipher. The problem of scan interpretation is even greater when other variables, such as the proposed heating rate, are considered. A sensible suggestion is to melt the ingredients prior to analysis at a temperature only 2–3°C higher than the final melting point of the blend and, following cooling, devitrify to a stable solid at a few degrees below the temperature of the eutectic horizontal. However the possibility of producing polymorphic materials different to the starting ingredient remains. When the ingredients are unstable the use of solvents to form an intimate blend coupled with faster heating rates than normally employed is favoured. Very unstable systems should be examined as physical mixes at fast heating rates. The use of resolidified melts is unacceptable if either of the two components sublime (Guillory *et al.* 1969)

Throughout the following sections, rather than discussing specific examples, reference is made to two hypothetical compounds, A and B, and to their interactive attraction (U) between each other (U_{AB}) or their own species (U_{AA} or U_{BB}).

2.3.1.1   Simple eutectic mixtures
Fig. 2.8 displays a phase diagram typical of a simple eutectic mixture of two

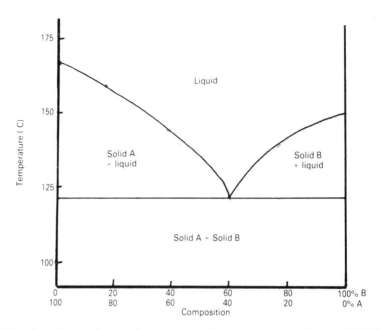

Fig. 2.8 — Typical phase diagram of two hypothetical components, A (melting point 167°C) and
B (melting point 150°C) showing simple eutectic formation.

hypothetical compounds derived from Fig. 2.9, which shows theoretical curves
produced by either DTA or DSC. Only three compositions display a single melting
endotherm and these are the pure compounds A and B and the composition which
represents the eutectic composition. The endotherm of the latter will be considerably
broader because of the depression of melting point caused by the intermixed nature
of the two components. Since trace impurities tend to broaden the melting range of a
major component it also follows that the endotherms of A and B will be narrow
unless they themselves are impure. This broadening induced by impurities forms the
basis of purity determination (Chapter 5). The depression of melting point is
increased by the increasing concentration of impurity and this, in essence, is the
observed effect of adding component A to B or B to A. The maximum depression of
melting point occurs at the eutectic concentration. As the composition of the binary
system is changed from either A or B to that of the euctectic the area of the
endotherm corresponding to the component in excess decreases in area as the
endotherm corresponding to the eutectic increases. This endotherm has its largest
area at the eutectic. This phenomenon may be used to characterize the eutectic
composition by DSC or quantitative DTA. The eutectic horizontal in Fig. 2.8 is
referred to as the solidus line whilst the final melting curve is the liquidus line. Solid

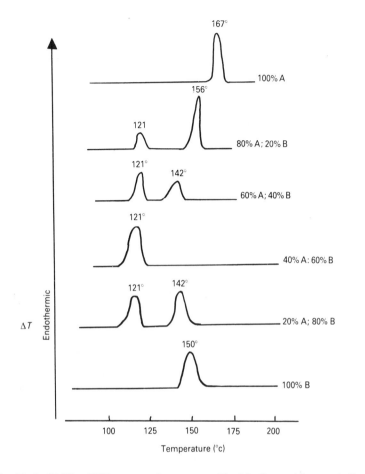

Fig. 2.9 — Typical DSC or DTA scans used to construct Fig. 2.8. Compositions are indicated.

solutions are not generally formed when the interactions between the molecules of the same species are less in the liquid state than the interactions between the different species and in the solid state are greater between the same species of molecules than between the differing species.

DSC or quantitative DTA provide an additional method of estimating the eutectic composition based on the heats of transition of the pure components, the eutectic and materials present in excess of the eutectic. The method was demonstrated by Ford & Francomb (1985) in a study on the physical mixes of sulphamethoxazole and mannitol. The peak melting temperatures (DSC) of the two materials were respectively 166–169°C and 162–168°C. HSM, a valuable technique for phase equilibria determination because of its ability to detect early melting, was used as a parallel technique. The eutectic composition was difficult to quantify because of the similarities in the melting points of the materials and their closeness to the eutectic temperature (determined as 159°C by DSC). Additionally the reproducibility of the peak temperatures (±1.5°C) represented some 30% of the difference between the melting points of the eutectic and the pure materials (Fig. 2.10) and compositions

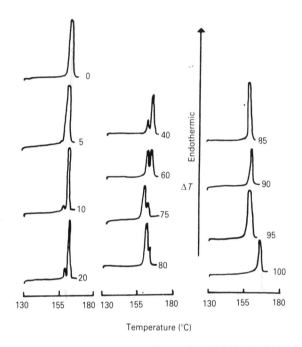

Fig. 2.10 — DSC scans of mannitol, sulphamethoxazole and their physical mixes. Numbers refer to percentage sulphamethoxazole content (reprinted from Ford & Francomb (1985) p. 1116, by courtesy of Marcel Dekker, Inc).

containing in excess of 80% sulphamethoxazole displayed only one endotherm. The phase diagram constructed from HSM data showed an eutectic at ≈85% sulpha-methoxazole. Heats of fusion for mannitol and sulphamethoxazole were 73.6 and 33.4 cal/g (Ford & Francomb, 1985) and were used, along with the heats of transition of the eutectic and excess component, mannitol, to produce Fig. 2.11. Straight-line relationships existed between the composition and the heats of fusion corresponding to mannitol and the eutectic component. Linear regression of the mannitol data indicated that the zero heat of transition, which corresponded to the eutectic, occurred at 90.3% sulphamethoxazole : 9.7% mannitol. The heat of fusion of the eutectic was predicted as 39.9 cal/g. The regresion coefficient (r) was above 0.99 for the 34 data points used to construct each line on Fig. 2.15 and was highly significant ($p < 0.001$).

Problems in examining the phase diagram of a system with a solidus line near to ambient were encoutnered by Brodin *et al.* (1984) in the lidocaine–prilocaine system. HSM of samples that were allowed to reach equilibrium following recrystalli-zation or examined in a non-equillibrium manner displayed phase diagrams which

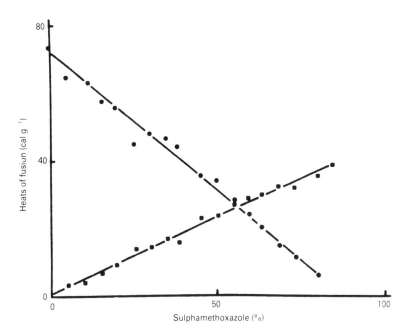

Fig. 2.11 — Heats of fusion of the eutectic component and excess mannitol components of sulphamethoxazole–mannitol physical mixes derived from DSC data plotted as a fraction of sulphamethoxazole. Key: ■, eutectic component; ●, excess mannitol component (reprinted from Ford & Francomb (1985) p. 1120, by courtesy of Marcel Dekker, Inc).

failed to reveal the nature of the interaction between the two materials. The extremes of the phase diagram could not be analysed to confirm the absence or existence of a solid solution. DTA was used to analyse samples which were mixed at room temperature and stored for various periods of time at 4–5°C to prevent ice formation and to establish equilibrium. Samples not equilibrated displayed only a single endotherm (Fig. 2.12) but samples stored at 5°C for three weeks displayed an additional endotherm corresponding to the eutectic temperature and confirming the absence of solid solution formation which was further substantiated by X-ray diffraction.

2.3.1.2 Monotectic systems
Monotectics are commonly found in metallurgy but have been rarely reported in the pharmaceutical literature. The phase diagrams are similar to those of the simple eutectic mixtures but with one of the liquidus arms absent and the other arm approaching the solidus horizontal at zero concentration of one of the components (Fig.2.13). The net effect is that the freezing point of the lower melting point substance replaces the eutectic point and this substance's melting point is unchanged with increased concentrations of the higher melting point substance. Conditions favouring a monotectic (Grant & Abougela, 1982) are interactions $U_{AB} \simeq U_{BB} > U_{AA}$ in the liquid but $U_{AB} \simeq 0$ in the solid state. The liquidus curve represents the solubility

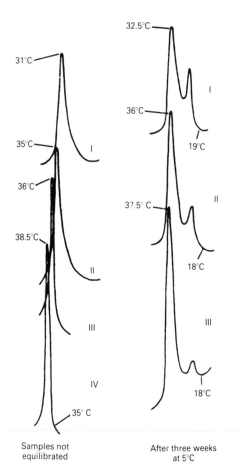

Fig. 2.12 — Some typical examples of the DTA registrations of ≈20 mg lidocaine plus prilocaine on heating showing the endothermic peaks. Prilocaine concentration (mol%). I, 85; II, 90; III, 95; IV, 100 (reproduced with permission from Brodin *et al.*, 1984).

of the higher melting point material in the lower melting point substance. Examples include griseofulvin with glyceryl monostearate (Grant & Abougela, 1982), polyethylene glycol 2000 (Kaur *et al.*, 1980) or polyoxyethylene 40 stearate (Grant & Abougela, 1982).

2.3.1.3 Solid solutions

These represent systems in which there is mutual solubility in both the liquid and solid states. They may be either substitutional, when the molecules of one compound replace the molecules of the other in the crystal lattice, or interstitial, where the molecules of one compound sit within the crystal lattice of the other without disrupting it. For either type of solid solution to form there must be an interaction between the ingredients in the solid state (Grant & Abougela, 1982). In substitutional solid solutions in the solid state $U_{AB}>0$ and in the liquid state $U_{AB}>U_{AA}$ or

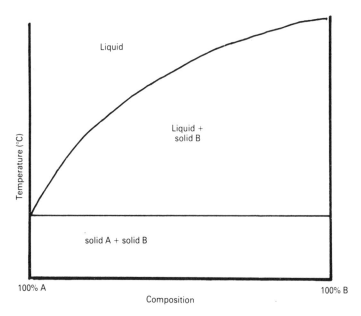

Fig. 2.13 — Typical phase diagram of two hypothetical components, A and B, showing monotectic formation.

U_{BB} and they occur when the molecular sizes and shapes of A and B are similar and this tends to lead to the formation of continuous solid solutions. Discontinuous solid solutions occur when there is limited solid solubility and occur at the extremes of phase diagrams separated by a eutectic. The molecules of A and B differ significantly in size and shape such that substitution into each other's crystal lattice only occurs to a limited extent (Grant & Abougela, 1982). There is only limited interaction in the solid state, i.e. U_{BB} and $U_{AA} > U_{AB} > 0$. The typical phase diagram is illustrated in Fig. 2.14.

The precise mode of incorporating two materials prior to examination for solid solutions is very important. Solid solutions may remain unidentified in the analysis of physical mixtures. The suphathiazole–urea system could have been mistakenly regarded as a simple binary mixture due to thaw points at the eutectic temperature had not more intimately blended systems shown a thawing temperature higher than the eutectic temperature which is indicative of solid solutions (Chiou & Niazi, 1971). Similarly preparative techniques lead to differences of interpretation of the chlorpropamide–urea system (Ford & Rubinstein, 1977). Three methods were used to prepare the blends. Physical mixes were prepared by trituration using mortar and pestle, other samples were premelted at 135–140°C and stored either at room temperature for up to four weeks or at 60°C for 96 hours. DSC of physical mixes indicated a eutectic composition at 89% chlorpropamide : 11% urea where only one endotherm was apparent. However for systems containing more than 89% chlorpropamide three endothermic peaks were discerned, originating from the eutectic and

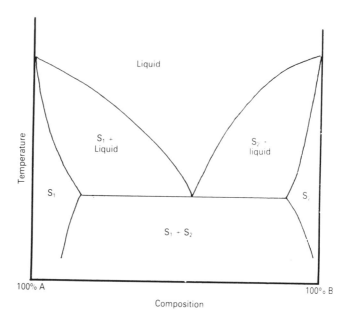

Fig. 2.14 — Typical phase diagram of two hypothetical components, A and B, showing partially miscible solid solutions with eutectic formation. S_1 and S_2 are solid solutions of B in A and A in B respectively.

the two polymorphic forms apparent in the drug itself. The DSC peaks for the pure drug were at 392.5 and 397°C for chlorpropamide forms II and I respectively. On grinding chlropropamide, the endotherm due to form II was lost and only form I recrystallized from its melt (Ford & Rubinstein, 1977). The solidus line, equivalent to melting of the eutectic, was horizontal for compositions containing 1–89% drug indicative of the absence of solid solutions. However in mixes containing more than 89% chlorpropamide the solidus increased in temperature which is indicative of solid solutions of urea in chlorpropamide despite occurring in physical mixes (Ford & Rubinstein, 1977). The use of the melted resolidified systems created problems due to the chlorpropamide undercooling to a glass which showed no typical melting endotherms. This supercooling continued even into melts containing as much as 60% urea which did not display any endotherms corresponding to the solidus line or fully crystallize even after storage at room temperatures or 60°C. Endotherms equivalent to the liquidus transition were apparent in 1-h-old melts containing 0–40 and 94–100% chlorpropamide. Their peak temperatures were 2–3°C lower than those of the physical mixes but increased by 1–2°C on storage. Small endothermic inflections, possibly second-order glass transitions, occurred in melts containing 50–92% chlorpropamide.

    The approach of using enthalpies of fusion to analyse phase diagrams may be used to determine the degree of solid solubility. Yang & Swarbrick (1986) used enthalpies to evaluate the phase diagrams of dapsone–monolauryl dapsone. The DSC scans of some comelts are given in Fig. 2.15 and their eutectic enthalpies were

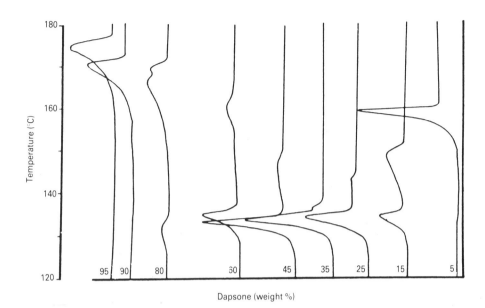

Fig. 2.15 — DSC scans of dapsone : monolauryl dapsone comelts. Numbers on the horizontal axis represent the percentage by weight of dapsone present in comelts (reproduced with permission from Yang & Swarbrick, 1986).

used to produce Fig. 2.16. Yang & Swarbrick (1986) tentatively proposed that this heat could be used to determine comelt composition but in reality two compositions either side of the eutectic would have the same value. The intersect of the straight lines corresponding to the heats of fusion gives the eutectic composition at 36.5% dapsone. The intersection of these lines with the composition axes did not correspond to the origins but zero enthalpy values gave terminal solid solubilities at the eutectic temperature of 11% monolauryl dapsone in dapsone and 2.8% dapsone in monolauryl dapsone. The eutectic composition contained equimolar amounts of each ingredient and is a mixture of two solid solutions. The phase diagram of the system (Fig. 2.17) was constructed from the DSC scans of the melts using peak temperatures to represent the liquidus and solidus lines. Compositions to the dapsone-rich side of the eutectic formed glassy solutions and on reheating failed to show any melting endotherms. Glass transition temperatures were between 30 and 40°C. The phase diagram obtained from the physical mixes was somewhat more complicated due to the existence of the various polymorphic forms of dapsone.

2.3.1.4 Solid complexes

When two materials interact significantly with each other they may form separate stable complexes. In the typical phase diagram (Fig. 2.18) the existence of a molecular compound with its own melting point is easily distinguished. This melting point may be said to be congruous, giving a congruent melting point at the exact stoichiometric ratio of the complex. Grant & Abougela (1982) considered that

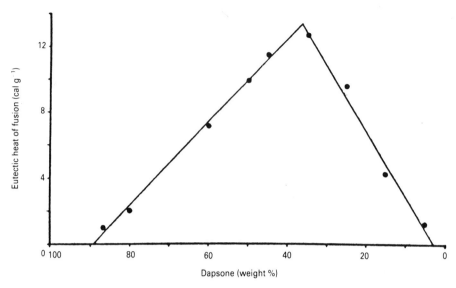

Fig. 2.16 — Enthalpy of fusion of the eutectic portion of dapsone : monolauryl dapsone comelts as a function of composition (reproduced with permission from Yang & Swarbrick, 1986).

griseofulvin (G) and phenobarbitone (P) formed two congruent complexes of formulae PG_3 and P_3G. Generally other techniques such as X-ray diffraction and IR spectroscopy will help identify the molecular compound.

The complex however may be too unstable to have its own recognized melting point and dissociation may occur before this temperature is reached. Such a system may be termed a peritectic, e.g. the paracetamol–phenazone system (Grant *et al.*, 1980). This system was examined using HSM and DTA. Physical mixes of the two materials were prepared in an agate mortar and fused samples were prepared at 2°C above the temperature of complete fusion. DTA and HSM of the pure materials gave sharp melting points at 111 and 169.5°C for phenazone and paracetamol respectively. The physical mixes (Fig. 2.19) of paracetamol and phenazone gave eutectic endotherms (a) at 83°C. Samples containing 5–50 mole% phenazone gave exotherms (b) at about 90°C due to complex formation. The exotherm (b) was not detected at above 50 mole% phenazone, due to the increase in size of the initial endotherm (a). Physical mixes containing 2–60% phenazone displayed a large endotherm (c) at about 107°C representing fusion of the complex. The area of this endotherm decreased in the range 55–70 mole% phenazone at the expense of an increase in size of the eutectic endotherm (a) at 84°C. Above these peaks a further endotherm (d) was apparent as a wedge-shaped transition and was attributed to the melting of the excess paracetamol in physical mixes containing 2–35 mole% phenazone (Grant *et al.*, 1980). Its area decreased with an increase in the phenazone content of the mixes. The melting point of the complex, which dominated the melting of the mixture in the range 40–60%, was itself depressed by the excess phenazone at concentrations above

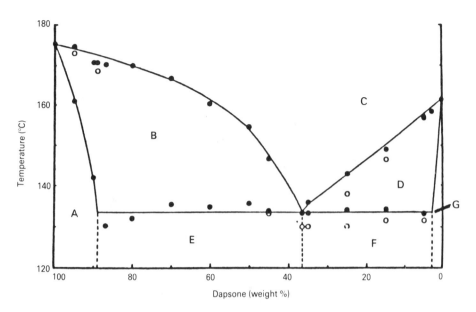

Fig. 2.17 — Phase diagram of dapsone : monolauryl dapsone comelts (●) first heating; (○) second heating. Solid solution α contains monolauryl dapsone dispersed in dapsone. Solid solution β contains dapsone dispersed in monolauryl dapsone. Key: A, solid solution α; B, solid solution α plus liquid; C, liquid; D, solid solution β plus liquid; E, solid solution α plus solid solution α : solid solution β; F, solid solution β plus solid solution α : solid solution β; G, solid solution β (reproduced with permission from Yang & Swarbrick, 1986).

60%. Mixes containing more than 75% phenazone displayed a further endotherm (f) which corresponded to the melting of the excess phenazone. The eutectic system contained 70 mole% phenazone and HSM was used to confirm that all the sample melted at this temperature. This phase diagram (Grant *et al.*, 1980) derived from physical mixes (Fig. 2.20) predicts that in an excess of paracetamol the eutectic temperature is followed by the exothermic formation of the complex at 90°C and then its melting or dissociation of 107°C. Point m represents the congruent melting point of the complex, given a composition of 1 : 1. The flatness of the melting point represents dissociation of the complex at or near its melting point, indicating a poor thermal stability that produces paracetamol which subsequently melts. Point p represents a type of peritectic point.

The data for resolidified melts (Grant *et al.* 1980) was similar but with some crucial differences. The exothermic transition was absent and the eutectic composition, at 76% phenazone, represented the eutectic of phenazone and the complex. Fig. 2.21 gives the true equilibrium phase diagram whereas Fig. 2.20 represents the diagram for a series of non-equilibrium events and reflects poor mixing of the ingredients at the molecular level.

Even from the brief description of the above systems it should be obvious that the appearance of DTA or DSC scans will depend on the history of the sample prior to analysis and that the physical mixtures will give different scans to those of systems

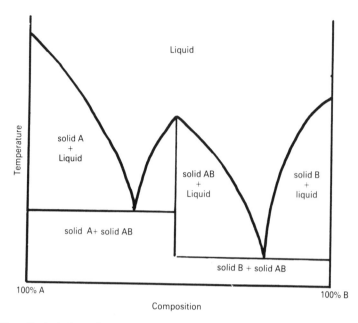

Fig. 2.18 — Typical phase diagram of two hypothetical components, A and B, displaying complex formation with a congruent melting point.

where intimate blending has been achieved. Other systems that created similar problems are griseofulvin–stearic acid systems (Abougela & Grant, 1979; Grant & Abougela, 1981, 1982) and the deoxycholic acid–menandione, caffeine–phenobarbitone and quinine–phenobarbitone systems (Guillory *et al.*, 1969).

Guillory *et al.* (1969) were amongst the first researchers to rationally examine the phase diagrams of mixtures showing complex behaviour using DTA. Their data emphasized that the resolidified melts of the pure materials should be scanned in addition to composites. This represents the minimum type of study expected to yield data considering the possibility of polymorphic modifications. Problems may be expected when the eutectic horizontal is close to the liquidus peak. Examination of cooling curves may result in difficulties in phase diagram determination due to the ability of ingredients to undercool to glasses, resulting in the absence of data due to the solidification. Menandione created such problems in the presence of deoxycholic acid. Guillory *et al.* (1969) further reported that experimental conditions such as agitation during cooling caused some peaks to disappear and others to change in size in the theophylline–phenobarbitone system. It is possible that these changes were due to polymorphic changes within the sample. When cooling curves are examined some degree of agitation may be required to avoid the problems of passing through the glassy state.

2.3.1.5 *Glass transition temperatures of mixtures*
Although this subject is more fully dealt with in Chapter 8 it is perhaps relevant to introduce the concept of its determination here. It is possible in a binary system that

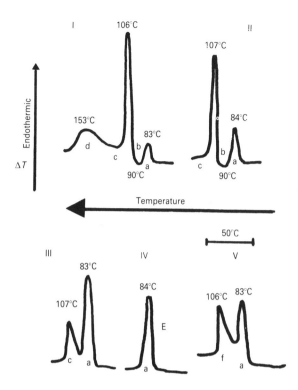

Fig. 2.19 — DTA scans of physical mixtures of paracetamol (A) and phenazone (Z). E refers to the eutectic mixture. The letters a,b,c,d and f refer to peaks discussed in text. Physical mixes contain: I, 20% phenazone; II, 40% phenazone; III, 60% phenazone; IV, 70% phenazone; and V, 90% phenazone (reproduced with permission from Grant *et al.*, 1980).

the glass-forming ability of one ingredient may exert its influence throughout that part of the phase diagram where such a component is present at high levels, e.g. in melts of chlorpropamide and urea (Ford & Rubinstein, 1977) or the eutectic or other specific composition may be the glass-forming entity. Where such binary systems are studied inevitably either DSC or DTA will be the method of analysis. There are several methods of defining the T_g (see section 2.1.1.6) but unfortunately many publications fail to state the precise method used, resulting in divergent values of what, although a property exceptionally dependent on previous thermal history, should be relatively easy to define. Yang & Swarbrick (1986) clearly stated that the T_g was derived as the midpoint of the change as measured from the extensions of the pre- and post-transition baseline. The thermal history of the sample was standardized by shock-cooling from a standard temperature to another standard temperature, less than ambient, in a controlled cooling fashion. Such treatment is imperative when determining the T_g of a mixture.

2.3.1.6 *Miscibility gaps*
Even for fully miscible systems one of the great problems in thermal analysis is that materials of vastly different melt viscosity will not readily mix. Since agitation during

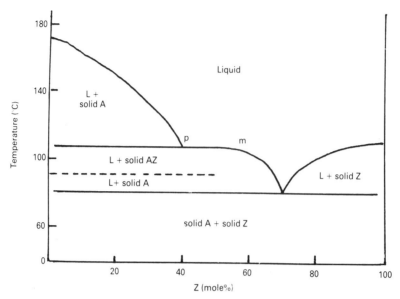

Fig. 2.20 — Phase diagram obtained on heating physical mixtures of paracetamol (A) and phenazone (Z). AZ, 1: complex; L, liquid; p, peritectic point; − − − exothermic change; m, congruent melting point of AZ (reproduced with permission from Grant *et al.*, 1980).

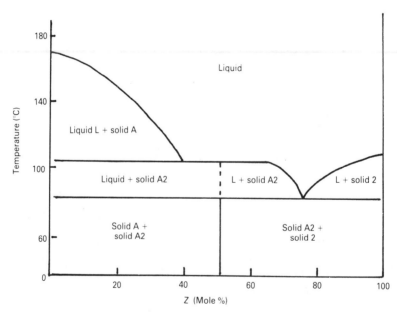

Fig. 2.21 — Phase diagram obtained on heating fused–cooled mixtures of paracetamol (A) and phenazone (Z). AZ, 1: complex; L, liquid; p, peritectic point; m, congruent melting point of AZ (reproduced with permission from Grant *et al.*, 1980).

a DSC or a DTA run is impossible, the slowness of the materials to melt and mix may lead to a phase diagram only poorly representative of the equilibrium state. Many binary systems however fail to show complete miscibility in the solid or liquid states and consequently show miscibility gaps in their phase diagram. As an example, Fig. 2.22 is taken from the study of Yang & Swarbrick (1986). In the composition 10–20%

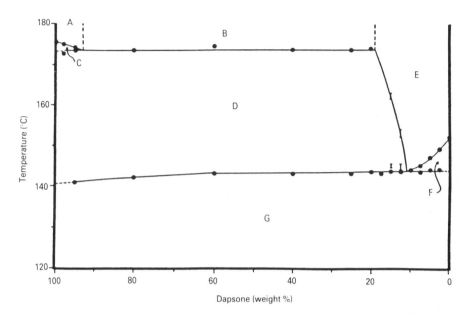

Fig. 2.22 — Phase diagram of dapsone : dilauryl dapsone physical mixtures: ●, from DSC; I, from HSM. Liquid 1 contains 19% w/w dapsone in dilauryl dapsone; liquid 2 contains 7% w/w dilauryl dapsone in dapsone. Key: A, liquid; B, liquid 1 plus liquid 2; C, solid dapsone plus liquid 2; D, solid dapsone plus liquid; E, liquid; F, solid dilauryl dapsone plus liquid; G, solid dapsone plus solid dilauryl dapsone. (Redrawn with permission after Yang & Swarbrick, 1986).

dapsone, because of the small amount of dapsone present, no melting endotherm corresponding to it was differentiated from the baseline. Consequently the melting ranges were determined by HSM. A eutectic containing 11% dapsone melting at about 143°C was apparent and the excess of dapsone melted generally at about 173.5°C. The liquids were considered to be liquid 1 (19% dapsone, 81% dilauryl dapsone) and liquid 2 (93% dapsone, 7% dilauryl dapsone). At 173.5°C samples containing more than 19% dapsone existed in three phases, i.e. solid dapsone and two liquids. Thus for systems showing miscibility gaps a combination of either DTA or DSC with HSM is required to distinguish ideally the phase diagram.

2.4   MISCELLANEOUS THERMAL ANALYTICAL METHODS

Certain, more recently developed, techniques have a limited although expanding pharmaceutical relevance. These are now introduced as techniques rather than for the results and information that they generate.

2.4.1 Evolved gas analysis

The practice of analysing gaseous products evolved during a scan is well recognized as an important technique that gives valuable information on the decomposition and stability of materials. It has not, as yet, been extensively used in pharmaceutical studies. The methods used to detect or analyse any gaseous effluent include mass spectroscopy and IR techniques. Two separate techniques are recognized: evolved gas detection is a qualitative method designed to detect one or more specific byproducts, evolved gas analysis is a method designed to analyse the gaseous byproducts.

The examination of volatiles from minerals by non-dispersive IR detectors has been described (Morgan, 1984). Non-dispersive IR was acceptable for evolved gases including water vapour, carbon dioxide and carbon monoxide, and electrochemical cells were considered acceptable for sulphur dioxide and oxygen.

Evolved gas analysis was used to examine the pyrolysis of tobacco (Baker, 1984). Evolved carbon dioxide was detected using mass spectrometry (MS), gas chromatography (GC) or Fourier transformation IR spectroscopy (FTIR). GC was the poorest method of detection since frequent sampling was required throughout whereas the other methods allowed the examination of different pyrolysis products from a single run. Baker (1984) considered the method suitable for examining the breakdown products of sugars, lignins, polysaccharides, amino acids, esters, celluloses and even inorganic carbonates.

Nachbaur *et al.* (1981) used evolved gas analysis in conjunction with DTA and TG (DTG) to quantify the degradation of urea. TG showed that two decomposition steps occurred: the first at 132–250°C equivalent to a 50.7% weight loss and corresponding to the formation of cyanuric acid, and the second weight loss between 300 and 430°C and equated to its depolymerization. Fig. 2.23 confirms the degradation mechanisms. HNCO (mass 43) and evolved ammonia (mass 17) are apparent in the spectra before the melting point of urea (132°C). The spectrum may be explained by the following reactions (schemes (2.1) to (2.4))):

$$H_2N-CO-NH_2 \qquad\qquad \rightarrow HNCO + NH_3 \qquad\qquad \text{Scheme (2.1)}$$

$$H_2N-CO-NH_2 + HNCO \rightarrow H_2N-CO-NH-CO-NH_2 \quad \text{Scheme (2.2)}$$

$$3\ HNCO \qquad\qquad\qquad \rightarrow (HNCO)_3 \qquad\qquad\quad \text{Scheme (2.3)}$$

$$HNCO + H_2O \qquad\qquad \rightarrow CO_2 + NH_3 \qquad\qquad\quad \text{Scheme (2.4)}$$

The mass species at 44 is due to the formation of carbon dioxide from the hydrolysis of HNCO. The double structure of the curves is due to the decomposition of biuret. The decomposition of cyanuric acid above 300°C is represented by the formation of the mass species at 43. The presence of a peak at mass 44 under these conditions is due to the formation of the ion H_2CNO^+.

The use of mass spectroscopy coupled with TG may involve continual monitoring of a particular mass species throughout the scan or the examination of gaseous residue that has been trapped during particular thermal events, e.g. weight losses. Suitable systems were described by Chiu & Beattie (1980). Derivative TG defined the temperature ranges for sample collection. Mass spectroscopy identified the

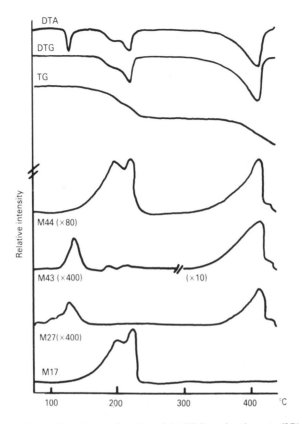

Fig. 2.23 — Thermal behaviour of urea. Sample weight: 20.5 mg; heating rate 6°C/min; reagent gas argon, 75 ml/min; ion source pressure 0.1 torr. (from Nachbaur, Baumgartner & Schober. *Proceedings of the 2nd. European Symposium on Thermal Analysis*. Editor D. Dollimore. Copyright 1982; Heyden. Reprinted by permission of John Wiley & Sons, Ltd).

major mass species for the degradation of calcium acetate monohydrate as 18 (water), 43 (acetone) and 44 (carbon dioxide) for the temperature cuts of around 100, 400 and 650°C. Continuous monitoring of the relative ion intensities at M/e of 18 and 31 were used to monitor the release of water and ethanol respectively from a polyimide prepreg and evaluate the degree of curing (Chiu & Beattie, 1980).

TG provides only quantitative information on the weight loss, not information on the nature of the specific component lost. Mass spectroscopy adds the advantage of identifying and determining the relative concentration of released gases (Chiu & Beattie, 1981). This latter requires the determination of the absolute weight of the component from the mass spectroscopic data. Chiu & Beattie (1981) described the use of a constant volume sampler to this aim. For a series of ethylene–vinyl acetate copolymers the first major weight loss occurred at 250°C and corresponded to a release of a species of $M/e=60$ (acetic acid). The relative intensity of this mass peak was plotted as a function of temperature and the integrated peak was a function of the acetic acid released. This was utilized to give the composition of the copolymer blends.

Other means of analysing evolved gas include FTIR. Wiedbolt *et al.* (undated) interfaced a Nicolet SXC FTIR spectrometer with a DuPont 9900 TGA system and examined the degradation products of polyvinylchloride. The FTIR we used in two modes, either the spectra of the gaseous products were analysed across the IR wavelengths or selected windows monitored the generation of functional groups or molecules. Thus wavelength windows at 2831–2785, 1760–1740 and 674–664 cm^{-1} were used to monitor hydrogen chloride, carbonyl absorption and benzene respectively. Compton (undated) used similar windows at 1780–1900, 2050–2200 and 2300–2380 cm^{-1} to monitor respectively water vapour, carbon monoxide and carbon dioxide during analysis of calcium oxalate.

2.4.2　Thermosonimetry and thermoacoustimetry

Thermoacoustimetry is a technique in which the characteristics of imposed acoustic waves are measured as a function of temperature after passing through a substance which is subjected to a controlled temperature programme. It has been used to characterize fast chemical reactions, chemical relaxation processes of molecules and transitions in solids including polymers. The method has not been used within the pharmaceutical arena.

Thermosonimetry has, however, found a variety of pharmaceutical applications. In thermosonimetry the sound which a substance, subjected to a controlled temperature programme, emits is measured as a function of temperature. The technique detects thermally induced acoustic emission (Clark, 1978). All thermally induced physical and chemical changes generate acoustic emissions but those derived from motions of lattice imperfections are especially important because they provide information about the defect centre themselves (Clark, 1986). The potential of thermosonimetry to pharmaceuticals and organic systems was discussed by Clark (1981, 1986) and Clark & Fairbrother (1981). Clark (1981) highlighted the potential of the system, using inorganic salts and their hydrates as model substances. Potassium sulphate and pottasium chlorate are acoustically active below their transition temperatures of 299 and 582°C respectively. These activities were respectively attributed to microcrack generation and fluid inclusion release (Clark, 1981). The frequency distribution for potassium chlorate showed a higher proportion of lower frequency emissions than the distribution for potassium sulphate. The resonance at 15 kHz was particularly active for potassium chlorate whereas potassium sulphate showed enhanced high frequency (50–100 kHz) with less active resonances at lower frequencies. Similar differences were noted when scanning copper sulphate pentahydrate between ambient and 300°C to produce the monohydrate and between ambient and 120°C to produce the trihydrate. The increase in height of the 15 kHz peak was particularly noticeable between 140 and 160°C and was attributed to the loss of two water molecules to form the monohydrate. Loss of liquid was registered by an increase in the frequency distribution at low resonances (Clark, 1981).

The full application of thermosonimetry to pharmaceutical research was demonstrated by Clark & Fairbrother (1981). Dislocations of adipic acid were induced using impurities such as oleic acid, linoleic acid, and ricinoleic acid. The molecular fit was not perfect for any of the impurities producing lattice distortion. The distortion was relieved by dislocation formation whose density was thought to be dependent on the amount of impurity included. The particle size of the crystals under examination must be carefully controlled by sieving prior to analysis because particle size influences the total acoustic emission (Clark & Fairbrother, 1981). Representative thermosonimetry curves of adipic acid and an oleic-acid-doped sample in which three regions of activity were seen is given in Fig. 2.24. Premelting activity at 60–120°C was

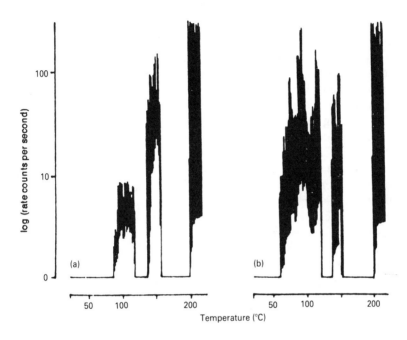

Fig. 2.24 — TS curves of adipic acid samples. (a) adipic acid plus 10.2 ppm oleic acid; (b) pure adipic acid. Heating rate 20°/min (from Clark & Fairbrother. *Proceedings of the 2nd European Symposium on Thermal Analysis*. Editor D. Dollimore, Copyright 1982; Heyden. Reprinted by permission of John Wiley & Sons Ltd).

due to microcrack production and propagation under the thermal stress of heating. The activity at 135–150°C was due to melting and that above 200°C was due to decomposition. Microscopically the impure sample possessed considerable crystal pitting and therefore there was little need to create new microcracks during heating to accommodate thermal stress. Consequently the pre-melting peak of this sample was small. Pure adipic acid microscopically exhibited low damage and consequently many microcracks would form during heating to relieve stress resulting in the large pre-melting peak. The peaks due to melting were similar although slightly bigger for the impure sample. This may have been due to the high pre-melting activity resulting in a less structurally strained sample at the onset of melting. Clark & Fairbrother (1981) determined a total count for 200 mg samples over the range from ambient to melt using a high heating rate of 60°C/min to obtain a significant total count. A plot of total acoustic emission versus impurity content (ppm) enabled the total count (c) to be related to the impurity level (x) by equation (2.8)

$$10^8/c^2 = 3.7x + 7.2 \tag{2.8}$$

The equation applied to all the impurities examined except undecylenic acid which was atypical and produced a considerably greater level of acoustic emission. This indicated that the induced dislocation density was dependent on the fatty acid chain length. Clark (1986) considered that thermosonimetry provides a measure of the integrity index or disruption index. Well-annealed samples of high crystallinity showed more pre-melting thermosonimetry than samples of low crystallinity and integrity. Clark (1981) proposed that thermosonimetry could be used to assess the condition of a substance, e.g. assessment of radiation damage, dislocation density and the degree of annealment.

Additionally crystallization involves acoustic activity. Clark (1986) found intense activity at the freezing point due to the initial nucleation of crystallites during cooling at 5–20°C/min. A broader, less intense peak some 10–30°C lower was also noted due to crystallization processes relieving lattice strain. During slow cooling (<1°C/min) the initial intense peak was followed by a series of active periods interdispersed between quiescent periods due to the slow build-up of lattice strain and strain relief due to microcrack formation (Clark, 1986). Different frequency spectra were obtained when adipic and benzoic acid were crystallized (Clark, 1986) which were not easily attributed to specific processes. The crystallization of adipic acid in the presence of impurities also produced a different acoustic spectra especially at 20–50 kHz, these peaks were less resolved at 10 ppm and had almost disappeared at an impurity level at 25 ppm, indicating that certain crystallization processes did not occur in the presence of impurities.

Clark (1986) also considered that thermosonimetry has potential in polymorphism studies. An unnamed pharmaceutical possessed a stable form melting at 81°C, a metastable form melting at 65°C and a glassy form with a T_g at 50°C. X-ray diffraction indicated that the stable form underwent some changes below its melting point that were undetected by DSC or HSM. However a thermosonimetric emission occured at a heating rate of 5°C/min which showed the melting of annealed samples as a twinned peak with a lower component at 70°C (Clarke, 1986). No explanation was forwarded for the phenomenon but thermosonimetry was the only technique to show the existence of the apparent transition.

REFERENCES

⁎Abougela, I. K. A. & Grant, D. J. W. (1979) *J. Pharm. Pharmac.*, **31**, 49p.

ASTM D3417-83 (1983) *Heats of fusion and crystallization of polymers by thermal analysis*, 107–111.

ASTM D3418-82 (1982) *Transition temperatures of polymers by thermal analysis*, 112–117.

Baker, K. F. (undated) DuPont Application Brief E-04240 *Measurement of glass transition temperature of films by thermomechanical analysis*.

Baker, R. R. (1984) *Anal. Proc.*, **21**, 12–13.

Barrall, E. M. (1973) *Thermochim. Acta*, **5**, 377–389.

Barrall, E. M. & Johnson, J. F, (1970) *Techn. Methods Polym. Eval.*, **2**, 1–39.

Brancone, L. M. & Ferrari, H. J. (1966) *Microchem. J.,* **10**, 370–392.

Brennan, W. P. (1973) 'What is a T_g?' A review of the scanning calorimetry of the glass transition, *Thermal Analysis Application Study TAAS-7*, Perkin-Elmer Corp., Norwalk, Connecticut.

Brennan, W. P. (1977) Thermomechanical analysis of polyethylene film, *Thermal Analysis Application Study TAAS-23,* Perkin-Elmer Corp., Norwalk, Connecticut,

Brennan, W. P. & Gray, A. P. (1974) Liquid crystals: the mesomorphic state,, *Thermal Analysis Application Study TAAS-13*, Perkin-Elmer Corp., Norwalk, Connecticut.

*Brodin, A., Nyqvist-Mayer, A., Wadsten, T., Forslund, B. & Broberg, F. (1984) *J. Pharm. Sci.,* **73**, 481–484.

Chen, F. C., Choy, C. L., Wong, S. P. & Young, K. (1980) *Polymer,* **21**, 1139–1147.

*Chiou, W. L. & Niazi, S. (1971) *J. Pharm. Sci.,* **60**, 1333–1338.

Chiou, W. L. & Riegelman, S. (1971) *J. Pharm. Sci.,* **60**, 1281–1302.

Chiu, J. & Beattie, A. J. (1980) *Thermochim. Acta,* **40**, 251–259.

Chiu, J. & Beattie, A. J. (1981) *Thermochim. Acta,* **50**, 49–56.

Clark, G. M. (1978) *Thermochim. Acta,* **27**, 19–25.

Clark, G. M. (1981) *Proceedings of the 2nd European Symposium on Thermal Analysis,* Dollimore, D. (ed.) Heyden, London, 85–88.

Clark, G. M. (1986) *Anal. Proc.,* **23**, 393–394.

Clark, G. M. & Fairbrother, J. E. (1981) *Proceedings of the 2nd European Symposium on Thermal Analysis,* Dollimore, D. (ed.) Heyden, London, 255–258.

Compton, D. A. C. (undated) *On-line FT-IR analysis of the gaseous effluent from a thermogravimetric analyzer: applications of an integrated TGA/FT-IR System,* Bi-Rad, Digilab Division, Cambridge, MA.

Daniels, T. (1981) *Anal. Proc.,* **18**, 412–416.

DuPont Brochure E-42397, *DuPont 982 Dynamic Mechanical Analysis System.*

Ellerstein, S. M. (1966) *Appli. Polym. Symposia,* No. 2, 111–119.

Ford, J. L. (1987) *Drug Dev. Ind. Pharm.,* **13**, 1741–1777.

Ford, J. L. & Francomb, M. M. (1985) *Drug Dev. Ind. Pharm.,* **11**, 1111–1122.

Ford, J. L. & Rubinstein, M. H. (1977) *J. Pharm. Pharmac.,* **29**, 209–211.

Fukuoka, E., Makita, M. & Yamamura, S. (1986) *Chem. Pharm. Bull.,* **34**, 4314–4321.

*Gaur, U., Mehta, A. & Wunderlich, B. (1978) *J. Therm. Anal.,* **13**, 71–84.

Gill, P. S. (1984) *Am Lab.,* **16**, 39–49.

Gillham, J. K. (1972) *Critical Rev. Macrom. Sci.,* **1**, 83–172.

Gillham, J. K. (1974) *Am. Inst. Chem. Eng. J.,* **20**, 1066–1079.

Gillham, J. K. & Boyer, R. F. (1977) *J. Macromol. Sci. Phys.,* **B13**, 497–535.

Gillham, J. K., Stadnicki, S. J. & Hazony, Y. (1974) *Am. Chem. Soc. Polym. Preprints.,* **15**, 562–569.

Grant, D. J. W. & Abougela, I. K. A. (1981) *J. Pharm. Pharmac.,* **33**, 619–620.

Grant, D. J. W. & Abougela, I. K. A. (1982) *Anal Proc.,* **19**, 545–549.

*Grant, D. J. W., Jacobson, H., Fairbrother, J. E. & Patel, C. G. (1980) *Int. J. Pharm.,* **5**, 109–116.

Gray, A. P. (1968) in Porter, R. F. & Johnson, J. M. (eds.) *Analytical Calorimetry,* Plenum Press, New York, 209–218.

Gray, A. P. & Brenner, N. (1965) *Am. Chem. Soc. Div. Polymer Chem.,* **61**, 956–957.

Guillory, J. K., Hwang, S. C. & Lach, J. L. (1969) *J. Pharm. Sci.*, **58**, 301–308.

Haida, O., Suga, H. & Seki, S. (1977) *Bull. Chem. Soc. Japan*, **50**, 802–809.

Kambe, H., Horie, K. & Suzuki, T. (1972) *J. Therm. Anal.*, **4**, 461–469.

Kaur, R., Grant, D. J. W. & Eaves, T. (1980) *J. Pharm. Sci.*, **69**, 1317–1321.

Koo, G. P. (1974) *Plastics Eng.*, **30**, 33–38.

Laye, P. G. (1980) *Anal. Proc.*, **17**, 226–228.

Lear, J. D. & Gill, P. S. (undated) DuPont Background Paper E-42400, *Theory of operation of the DuPont 982 Dynamic Mechanical Analyzer*.

Maurer, J. J. (1978) *Thermal methods in polymer analysis*, Franklin Inst. Press, USA, 129–162.

Morgan, D. J. (1984) *Anal. Proc.*, **21**, 3–4.

Nachbaur, E., Baumgartner, E. & Schober, J. (1981) *Proceedings of the 2nd European Symposium on Thermal Analysis*, Dollimore, D. (ed.) Heyden, London, 417–421.

Perkin-Elmer Thermal Analysis Newsletter TAN-3 (1966). Specific Heats by DSC. Perkin-Elmer Corp., Norwalk, Connecticut.

Rak, J., Ford, J. L., Rostron, C. & Walters, V. (1985) *Pharm. Acta Helv.*, **60**, 162–169.

Roller, M. B. (1981) *Proceedings 11th North American Thermal Analysis Society Conference*, Schelz, J. P. (ed.) New Orleans, 519–529.

Serajuddin, A., T. M., Rosoff, M. & Mufson, D. (1986) *J. Pharm. Pharmac.*, **38**, 219–220.

Stadnicki, S. J. & Gillham, J. K. (1976) *J. Appl. Polym. Sci.*, **20**, 1245–1275.

Staub, H. & Schnyder, M. (1974) *Thermochim. Acta*, **10**, 237–243.

Suga, H. & Seki, S. (1974) *J. Non-crystalline Solids*, **16**, 171–194.

Takahashi, Y. (1985) *Thermochim. Acta.*, **88**, 199–204.

Wetton, R. E. (1981) *Anal. Proc.*, 416–421.

Wieboldt, R. C., Adams, G. E., Lowry, S. R. & Rosenthal, R. J. (undated), *Analysis of a vinylchloride polymer by TGA-FTIR*, Nicolet Spectroscopy Center, Maddison.

Yang, T-T. & Swarbrick, J. (1986) *J. Pharm. Sci.*, **75**, 53–56.

Zordan, T. A., Hurkot, D. G., Peterson, M. & Hepler, L. G. (1972) *Thermochim. Acta*, **5**, 21–24.

3

Practical considerations for optimizing and improving the performance and quality of results obtained from thermal analysers

This chapter is concerned with factors, either sample- or instrument-related, that may affect the accuracy and precision of thermo-analytical results and is primarily concerned with DSC, DTA and TG since these are the methods most frequently used by the pharmaceutical scientist. The reader is reminded that not all the controls and precautions cited need be utilized to produce acceptable results. Their quality, however, will be considerably enhanced if the subtleties behind the methods are fully understood.

3.1 REFERENCE MATERIALS

In DSC and DTA, a suitable inert material should be used to balance the reference container and the sample container with respect to weight and/or specific heat. The material used should not undergo any interfering transition or absorb water but ideally should be stable and unaffected by repeated heating and cooling. ASTM D3418–82 recommends glass beads, alumina powder and silicon carbide as suitable reference materials. This standard also suggests that the thermal diffusivity of the reference should be as near as possible to that of the sample.

3.2 INSTRUMENT CALIBRATION

Both DTA and DSC need instrumental calibration of their temperature scale and, when used for calorimetric analysis, their calorimetric response. Variations in the enthalpies of fusion of tin, lead, zinc, bismuth and selenium were reported as 3, 8.9, 10.2, 7.5 and 6.7% respectively (Barnes *et al.*, 1984), indium being perhaps the most widely used calibrant. The development of temperature standards for thermal analysis equipment was reviewed by Charsley (1980) who described the standardization rationale undertaken by the International Confederation for Thermal Analysis

(ICTA). The objective of these standards are to provide a common basis for relating independently acquired data, to enable comparison of all thermo-analytical equipment independent of design and to provide a means of relating thermo-analytical data to physical and chemical properties determined by conventional isothermal procedures. The ICTA has developed standards for use under dynamic conditions. The temperatures relate only to the batches of material supplied as standards and not to the materials themselves.

Sarge & Cammenga (1985) listed the ideal properties of references intended for temperature and calorimetric calibration. Calibrants should be easily purified, chemically stable in the solid and liquid phases, non-toxic and non-hygroscopic, possess only one phase transition at the temperature of interest, have transitions with exactly determined enthalpy and temperature, be inert in the containers used and have small vapour pressures.

3.2.1   Temperature

The precise transition temperature may be more easily determined by using a drop of a non-volatile liquid on the sensor and pressing to a thin layer by setting the encapsulated sample on it firmly. This will reduce the instrumental lag by removing the air gap (Barrall & Dawson, 1974). Suitable liquids are silicone oils which have an effective range of -40 to $320°C$.

Langier-Kuzniarowa (1984) studied reference materials. Although the glass transition of polystyrene at $\simeq 100°C$ is often used for calibration of DSC and DTA its onset temperature is not reproducible enough to be used as a calibrant because it is operator-subjective. The extrapolated temperature and the midpoint of the glass transition were acceptable with values of $104.4 \pm 1.5°C$ and $107.5 \pm 1.7°C$ respectively (Langier-Kuzniarowa, 1984). This material is supplied as NBS-ICTA GM 754 reference material issued by the United States National Bureau of Standards. The range of temperatures below 350 K are covered by references NBS-ASTM GM 757 and consist of four materials: o-terphenyl, phenyl ether, cyclohexane and 1,2-dichloroethane. The extrapolated onset temperature and peak temperatures of the melting transitions are used to provide the temperature calibration and for cyclohexane a phase transition at 187.1 K (onset) and 190.9 K (peak) is also recommended. Hexachloroethane and hexamethylbenzene could be use as calibrants at 71 and $110°C$ respectively (Kambe *et al.*, 1972).

The temperature range $125–940°C$ is covered by three sets of references (NBS-ICTA GM 758, GM 759 and GM 760) each consisting of five reference materials. Solid–solid first-order transitions represent the main monitoring events (e.g. potassium nitrate and barium carbonate transitions) but two melting transitions (indium and tin) are also included. Certain special treatments need to be carried out before certain materials are used as references, e.g. potassium nitrate must be heated *in situ* to $150–160°C$. The results of inter-laboratory findings on the use of these standards have been described (Langier-Kuzniarowa, 1984). ASTM E474-80 provides a description of the methods of calibration. The use of solid–solid transitions is preferred because some instruments are unable to cope with melting materials and the thermocouples used in some apparatus may form alloys with molten metals. Additionally these transitions involve less change in specific heat, thus avoiding problems with baseline changes, and they are less susceptible to changes induced by impurities.

It must not be assumed that the instrumental readout of temperature either directly reflects the true temperature or is linearly related to true temperatures. Sarge & Cammenga (1985) found that the temperature calibration for the Mettler FP 800/84 heat-flux calorimeter was parabolic over the temperature range -60 to 300°C.

The peak temperature may not correspond to the true melting temperatures because of thermal resistance (R_o) between the sample and the sample holder. Barrell & Johnson (1970) described how allowances were made for this small but in some cases significant error in early DSC instruments. For accurate temperature determination the lag T_p/R_o should be evaluated where T_p is the scanning rate. The lag can be experimentally determined by heating a very pure material, such as semiconductor grade indium, encapsulated in the same manner as the sample and heating at the same rate. The slope of the landing edge of the melting endotherm will be T_p/R_o (Perkin-Elmer, 1967). Such a calibration is very sensitive to heating rate and encapsulation and should these variables be altered the instrument will need calibration at the new settings. The true temperature may be determined from the intersection of T_p/R_o with the true baseline as indicated in Fig. 3.1 where the

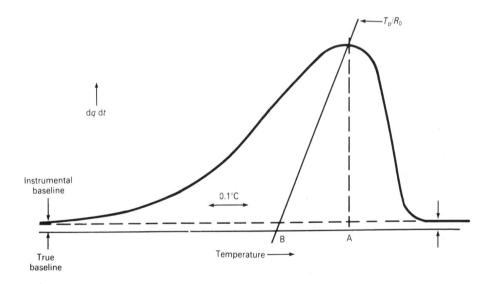

Fig. 3.1 — The use of an indium melting curve to compensate for thermal lag in the melting of triphenylmethane. The line of slope T_p/R_o represents the slope of the leading edge of an indium fusion endotherm (modified with permission from Perkin-Elmer, 1967).

determination of a vertex temperature with the aid of an indium melting curve and the true baseline is shown. The actual melting temperature is B and not A. In most cases the projected scanning baseline is close enough to the isothermal baseline but a severe error is not introduced if it is substituted for the reading at B. A high-purity standard should be used whose melting temperature is close to the material of interest. Additionally the dial readings of early DSC or DTA models were not linearly related to the real temperature because of variation of R_o with temperature, and a somewhat parabolic relationship was often obtained between corrected temperature and instrumental readout. Generally the isothermal baseline is not constructed and the intercept of this imposed slope from the position of interest to the scanning baseline is taken as temperature. R_o may be altered by unclean sample pans, distortion of the bases of the pans and the type of encapsulation used. Therefore the temperature correction used must have been derived from the pan geometry used for the reference. For accurate temperature determination the following should be used: (1) a standard position for the pan in the holder, (2) nitrogen and not helium as the inert flushing gas, (3) a small and constant sample weight, (4) a constant sample geometry, and (5) in the study of organic materials, organic calibrants due to the higher thermal conductivity of metals such as indium.

3.2.2 Calorimetric
Barnes *et al.* (1984) used a combination of quantitative DTA and TG to assess suitable calibrants. TG revealed that organic compounds (e.g. naphthalene, benzil, acetanilide, benzoic acid and diphenylacetic acid) sublimed to some extent prior to fusion. Therefore to obtain accurate results these samples were placed in hermetically sealed sample pans. It should be automatic to weigh such materials after scanning to avoid excessive sample loss. It is essential to calibrate at the temperature range of interest. Calibration constants, relating energy conversion factors at different temperatures, were shown by Barnes *et al.* (1984) to vary linearly with temperature between approximately 25 and 240°C for a Stanton-Redcroft DTA, type 671B. Sarge & Cammenga (1985) found that the calibration factors for heat-flux calorimeters were parabolic over the temperature range -60 to 300°C (Mettler FP 800/84) and -100 to 500°C (Heraeus TA 500) although being relatively independent on heating rate.

Indium is probably the most used calibrant for both DSC and DTA instruments. This is because it is available as a very pure (greater than 99.9999%) material. However, the normally accepted value for its heat of fusion is 6.80 cal/g although Richardson & Savill (1975) obtained a value of 29.2 J/g (7.04 cal/g). McNaughton & Mortimer (1975) considered 5 to 10 mg acceptable as reference size. The calibration constant K in mcal/unit area was determined from the relationship of (3.1)

$$K = \frac{[\Delta H \, (\text{fusion}) \times M_c]}{A_c} \tag{3.1}$$

where ΔH (fusion) is the enthalpy of fusion of calibrant (mcal/mg), M_c is the mass of calibrant in mg and A_c is the peak area of the calibration endotherm. K may be used to convert the area of transitions of unknown energies to specific calorific values provided the sample weight is known.

Calibrants suggested by ASTM D3417-83 and ASTM D3418-82 are benzoic acid (melting point 122.4°C, heat of fusion 142.04 kJ/kg), indium (156.4°C, 28.45 kJ/kg), tin (231.9°C, 59.5 kJ/kg), lead (327.4°C, 22.92 kJ/kg) and zinc (419.5°C, 102.24 kJ/kg).

3.3 ENTHALPY DETERMINATION

The determination of the energy of a transition may be measured by the area under the curve describing that transition, using either DSC or quantititative DTA. The major requirements in measuring heats of transition are accurate weighing and accurate determination of the area under the curve. The former requires accurate balances but the latter, unless dedicated computation is used, becomes somewhat subjective when determined by paper weighing and planimetry.

The problem of deciding what constitutes the precise area has been a thorny problem for many years. Barrall & Johnson (1970) described the types of corrections and baseline construction available to the operator. These baselines may be described as an isothermal baseline and the scanning baseline. For the isothermal baseline technique the line is drawn between isothermal points before and after the transition. These are obtained by allowing the instrument to come to a balance at some isothermal point below the transition temperature to obtain a baseline, then scanning through the transition and, once a temperature well above the transition has been reached, stopping the scanning and running the instrument isothermally to provide the new baseline. If this method is used the normal specific heat or enthalpy change in the sample is integrated into the enthalpy change of the sample from the specific heat.

The scanning baseline approach is the easier to deal with experimentally. The area is constructed from the first sensible departure from the program line of the instrument to the last sensible departure of the endotherm or exotherm from the program line of the instrument. With this method there is a high degree of operator bias. Where the transitions are relatively sharp, i.e. for pure materials, the area is easy to determine. In sharp transitions with little change in specific heat (heat capacity) this is a very simple procedure and provides an accurate interpretation of the transition energy. However, for impure materials the departure temperatures are less easily defined and areas that should have been ascribed to the transitions are neglected, giving lower estimates of the transition energy.

Corrections to the transition energy and the apparent energy lost are described in the chapter on purity determination. However, in broad transitions and those in which there are either a loss of volatile materials, e.g. decomposition reactions, or great changes in the specific heat, alternative corrections must be used in an attempt to define the correct baseline. It should be remembered that the abscissa is in fact a time axis and its conversion to a temperature scale is dependent on uniformity in heating rate. Therefore in isothermal crystallization reactions the area under the curve is still a measure of the heat of crystallization. Since the ordinate of these curves is a power function integration of the curve (power versus time) will give an energy that corresponds to an enthalpy change.

The various methods of deriving these energies were described by Richardson (1980). A baseline may be constructed by connecting the two points at which the endotherm or exotherm deviates from the relatively straight baseline (ASTM D3417-83). Although modern facilities would use a microcomputer to determine the peak area, methods such as paper weighing and planimetry may be used to determine this area. When the baselines before and after transition are different it may be possible to draw a line to join the points where the peak leaves one established baseline and joins another. Brennan *et al.* (1969) also showed that the area can be interpolated by extending both pre-event and post-event baselines until they connect with a line dropped vertically from the transition peak. The rationale behind the respective approaches is that the effective heat capacity changes linearly with the course of the event or changes abruptly at the peak temperature. Neither is likely to be correct. Brennan *et al.* (1969) demonstrated correction methods that may be used to produce the correct baseline. The method is based on the contribution of the reactant and product to the specific heat of the system and the function melted at any given temperature. The baseline is not straight but is sigmoidal in shape. The interested reader is invited to study references such as Guttman & Flynn (1973) who considered the theory behind baseline construction. Amongst graphical methods the method of Laye (1980) was considered the soundest by Richardson (1980) in that it gives heat capacity changes versus temperature. However, without computer facilities the method is tedious. The method is best described with reference to Fig. 3.2.

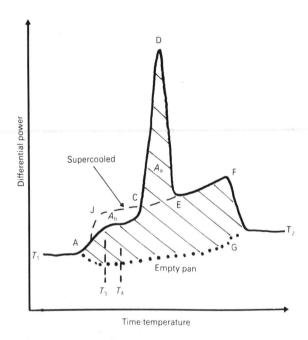

Fig. 3.2 — The use of a supercooled liquid to define an enthalpy of fusion (reproduced with permission from Richardson, 1980).

This represents a sample heated through its melting transition in a poorly set-up instrument which showed baseline curvature and disparity in the isothermal baseline. In these circumstances the estimation of the baseline has little significance but a meaningful baseline may be obtained by returning the sample to the starting temperature T_1 and rescanning through to T_2. Provided that the sample has supercooled rather than recrystallized, the curve indicated by the broken line will be obtained. The shaded area ABDG (termed A_1) is proportional to $H_1(T_2) - H_s(T_1)$ and $H_1(T_2) - H_1(T_1)$ is proportional to the dotted area AJFG (A_2). Equation (3.2) was derived:

$$\Delta H(T_1) = H_1(T_1) - H_s(T_1) = K(A_1 - A_2) = K(A_a - A_b) \tag{3.2}$$

where A_a and A_b are the areas CDE and AJCB respectively in the figure. K is the area-to-enthalpy conversion factor (equation (3.1)). The resultant quantity is $\Delta H(T_1)$ but there is no reason why it cannot be determined from steady-state conditions to define the course of the $\Delta H(T)$ versus T curve which may be extrapolated to T_m (Richardson, 1980).

Although the ordinate is a power differential or a temperature differential in power-compensation and heat-flux DSC instruments respectively, the magnitude is proportional to the difference in heat capacity between the sample and the reference (Richardson, 1988). This so-called capacity includes all reactions giving in or taking in heat. Modern instruments are capable of giving heat capacities accurate to $\pm 1\%$ over the temperature range. Integration of such curves will give enthalpy–temperature curves. These should be referred to a reproducible and well-defined state (Richardson, 1988).

Figs 3.3 and 3.4 illustrate the principle behind the process for 4,4′-ethoxycyanobiphenyl (2.OCB) where the defined state is the isotropic liquid at 390 K. Heats of fusion [$\Delta H(T)$] may be determined from extrapolation of the data for the liquid. More often than not a linear C_p versus temperature relationship exists. A similar form of plot occurs for the solid state and the broken lines indicate how an idealized enthalpy–temperature curve may occur in the absence of pre-melting and instrumental rate effects. In a correctly calibrated DSC the variation caused by differing heating rates would be an apparent displacement in the enthalpy step although the magnitude at the melting temperature would remain unchanged (Richardson, 1988). This is only an apparent displacement because it is due to the finite time required to transfer heat of fusion to the sample.

When determining the energies for rapid exothermic transitions the instantaneous heat flow rate into or out of the sample should not exceed the maximum rating of the differential power measuring system. Relatively large samples of, for example, water may cause problems and O'Neill (1985) suggested that for water the sample should perhaps be as low as 0.8 mg. An alternative solution is to increase the resistance between the sample and holder, thereby reducing the heat flow. This may be achieved by placing an inert thin disc of thermal insulating material between the sample pan and the sample holder. This will not affect the measurement of the energies of the transitions.

Barrall & Johnson (1970) outlined the importance of determining the calibration factor K (equation (3.1)) at several temperatures throughout the temperature range of interest. Differences in scanning speed of less than 2°C/minute were unimportant (Barrall & Johnson, 1970). Neglecting early melting areas and poor weighing were

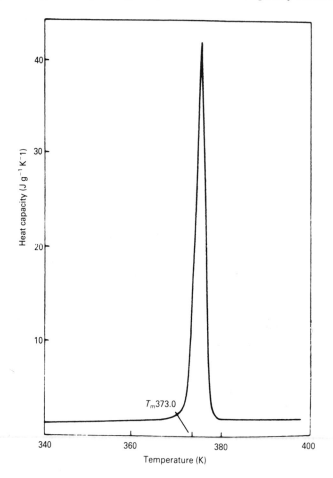

Fig. 3.3 — Heat capacity of solution-grown crystals of 4,4′-ethoxycyanobiphenyl. Heating rate 10 K/minute (reproduced with permission from Richardson, 1988).

also considered as important errors. Small sample sizes were required for materials with large heats of transitions. When gaseous products are formed these must be entrapped in the sample pans otherwise their contribution to heat capacity will be lost. Sublimation of materials may also occur in standard sample pans and may result in an apparent decrease in the heat of melting. Solvent evaporation may cause a drifitng baseline.

3.4 FACTORS INFLUENCING DSC AND DTA SCANS

van Dooren (1980) reviewed some factors affecting DSC curves. On transfer from isothermal mode to the dynamic mode at the start of a scan, the initial baseline deflection depends on the masses used, specific heat of samples, heating rate and instrument sensitivity. Particle size appeared not to affect this deflection (van

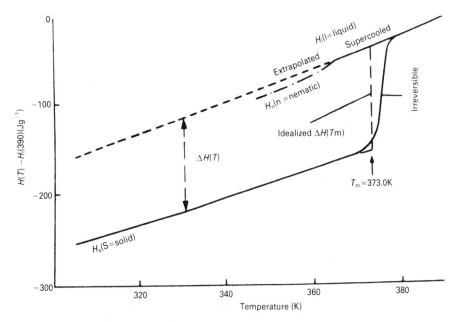

Fig. 3.4 — Enthalpy changes calculated from Fig. 2.3 using H_1 (390 K) as the reference state; the monotropic liquid crystal phase is shown ———, idealized curve ---- (reproduced with permission from Richardson, 1988).

Dooren, 1980). The onset temperature is often dependent on the experimental setting. van Dooren & Muller (1981) considered the most important factors influencing DSC curves as:

(1) The adjustment of the apparatus: calorimetric sensitivity and heating rate.
(2) The sample: its nature and mass, particle size, packing and porosity, pretreatment and dilution.
(3) The reference material: nature, mass and pretreatment.
(4) The atmosphere: oxidizing or inert, thermal conductivity, flowing or static conditions.

This section briefly outlines the importance of some of the factors, classing them as instrument- or sample-controlled.

3.4.1   Instrument-controlled factors

3.4.1.1   Heating rates
The choice of a correct heating rate is of vital importance in ascribing the correct temperature to a transition. Brancone & Ferrari (1966) recognized that, for DTA, rapid heating rates give large sharp peaks but small details may be lost, whereas at slower rates the temperature difference between the sample and reference becomes too small and some transitions may not occur on the scan.

The accuracy and precision anticipated from DSC data may be judged from values obtained by van Dooren (1982) for transition enthalpies and their coefficients of variation (%) for the fusion of adipic acid (255.4 J/g, 1.24%), a solid–state transition in potassium nitrate (50.2 J/g, 1.75%), fusion of naphazoline nitrate (93.5 J/g, 2.32%) and dehydration of sodium citrate dihydrate (372.7 J/g, 4.16%). The greater coefficients of variation for the latter two substances were attributed to decomposition but may be reduced by careful manipulation of heating rate. Additionally the fusion enthalpy of adipic acid was higher at 0.08 K/s than at 0.01 and 0.32 K/s, the differences being due to sublimation and inaccuracies in drawing the baseline at the two heating rates respectively. At the higher rates the peaks became wider and less sharp. van Dooren (1980) found that the enthalpies of transition for potassium nitrate and sodium citrate dihydrate were independent of the heating rate at rates of less than 0.32 K/s.

However, the determined transition temperatures of these transitions depend to a certain extent on the heating rates employed and the properties of the materials under test. Various phenomena accounted for the variations in response to heating rate for each of the transitions. Adipic acid is stable at its melting temperature yet an increase in heating rate resulted in a decrease in the peak temperature (van Dooren, 1982). This indicated a failure of the system to maintain equilibrium and the process was not recorded accurately. When accurate transition temperatures are required the heating rate should be as low as possible. Because naphazoline nitrate melts with decomposition, slow rates increased the amount of degradation prior to melting, resulting in a decreased area due to the evolution of heat, thereby reducing the endothermic peak area. The impurities decreased the peak temperature (van Dooren, 1982).

van Dooren (1982) also examined the solid–solid transformation in potassium nitrate. Both the extrapolated onset temperature and the peak temperature increased with an increase in heating rate. Use may be made of these in determining activation energies (see Chapter 4). The dehydration of sodium citrate dihydrate did not display the same temperature response probably because the mechanisms are governed by the formation and growth of nuclei which usually occur along the lines of strain within the crystal lattice (van Dooren, 1982).

Other data confirm the complexity of heating rate effects. The area of the fusion endotherm of indium is independent of heating rates in the range 1–20°C/min (Barnes *et al.* 1984) but an increase of 5.1 K occurred in the melting transition of polyethylene when the rate was increased from 2.5 to 80°C/min (Harrison & Varnell, 1982). This reflected the problem that, compared to indium, polyethylene is a poor conductor and consequently as the heating rate was increased the temperature differential between the top and bottom of the sample also increased. This was apparent as the increase in the melting temperature but was not observed with small samples.

As a further example of the problems caused by the selection of heating rates, variation in the heating rate from 1 to 15°C/min for tristearin resulted in the loss of whole transitions and a reduction in the melting point by 13°C (Liversidge *et al.* 1982).

A further problem encountered is that a decomposition reaction is often preceded by the melting endotherm. Such problems were reported by ASTM E537-

76. This make resolution of the onset temperature almost impossible. The standard recommends in this case that the heating rate be reduced from 20–30'C/min to 2°C/min.

3.4.1.2 Atmosphere

The atmosphere surrounding the cells, although usually inert, often plays an important role in controlling the shape and size of any transition. When the material under test is liable to oxidative degradation the peak will commence at a higher temperature in an inert atmosphere such as nitrogen (van Dooren, 1980). This was the case for naphazoline nitrate (van Dooren, 1980, 1982). If the atmosphere is air or oxygen, degradation will occur in the solid state and the presence of these impurities will lower the peak temperature and increases the increment between the peak temperature and the onset temperature. Helium as an inert gas also gives problems (van Dooren, 1980) causing a decrease in the enthalpy of transition to about 40% of its original value due to its having a thermal conductivity 5.5 times greater than that of air.

3.4.2 Sample-controlled effects

3.4.2.1 Peak shape

Care should be paid to the peak shape of any transition. Samples which display solid solution formation throughout the whole of the composition range normally give single broad peaks whereas eutectic systems will give double peaks for compositions within the miscibility gap and single peaks only for the terminal solid solutions. Robinson & Scott (1972) pointed out that a small peak may be present at the start of the initial melting endotherm. This may be due to the presence of a small amount of eutectic solid. Conditions favouring double peaks are a wide separation of the solidus and liquidus temperatures, a small sample and a slow scanning speed. Double peaks may occur even in solid solutions and are spurious in that they may be artifacts arising from the shape of the phase boundaries. They do not therefore correspond to phase changes within the sample.

Radomski & Radomska (1982) reviewed the determination of solidus and liquid temperatures by DSC. The types of transitions studied were classed as isothermal or sharp (e.g. melting of pure compounds), non-isothermal or diffusion (e.g. melting of solid solutions) and complex (e.g. melting of binary systems containing some eutectic). For each transition the onset temperature and the peak temperature (in the binary system this corresponds to the second endotherm) were determined. They suggested that the solidus and liquidus temperatures should be determined by DSC and HSM respectively. The use of these independent measurements removed some of the problems caused by the dependence of these temperatures on the heating rates employed in their determination. Radomski & Radomska (1982) developed the theory of how the interval of a sharp transition varies with scanning rates, the melting interval (ΔT_{iso}) being defined as the difference between the peak and onset temperatures. ΔT_{iso} increased linearly with the square root of the scanning rate but did not disappear when extrapolated to zero heating rate when the material was a solid solution or decomposed on heating. Such a plot was suggested as a potential test for purity.

3.4.2.2 Sample encapsulation

The most important factor in obtaining good accurate results from DSC and DTA is proper encapsulation of the sample, coupled with good maintenance of contact of the sample with the bottom of the pan to ensure uniform contact with the temperature sensing device. These all tend to reduce the thermal lag and the thermal resistance of the system. The aim is to immobilize the position of the sample such that movement is kept to a minimum during melting and recrystallization. When the areas of endotherms or exotherms are used quantitatively these must be equivalent to the whole of the sample weighed into the container prior to testing. Any sample loss prior to a transition will reduce this area and therefore volatile samples must be encapsulated in suitable pans such as volatile sample pans. Additionally a suitable metal insert should be placed onto the sample, especially if a liquid, to ensure uniform sample contact with the pan base (Barrall & Dawson, 1974). Samples may also be degassed, and, if unstable or susceptible to oxidative degradation, encapsulated under a nitrogen atmosphere in order to exclude oxygen. Even solid samples should be pressed into position. Barnes *et al.* (1984) suggested flattening the base of sample pans against a glass plate. Careful positioning of the pans on the sensor and replacement of the cell covers and thermostat lid is required to avoid jarring the encapsulated sample out of place (Barrall & Dawson, 1974).

Sealed cells are useful in the analysis of volatile liquids and solutions since they ensure that volatiles are not lost and consequently the latent heat of volatilization will not interfere with other transitions of interest. Freeberg & Alleman (1966) described cells with brass caps and cell bottoms made of stainless steel, aluminium and brass. A removable Teflon disc was used to seal the cell and special wrenches were used to close the cell firmly by use of the screw threads. When uncapped cells are used, problems such as thermal emissivity and change of sample shape complicate interpretation of the curves (Barrall & Johnson, 1970). The former may lead to a more than full-scale recorder deflection with instruments set at the highest sensitivity. Unless unavoidable, sealed pans should always be used. Volatile samples or oxygen-labile samples should ideally be placed in volatile sample pans that have been cold-welded under nitrogen. More reliable results may also be obtained if the sample and reference pans are covered with aluminium domes. This may add some thermal lag to the system but equalizes temperature gradients over the sample.

Care should be taken when the sample is subject to sublimation. Thus van Dooren (1980) found that under vacuum most of a subliming sample (adipic acid) disappeared following DSC treatment. Materials that normally undergo sublimation require encapsulation in volatile sample pans. When sublimation occurs a melting peak may develop a rounded leading edge after a first run in a sealed sample pan (Barnes *et al.* 1984). The pan would be coated with the sample. The reason postulated by Barnes *et al.* (1984) was that the sample would melt at slightly different temperatures due to temperature gradients across the pan and its hermetically sealed lid. Provided no sample is lost the peak area would, however, be unchanged.

3.4.2.3 Sample handling

ASTM D3418-82 gives some procedures to be followed in the handling of polymers prior to DSC or DTA study. Grinding should be avoided in order to prevent changes in the crystal order. Polymeric samples could be cut into thin slices using a microtome, razor blade, hypodermic punch, paper punch or cork borer. Film sheets could be cut into slivers. The sample should be carefully packed to maintain intimate thermal contact between the sample and its container. Crimping the sample into the container assures firm contact of the sample with the container and may also induce a flattening of the sample. The addition of a pan lid above the sample ensures better thermal heat transfer. The pan bottoms should be flat. The effect of differences in sample size, e.g. between granular and powdered, were recognized as long ago as 1964 when Watson *et al.* showed that a granular sample of bismuth produced a broad transition whereas a flat sample gave a relatively sharp endotherm. Poor heat transfer from the granules gives an irregular curve shape. It is not uncommon in DSC for lumps of samples when encapsulated to give two, almost separate, transitions. Movement of the partially molten material may contribute to these discrepancies. It should be emphasized that such occurrences generally do not influence the determined heats of transition but make estimation of the true melting point difficult.

3.4.2.4 Sample size

van Dooren (1980) considered that determined enthalpies were independent of mass size provided it exceeded 0.2 mg. Below this value low values of enthalpies were determined. Radomski & Radomska (1982) suggested that for power-compensation DSC the sample size should be ideally 5 to 10 mg. For samples smaller then 5 mg which are also impure a large proportion of the sample may have melted before deviation from the baseline is observed. Corrections for this are described in Chapter 5.

In assessing the use of indium as a calibrant Barnes *et al.* (1984) found a linear relationship between peak area and sample weight for up to a maximum weight of 100 mg. In the DTA of tristearin, sample size played a small but significant role in the appearance of the scans (Liversidge *et al.*, 1982). Increasing the sample mass from 2 to 10 mg resulted in the loss of one exotherm and an increase in peak temperature of an endotherm from 32.7 to 35.8°C. This indicates that when several samples of material are to be analysed an optimum sample size or constant sample size should be used.

3.4.2.5 Particle size

Again the study of van Dooren (1982) is important to our understanding of the influence of particle size on DSC and DTA scans. Although no precise determination of the particle size was made in this study the trends observed have vast implications for results obtained from different particle size fractions of materials. Large particles of adipic acid gave a decrease in onset temperature, an increase in peak temperature, a decrease in fusion enthalpy and an increase in the differences between the onset temperature and both the extrapolated onset and peak temperatures (van Dooren, 1982). Such changes were explained as heat transfer problems. The transfer between large particles is less and therefore there was a reduction in the thermal effect recorded and a decrease in the peak area. The temperature difference between the hot and cold sides of the bed became greater with the result that the hotter side melted and released its energy to other particles. The net effect was to lower the onset temperature and increase the onset interval and total transition interval (van Dooren, 1982). For unstable samples, such as naphazoline nitrate (van Dooren, 1982), which are sensitive to oxidation that starts in the solid phase, smaller particles displayed a lower peak temperature and a larger differential between the onset temperature and the peak temperature due to the increase in impurity during heating. For solid–solid transitions, such as in potassium nitrate, the predicted effects are difficult to quantify (van Dooren, 1982). This may be due to two counteracting phenomena. Grinding leads to a reduction in particle size inducing lattice strain reducing peak area and shifting temperatures to lower values. However, in larger particles the thermal lags will be greater. Each particle separately contributes to the total thermal effect and may transform at slightly different temperatures (van Dooren, 1982). Due to the lags the peaks become wider which means that either the peak temperature is increased or the onset temperature is decreased.

3.4.2.6   Sample shape
The actual position and shape of the sample within the pan importantly influences the DSC or DTA curve shape. Harrison & Varnell (1982) examined the effect of sample geometry on the melting point of polyethylene. The work outlined some important factors which should always be considered. For instance the fusion peak temperature, increased by 1.7 K when the sample thickness was increased from 1 to 8 μm when a constant weight of 0.05 mg was used. A chunk sample of 1.17 mg displayed a peak temperature some 7.5 K above that of the 1 μm sample. Harrison & Varnell (1982) considered that provided the sample thickness is kept constant then weight should not effect the peak position. Obviously, however, the problems of low sample weight, such as poor discrimination of onset temperature and possible superheating in polymeric samples may be heightened by small sample size.

3.5   THERMOGRAVIMETRY (TG)
The successful calibration of TG is more difficult to obtain than for DTA and DSC. The problem in temperature calibration is that weight loss is often not a discrete process but occurs over a wide range of temperature. Additionally it is difficult to obtain a reliable and repeatable weight loss. Langier-Kuzniarowa (1984) outlined the development of suitable standards. Originally organic and inorganic materials were selected and then the thermal decomposition of carbonates and oxalates was used. Nowadays the basis is the Currie point of ferromagnetic materials. The temperature point of the magnetic transition measured in the TG curve is not the exact Currie point and is termed the magnetic reference point. The principle of the magenetic transitions is that the references are heated under an imposed magnetic field and the magnetic properties are lost at a very narrow temperature range. This will induce an apparent weight sample loss or gain depending on the position relative to the imposed magnetic field. This temperature is the Currie temperature. Accuracies as high as $\pm 4^\circ$C have been claimed. Concurrently the precise temperature is

determined with a temperature sensor which compares its temperatures with that recorded during TG to detect any possible systematic error.

The ICTA has certified a set of references for TG which are Permanorm 3, nickel, Mumetal, Permanorm 5 and Trafoperm (NBS-ICTA GM 761). Garn *et al.*, (1981) highlighted the problem of comparing interlaboratory data. Several models of the same instrument gave spans of measured temperatures from 3 to 15°C whereas when 18 instruments were compared the range became 17–39°C. Therefore calibration references are essential for absolute measurements and comparison. The interested reader will find the work of Garn *et al.* (1981) essential reading. TG may be used to determine the Currie point and crystallization temperatures of many amorphous glassy metals and alloys.

Alternative methods of calibration involve the use of weights connected to the TG arm by low melting point materials. The method is fully explained by McGhie *et al.* (1983) and utilizes a small platinum weight mounted on the TG apparatus by a small fusible link made of the calibration materials usually used for DSC or DTA calibration. As the temperature rises during the programmed heating the link will melt causing the weight to drop. If this is collected on the boat it will register as a blip in the instrument readout. However, if the boat has a hole placed in it to allow the weight to pass through the apparatus the weight loss will be monitored as a discrete weight loss. Calibration to within $\pm 2°C$ may be achieved by the method. The problem is that temperature calibration may be made at distances varying from millimetres to centimetres from the test specimen depending on the style of the apparatus (McGhie *et al.*, 1983).

Usually TG does not require highly accurate calibration and its use in decomposition kinetics only requires accurate temperature calibration when the reaction rate constants, and the pre-exponential factors are described (see Chapter 4). The activation energy does not need an accurate temperature calibration.

3.6 DERIVATIVE THERMAL ANALYSIS

The use of microprocessors in thermal analysis has advanced considerably the determination of energy changes and temperature changes in thermal analysis. The advantages that derivative data treatment gives include improved accuracy, improved peak resolution and quantitative determinations. Specifically Lever (1987) outlined that enhancement of resolution for single and overlapping peaks may be determined and that the odd derivative curves (1st, 3rd, 5th, etc.) are useful in evaluating second-order transactions such as the glass transition. The peak of the first derivative curve gives far more accurate determination of the T_g than more conventional methods. The second derivative curve may be used to determine the beginning and end of the glass transition. Derivative data is exceptionally useful for analysis of DMA and TMA data.

REFERENCES

A.S.T.M E474-80 (1980) *Evaluation of temperature scale for differential thermal analysis*, 412–414.

A.S.T.M. E537-76 (1976) *Assessing the thermal stability of chemicals by methods of differential thermal analysis*, 431–435.

A.S.T.M. D3417-83 (1983) *Heats of fusion and crystallization of polymers by thermal analysis*, 107–111.

A.S.T.M. D3418-82 (1982) *Transition temperatures of polymers by thermal analysis*, 112–117.

Barnes, P. A., Charsley, E. L., Rumsey, J. A. & Warrington, S. B. (1984) *Anal. Proc.*, **21**, 5-7.

Barrall, E. M. & Dawson, B. (1974) *Thermochim. Acta.* **8**, 83–92.

Barrall, E. M. & Johnson, J. F. (1970) *Technical methods of polymer*, **2**, 1–39.

Brancone, L. M. & Ferrari, H. J. (1966) *Microchem. J.*, **10**, 370–392.

Brennan, W. P., Miller, B. & Whitwell, J. C. (1969) *I. & E. C. Fundamentals*, **8**, 314–318.

Charsley, E. L. (1980) *Anal. Proc.*, **17**, 223–226.

van Dooren, A. A. (1980) *Thermal analysis, vol 1: theory and instrumentation, applied sciences, industrial applications*, Birkhauser Verlag, Basel, 93–98.

van Dooren, A. A. (1982) *Anal. Proc.*, **19**, 554–556.

van Dooren, A. A. & Muller, B. W. (1981) *Thermochim. Acta*, **49**, 151–161.

Freeberg, F. E. & Alleman, T. G. (1966) *Anal. Chem.*, **38**, 1806–1807.

Garn, P. D., Menis, O. & Wiedemann, H. G. (1981) *J. Therm. Anal.*, **20**, 185–204.

Guttman, C. M. & Flynn, J. H. (1973) *Anal. Chem.*, **45**, 408–410.

Harrison, I. R. & Varnell, W. D. (1982) *J. Therm. Anal.*, **25**, 391–397.

Kambe, H., Horie, K. & Suzuki, T. (1972) *J. Therm. Anal.*, **4**, 461–469.

Langier–Kuzniarowa, A. (1984) *J. Therm. Anal.*, **29**, 913–918.

Laye, P. G. (1980) *Anal. Proc.*, **17**, 226–228.

Lever, T. (1987) *Lab. Practice*, **38**(8), 17–18.

Liversidge, G. G., Grant, D. J. W. & Padfield, J. M. (1982) *Anal. Proc.*, **19**, 549–553.

McGhie, A. R., Chiu, J., Fair, P. G. & Blaine, R. L. (1983) *Thermochim. Acta*, **67**, 241–250.

McNaughton, J. L. & Mortimer, C. T. (1975) *Differential Scanning Calorimetry* Perkin-Elmer Corp., Norwalk, Connecticut, U.S.A.

O'Neill, M. J. (1985) *Anal. Chem.*, **57**, 2005–2007.

Perkin-Elmer Thermal Analysis Newsletter TAN-5 (1967) Determination of purity by Differential Scanning Calorimetry. Perkin-Elmer Corp., Norwalk, Connecticut.

Radomski, R. & Radomska, M. (1982) *J. Therm. Anal.*, **24**, 101–109.

Richardson, M. J. (1980) *Anal. Proc.*, **17**, 228–231.

Richardson, M. J. (1988) *Anal. Proc.*, **25**, 16–18.

Richardson, M. J. & Savill, N. G. (1975) *Thermochim. Acta*, **12**, 221–226.

Robinson, P. M. & Scott, H. G. (1972) *Nat. Phys. Sci.*, **238**, 14–15.

Sarge, S. & Cammenga, H. K. (1985) *Thermochim. Acta*, **94**, 17–31.

Watson, E. S., O'Neill, M. J., Justin, J. & Brenner, N. (1964) *Anal. Chem.*, **36**, 1233–1238.

4

Thermal analysis in reaction and decomposition kinetics

The aim of this chapter is to introduce some of the concepts involved in the assessment of stability, reaction and decomposition kinetics by thermal methods of analysis. The methods most suited are DTA, DSC and TG. The major limitation is that where parallel reactions occur these will not be detected without the aid of a form of evolved gas analysis, e.g. mass spectroscopy.

Li Wan Po (1986) reviewed the application of thermal analysis to stability determinations. Despite the plethora of information concerning the stability and degradation kinetics of inorganic salts, the information concerning pharmaceuticals is very limited. Li Wan Po (1986) considered that most DSC or quantitative DTA reports on stability were at best semi-quantitative. This was considered to be due in part to the decomposition of drugs needing an external agent, e.g. moisture or oxygen, rather than being a spontaneous process. The requirement to modify the DSC or DTA cell accordingly is usually deemed too expensive to contemplate and other methods are sought. Special hermetically sealed cells may be constructed but there are probably too many other variables even then to allow easy analysis of the results. The potential advantage of thermal methods still lies in the potential of calculating stability data from a single run. Nevertheless DSC and DTA runs can be used to assess stability in certain systems. Weight loss via either isothermal or non-isothermal TG testing is somewhat easier to quantify. The use of thermal analysis in screening for incompatability is discussed in Chapter 10.

4.1 KINETIC ANALYSIS AND TREATMENT OF DSC AND DTA DATA

The type of degradation reaction is a vital clue in assessing stability: if degradation proceeds by endothermic processes the drug or chemical is reasonably stable (Hardy, 1982). Exothermic degradation represents instability that could be analysed by the ASTM E698-79 method which is attractive in that stability may be estimated from a single determination without the recourse to other methods such as high-pressure liquid chromatography (HPLC) or gas–liquid chromatography (GLC).

The kinetic analysis of the stability of a drug is usually related to the Arrhenius equation (equation (4.1))

$$k = Z e^{-E/RT}$$
(4.1)

where k is a specific rate constant at a temperature T (K), Z is the Arrhenius frequency or pre-exponential factor, E is the activation energy and R is the gas constant. It is generally assumed that k is derived from first-order kinetics and at isothermal temperatures. However many solids decay by non-first-order reactions and the thermal data need careful mathematical manipulation before accurate and reliable interpretation may be made.

The sources of error in kinetic analysis of thermal analytical data (Sestak, 1980) include (a) the varying conditions of measurement due to incorrect temperature and pressure determination; (b) undefined sample preparation yielding different fractions of solid compacts, with concomitant inconsistancies in shape and size distribution, concentration profiles and structural defects; (c) the inappropriate choice of mathematical modelling of the process in question and the consequent treatment of the kinetic equation; (d) the possible inadequacy of the Arrhenius law in describing the solid-state reactions and invalid assumptions of the energy distribution; and (e) mathematical reasons hidden in the entire form of the correlation employed.

Many workers assume that the pre-exponential factor, the order of the reaction and the activation energy do not change during the course of a reaction. Gyulai & Greenhow (1974) pointed out that the methods where the three functions were treated as constants were less successful at curve interpretation than when at least one of the factors was considered variable during the course of a reaction. The method of Ozawa (1970) was considered to be the most reliable method by Gyulai & Greenhow (1974). Various approaches have been used to combat the problem and Petty *et al.* (1977) described methods to determine the temperature-dependence of the pre-exponential factor of the approaches of Coats & Redfern (1964) and Freeman & Carroll (1958). Torfs *et al.* (1984) claimed that a single temperature-programmed DSC experiment can be used to compute the Arrhenius pre-exponential factor, activation energy and the order of the reaction within a few hours.

Because there are many methods of analysing kinetic data from TG, DSC or DTA curves this chapter only introduces some of those methods which have potential interest to the pharmaceutical formulator.

There are several methods which may be used to estimate stability characteristics, including isothermal methods and determining reaction kinetics from either a single scan or from a number of scans derived at different heating rates. Not all kinetic approaches are equally reliable (Duswalt, 1974) and no single method is satisfactory for all reactions. However an isothermal run is an absolute requirement to confirm the findings of results obtained by dynamic methodologies. Discrepancies occur in the literature describing results from DSC analysis. These have been attributed (Torfs *et al.*, 1984) to the approximations inherent in the assumptions used to calculate the reaction constants, the incidence of side-reactions and phenomena such as boiling and melting giving different thermal properties to the material under examination. Care should be taken when using the variable heating-rate methods since underestimates of the activation energy by as much as 10 to 25% may occur (Torfs *et al.*, 1984). This was attributed to the strong dependence of the activation energy value on the small variations in measured peak temperature values.

4.1.1 Method of Borchardt & Daniels (1957)

This method makes several assumptions in the derivation of kinetic data including (a) the heat transfer coefficients for the sample and reference are identical, (b) the temperature is uniform throughout the reference and sample cells, (c) the heat capacity of the sample remains identical with that of the reference during the reaction and (d) the reaction is endothermic. Equation (4.2) was derived:

$$k = \left(\frac{VKA}{x_0}\right)^{n-1} \frac{C_p d(\Delta T)/dt + K\Delta T}{[K(A-a) - C_p\Delta T]^n} \tag{4.2}$$

where K is the heat transfer coefficient of the reactants, n is the order of reaction, x_0 is the initial number of moles of the reactant in volume V, A is the area of the endotherm and a is the area under the endotherm from time 0 to time t. C_p is the heat capacity of the sample and the reference, $d(\Delta T)/dt$ is the slope of the endotherm and ΔT the height at time t. Equation (4.2) may be simplified for first-order kinetics but for solids there is little justification for doing this (Pope, 1980). A plot of $\log K$ against $1/T$ may be drawn for several values of n and the one which gives the straight-line plot would give the correct value of n. The so-called energy of activation may be derived directly from the slope. The problems in justifying the reaction kinetics for solids (Pope, 1980) include (a) the thermal conductivity and capacity of the sample will change for a solid during the course of a chemical reaction; (b) during the uniform increase in temperature while scanning, a temperature gradient will develop between the surface and interior of the sample; (c) heat given out during the course of the reaction will further exaggerate this differential, thereby modifying the reaction rate; and (d) the reaction will probably occur at the surface of the solid.

The reaction within a solid proceeds in three stages (Pope, 1980). At a fraction reacted (α) of less than 0.05, nucleation and initiation of the new phase predominates with the nuclei merging, to form a continuous reaction interphase. At $0.01 < \alpha < 0.9$ a contracting interphase provides the basis of the experimental data. At $\alpha > 0.9$ diffusion of gases through the bed becomes rate-determining. Pope (1980) additionally interpreted values of n. A value of 0 corresponds to a reaction occurring over the surface of a uniform plane with no contraction of the reaction interface, a value of 0.5 is consistent with a disc contracting inwards at the circumference only, a value of 0.66 is the figure for a contracting cube or sphere where the reaction interface is advancing at a constant rate through the crystal planes, and a value of 1 equates to a reaction occuring in a liquid phase as when the sample is above its melting point or when a low melting eutectic point is formed. Only a value of 0.66 can be satisfactorily explained and occurs, for reacting solids, as a simple model. It is not acceptable to assign to n an arbitrary rate with our current state of knowledge (Pope, 1980). An additional criticism is that E should only be stated when the reaction order is known and this is unlikely for solids. Therefore estimates of E are only a measure of the slope.

The method of Borchardt & Daniels (1957) is a heat evolution method and allows the determination of the Arrhenius factors in equation (4.1) from a single DSC scan of an exothermic reaction. The principle uses equation (4.3).

$$\frac{d\alpha}{dt} = k(1-\alpha)^n \tag{4.3}$$

The DSC exotherm measures two parameters, enabling equations (4.1) and (4.3) to be solved (DuPont, undated a). The reaction rate $d\alpha/dt$ may be obtained by dividing the peak height at temperature T by the total peak area and the fraction unreacted $(1-\alpha)$ may be calculated by measuring the ratio of the partial area at temperature T to the total peak area to obtain α and then subtracting this value from 1. The information derived for exothermic reactions was useful for thermoset curing, chemical decomposition and thermoplastic polymerization. The method as described by Borchardt & Daniels (1957) was devised for solutions and later modified for solids. An accuracy in estimating the Arrhenius constant of 0.5 to 1 Kcal/mol was claimed (Duswalt, 1974).

On a DSC curve at any particular temperature the rate of heat change ($\delta H/\delta t$) may be assessed by the height above the extrapolated baseline, and the area of this peak up to portion (a) calculated. For a first-order reaction the reaction rate k may be determined from equation (4.4), which is derived from equation (4.3) from a knowledge of the total area A of the exotherm. Additionally the activation energy E may be determined from equation (4.5).

$$k = (\delta\alpha/\delta t)/(1-\alpha)^n = (\delta H/\delta t)/(A-a) \tag{4.4}$$

$$\delta\log k/\delta(1/T) = E/2.303R \tag{4.5}$$

The presence of small reaction endotherms or exotherms on the main reaction curve will give rise to errors. Ideally the mechanism of the reaction is known or has been correctly identified and then correct determination of E and Z can be made from a single DTA scan. To gain an insight into the complexities that may be encountered in the determination of kinetics, equation (4.6), a general form of equation (4.3), may be seen to hold,

$$f(\alpha) = k(T)t \tag{4.6}$$

Substitution into the Arrhenius equation gives equation (4.7)

$$f(\alpha) = Ze^{-E/RT}t \tag{4.7}$$

The $f(\alpha)$ term may be very complex (Brown & Phillpotts, 1978). Based on orders of reaction, $-\ln(1-\alpha)$, $1/(1-\alpha)$ and $[1/(1-\alpha)]^2$ are the function terms for first-, second- and third-order reactions respectively. For one- and two-dimensional diffusional models the dependencies are α^2 and $[(1-\alpha)\ln(1-\alpha)+\alpha]$ respectively and for a contracting-volume model the dependency becomes $1-(1-\alpha)^{1/3}$. These and many more complicated models may be required to describe reaction kinetics in the solid state. No method exists of predicting which model will describe the degradation of interest.

Torfs *et al.* (1984) modified the method to describe chemical reactions using heat-flux DSC where the heat flow signal was considered a reasonable approximation when small samples (less than 20 mg) and heating rates (not more than 20°C/min) were used. Under these conditions the instantaneous reaction rate can be derived from the measured heat flow. Thus any reaction rate k may be calculated at temperature T from equation (4.8)

$$k = \frac{dH}{dt} \frac{1}{\Delta H_{tot}(\Delta H_{rest}/\Delta H_{tot})^n} \tag{4.8}$$

where dH/dt is the heat flow deviation from the baseline, ΔH_{tot} is the total heat evolved from the reaction, ΔH_{rest} is the reaction heat evolved above a certain temperature T and n is the order of the reaction. The method of analysis consisted of calculating the data using the values of $n=0, 1, 2$ and 3 plotting as the Arrhenius plot. The nearest to the true reaction order should give a linear plot. Choosing the most nearly linear curve is a difficult process and acknowledges that no one curve may really describe the reaction order. Since all the kinetic determinants are interrelated, each will vary with the chosen reaction order and Torfs et al. (1984) suggested that determination of the reaction order is best done at a series of conditions. Following order assessment, the Arrhenius plot could be made and high and low values of reaction rates discarded until the correlation coefficient reaches a maximum. Discarding further results will decrease the coefficient due to the increased error caused by instrumental noise.

4.1.2 Method of Kissinger (1957)
Kissinger (1957) determined reaction kinetics by DTA using the temperature corresponding to maximum peak height. Reactions of the type solid → solid + gas were considered. If the reaction rate varies with temperature then the peak temperature will vary with heating rate. Kissinger (1957) presented theoretical considerations to demonstrate the validity of the assumption that the temperature of maximum deflection corresponded to the temperature at which the reaction rate was maximum. Reactions examined were described by equation (4.9)

$$d\alpha/dt = Z(1-\alpha)^n e^{-E/RT} \tag{4.9}$$

Usually $n \leqslant 1$ and does not vary during most of the course of the reaction. During scanning $d\alpha/dt$ will increase to a maximum value and then return to zero when the reactants have been exhausted. This maximum rate will occur when $d/dt(d\alpha/dt)$ is zero. If the temperature rise (heating rate, ϕ) is constant, then differentiation of equation (4.9) gives equation (4.10)

$$\frac{d}{dt}\frac{d\alpha}{dt} = \frac{d\alpha}{dt}\frac{E\phi}{RT^2} - Zn(1-\alpha)^{n-1}.e^{-E/RT} \tag{4.10}$$

As the maximum rate occurs at temperature T_m, this may be defined by setting equation (4.10) to zero to produce equation (4.11).

$$\frac{E\phi}{RT_m^2} = Zn(1-\alpha)_m^{n-1} e^{-E/RT_m} \tag{4.11}$$

The temperature T_m corresponds to the peak of the DTA scan deflection. The amount unreacted $(1-\alpha)$ cannot be readily determined from the DTA trace. Equation (4.9) may also be integrated and by substitution yields equation (4.12)

$$\frac{d(\ln \phi/T_m^2)}{d(1/T)} = -E/R \qquad (4.12)$$

This equation is independent of the reaction order. Thus a plot of $\ln\phi/T_m^2$ against $1/T$ will yield a straight line plot of slope $-E/R$ enabling the determination of E. The amount of unreacted product $(1-\alpha)$ at T_m decreased as n decreased and the DTA peak became more assymetric (Kissinger, 1957). This gives rise to the concept that the peak symmetry can be used to evaluate n, provided the peak shape is independent of heating rate and kinetic constants. Fig. 4.1 gives typical peak shapes at

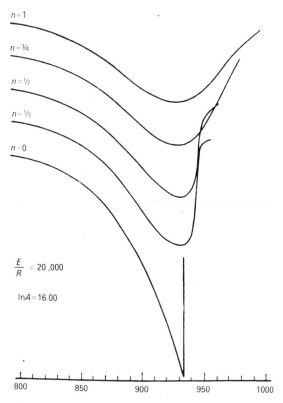

Fig. 4.1 — Effect of order of reaction (n) on plots of reaction rate versus temperature for constant heating rate, frequency factor and activation energy. Reprinted with permission from Kissinger, H. E., *Anal. Chem.*, **29**, 1702–1706. Copyright (1957) American Chemical Society.

different heating rates. A shape index (S) may be used to quantify the values of the pre- and post-peak slopes and is defined as the absolute value of the ratio of the gradients of the tangents of the curves at the inflection points. Fig. 4.2 illustrates the

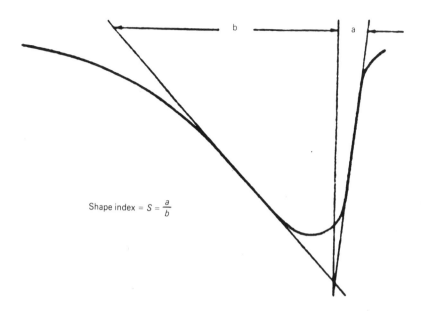

Fig. 4.2 — Method for measuring amount of asymmetry in an endothermic differential thermal analysis peak. Reprinted with permission from Kissinger, H. E., *Anal. Chem.*, **29**, 1702–1706. Copyright (1957) American Chemical Society.

derivation of such a ratio. The derived relationship between n and S is given in equation (4.13)

$$n = 1.26\, S^{1/2} \tag{4.13}$$

Several reservations have been expressed concerning Kissinger's method. The proposed straight-line relation may deviate at low heating rates (Anderson *et al.*, 1977) and hence low values of T_m vary from the theoretical curves. This resulted in a suggested $\simeq 6\%$ underestimation in the activation energy. Another problem is that the simple equation representing the decomposition of the solid is an oversimplification of a multistage process. The stages involved are the chemical act of breaking bonds, destruction of the initial crystal lattice, formation of the crystal lattice of the decomposition product, adsorption and desorption of the gaseous product and its diffusion, and finally heat transfer (Zsako, 1968). Taking pyrolysis as an example it is possible that the rate-determining stage at the start of the process is replaced by one of the other stages as the process proceeds. Instrumental design and sample size are other factors that might limit the rate of the reaction. Given the process complexity it is unlikely that a simple equation could describe the whole of the reaction.

4.1.3 Method of Piloyan *et al.* (1966)
This method derives activation energies from a single DTA scan. In the initial stages of a reaction the deviation from the baseline is described by the equation (4.14)

$$\Delta t = S \, d\alpha/dt \tag{4.14}$$

where Δt is the deviation from the baseline, S is the area of the thermal effect (s °C). The rate of a chemical reaction is described by the equation (4.15)

$$d\alpha/dt = A_0 f(\alpha) e^{-E/RT} \tag{4.15}$$

where A_0 is a constant, and $f(\alpha)$ is some function of the extent of the reaction. It is often thought that this is equivalent to $f(a) = (1-\alpha)^n$ (e.g. equation (4.9)) but Piloyan et al. (1966) considered a better approximation would be $f(\alpha) = \alpha^m (1-\alpha)^n$ where n and m are constants. By substituting equation (4.14) into equation (4.15) and taking logarithms, equation (4.16) may be produced:

$$\ln \Delta t = C - \ln f(\alpha) - E/RT \tag{4.16}$$

where C is a constant and combines the constants of equations (4.14) and (4.15). Under isothermal conditions the only variable is α, but for values of α in the range 0.05 to 0.8 at the heating rates 10–40°C/min the change in heating rates has a greater effect on Δt than the change in α. In equation (4.16) the term $\ln f(\alpha)$ can be neglected and the equation rewritten as equation (4.17)

$$\ln \Delta t = C - E/RT \tag{4.17}$$

Values of Δt may lead directly off the DTA scan. The data points usually considered were from the point of maximum curvature to the exothermic peak. Anderson et al. (1977) considered that the method tended to underestimate the activation energy because of negative deviations from the theoretical curve at low heating rates. The errors were greater than those derived at similar rates by Kissinger's method.

4.1.4   ASTM E698 method

The method examines exothermic peak maximum temperatures at different heating rates and is based on the method of Ozawa (1970) and modified by Duswalt (1974). Ozawa (1970) considered that the method was applicable to reactions which may be described by a single activation energy and described two methods, one using the linear relation between peak temperature and heating rate to estimate the activation energy. The frequency factor of the Arrhenius plot (equation (4.1)) was derived from a plot of $\log \phi/T_m^2$ and $1/T_m$ (see equation (4.20)). The other method requires information both on the conversion and the rate of conversion. The Arrhenius plot is then made on the basis of assumed reaction kinetics. Three assumptions were made in the determinations (DuPont, undated b) namely (a) the peak maximum represents a point of constant conversion for each heating rate, (b) the temperature dependence of the reaction rate constant obeys the Arrhenius equation and (c) the reaction is a first-order reaction.

Equations, including equation (4.1), relevant to this method include equation (4.18)

$$E \simeq -2.19R . \frac{d \log \phi}{d(1/T_m)} \tag{4.18}$$

The activation energy can be obtained within $\pm 3\%$ (Duswalt, 1974) by measuring the reaction peak maximum temperature as a function of the heating rate and plotting $\log \phi$ versus $1/T_m$. Materials that undergo isomerization at reaction temperatures and which decompose on melting are unsuitable for this type of analysis. Isothermal studies would prove successful for these types of materials. The experimental uncertainty in the slope was considered to be less than 5% (Cassel, 1979).

Because first-order kinetics are assumed, the first-order derivative of equation 4.3 is used to describe the reaction, i.e. equation (4.19)

$$\frac{d\alpha}{dt} = k(1 - \alpha) \tag{4.19}$$

and the frequency factor may be determined from equation (4.20)

$$Z = \frac{\phi E e^{E/RT}}{RT^2} \tag{4.20}$$

where T and ϕ are determined from a mid-range heating rate. This equation is considerably less complicated than the order-independent methods of Ozawa (1970) but the agreement was reckoned by Duswalt (1974) to be in the order $\pm 2\%$. No further measurements were required to effect determination of Z other than those required for equation (4.18).

It will be noticed that this predicts a similar relationship to that shown by Kissinger (1957). Ozawa (1970) stated that the height of the temperature difference from the baseline is not proportional to the rate of the process in DTA. Kissinger considered that the height was proportional to the rate according to the limitations that the temperature measured is a function of the time, the position of the temperature measurement and the rate of conversion, but Ozawa (1970) considered that it was also a function of the conversion.

The ASTM E698 method is used for exotherms obtained by either DTA or DSC. The method is unsuitable when (a) deviation from straight-line plots occurs, (b) disagreement with the isothermal test occurs, (c) the reactions are inhibited, (d) reactions have simultaneous or consecutive reaction steps, or (e) reactants display transitions at or near the reaction temperature.

The essential stages of the ASTM method may be summarized as (1) recording the peaks of any reaction exotherms when the sample temperature is increased at a linear rate, (2) repeating at various reaction temperatures between 1 and 20K/min and (3) plotting peak maximum temperatures against the heating rates at which they are recorded.

ASTM E698 suggests that the sample should be small enough to minimize temperature gradients within the sample — a sample weight resulting in the generation of less than 8 mJ/s is satisfactory — good thermal contact should be maintained between the sample and the container, the sample should be representative of the test material and the container should not react with the sample. Volatile samples should be sealed in hermetically sealed pans to prevent volatilization and the sample atmosphere should represent the conditions of usage. The standard suggests that the instrument is able to provide different atmospheres to the samples. Good conductivity between the sample and container may be accomplished by the insertion of an inert metal disk.

Ozawa (1970) obtained an approximate value of the activation energy (E) using equation (4.18). Once this value has been obtained it is refined with reference to calculated tables (ASTM E698). The pre-exponential factor may be calculated using equation (4.20). The equations basically predict that the peak temperature (T_m) will decrease (indicating a slower rate of reaction) as the scan speed is lowered. ASTM E698 refines E by calculating E/RT approximately and finding a corresponding value of D in equation (4.21) from provided tables. D may be defined as $=-\mathrm{d}\ln p(x)/\mathrm{d}x$ where $p(x) = (x+2)^{-1}(x^{-1})(e^{-x})$. Thus a new value of E can be calculated from equation (4.21).

$$E=(-2.303\,R/D)[d\log \phi/\mathrm{d}(1/T)] \qquad (4.21)$$

The value of E may be similarly refined several times, although the standard infers that a second correction will allow determination of E close to its actual value. The value of Z may be calculated from equation (4.20) using a value of ϕ in the middle of the heating range.

The ASTM E698 method also includes a quick check of the validity of the derived data. A sample of the compound under test may be weighed and held isothermally at the temperature used for its half-life ($t_{1/2}$) determination for a period equivalent to $t_{1/2}$. This involves prediction, using equation (4.1), of the reaction rate at a temperature lower than that examined by experimental analysis or at a temperature not examined experimentally and requiring ageing for at least one hour. This rate (k_{calc}) can then be used to determine the reaction half-life ($t_{1/2}$) via equation (4.22)

$$t_{1/2}=0.693/k_{calc} \qquad (4.22)$$

The area under the curve of the sample as it is subsequently scanned through its melting point can readily be determined and should be equivalent to half the area of a sample of identical weight and not subjected to partial decomposition. If the aged sample possesses half the area the kinetic values are confirmed. The isothermal hold should be followed by quenching the sample at least $50\,K$ below the ageing temperature. If the area of the sample is greater than 50% of the area of the undecomposed sample the material is less reactive at lower temperatures than predicted (Cassel, 1979) and indicates that the degradation pathways were different at the lower temperatures. Therefore the method may well be deemed unsuitable.

Instrumental capabilities are important (Cassel, 1979). The instrument should be capable of accurate and non-fluctuating isothermal hold. A temperature difference of 1°C between the theoretical and actual temperatures will produce a difference of

13% in the calculated half-life of the product. The instrument should be capable of controlling the temperature during the exothermic stage of the output otherwise higher temperatures than anticipated occur and over-estimations of the rate of the reaction will result. Therefore any heat generated should be rapidly dissipated and the instrument should have a large heat sink to dissipate the heat or alternatively smaller sample sizes should be used. Heat-flux DSC machines are considered more susceptible to this error than power-compensation DSC (Cassel, 1979). Lastly accurate instrumental temperature calibration is required. Instruments capable of high resolution at lower scanning speeds should be used (Cassel, 1979) because it is these slow heating rate data that are particularly important in extrapolation when predicting stability at lower temperatures than those studied.

   ASTM E698 also utilizes the method of Kissinger (1957) to determine the reaction constants. Thus a plot of $-\ln(\phi/T^2)$ versus $1/T$ may be made giving a slope of $d(-\ln[\phi/T^2])/d(1/T)$. This may be used to calculate E from equation (4.11).

4.1.5   Isothermal studies
Isothermal studies are used either as an alternative to dynamic methods or to check the results from such determinations (see section 4.1.4). Cell samples are held isothermally at different temperatures for varying amounts of time. Following isothermal hold the samples are subjected to heating at a standard rate (fast enough to avoid excessive further decomposition) and the areas of the reaction exotherm determined in the usual manner. The logarithms of these areas can then be plotted as a function of time (assuming first-order reactions) and the reaction rate constants determined.

4.2   KINETIC ANALYSIS AND TREATMENT OF TG DATA

Thermogravimetric methods are particularly useful in the study of pyrolysis, especially when volatile products are produced. Many of the equations described earlier in this chapter are relevant.

4.3.1   Method of Freeman & Carroll (1958)
This was one of the earliest methods to ascribe reaction kinetics to TG data. The method utilized the treatment of Borchardt & Daniels (1957). Equation (4.23) is used for kinetic analysis

$$\frac{-E/2.303R\Delta(1/T)}{\Delta \log W_r} = -n + \frac{\Delta \log dw/dt}{\Delta \log W_r} \tag{4.23}$$

where W_r is the difference between the weight loss at the completion of the reaction and the weight loss (w) up to time t. The equation predicts therefore that a plot of $\Delta\log(dw/dt)/\Delta\log W_r$ against $\Delta(1/dt)/\Delta\log W_r$ would give a straight-line relation of slope $E/2.303R$ and intercept n. Particle size variation may affect reaction kinetics.

   The method possessed the advantage that it did not pre-identify the value of n (Anderson et al., 1977). However small deviations from the model course result in large deviations from the straight line. This means that high measuring accuracy is required. The method tends to concentrate the data points near to the abscissa in the region of small values, which potentiates any errors.

4.2.2 Method of Coats & Redfern (1964)

The method of deriving kinetic interpretation from TG was in its infancy when Coats and Redfern described their method. Factors such as crucible geometry, heating rate, sample prehistory and particle size were yet to be investigated. However, accurate temperature determination was considered a prerequisite for the precision and detection of a departure from a true linear heating rate due to endothermic and exothermic reactions. Reaction kinetics were based on the basic equations (4.1) and (4.3). For TG, α may be redefined as the weight loss at time t divided by the total weight loss at completion of the reaction. Combination, integration and rearrangement of these equations lead to the derivation of equation (4.24) which is valid for all values on n except $n=1$ when equation (4.25) was derived

$$\log\frac{[1-(1-\alpha)^{1-n}]}{T^2(1-n)} = \log\frac{ZR}{\phi E}[1-2RT/E] - E/2.303RT \qquad (4.24)$$

Equation (4.14) predicts a plot of $\log[1-(1-\alpha)^{1-n}]/T^2(1-n)$ against $1/T$ should give a straight line of slope equivalent to $-E/2.303R$

$$\log[-\log(1-\alpha)/T^2] = ;og\,ZR/\phi E[1-2RT/E] - E/2.303RT \qquad (4.25)$$

Equation (4.25) predicts a straight-line plot of slope $-E/2.303R$ when the function $\log[-\log(1-\alpha)/T^2$ is plotted against $1/T$. Each of these assumes the correct value of n is used and that for most values of E and the term $\log ZR/\phi E.[1-2RT/E]$ is constant in the temperature range over which most reactions occur. For solid-state kinetics n has theoretical justification for values of $n = 0$, 1/2, 2/3 and 1 (Coats and Redfern, 1964). The value of n for the water loss from calcium oxalate monohydrate was 2/3 and the calculated activation energy was 21.7 kcal/mol. Anderson *et al.* (1977) considered that this was a good method and could be used for DTA data. The approximations used showed that the graph depended strongly on the reaction order near to the DTA maximum and at high temperatures but only slightly on the leading edge of the DTA trace. The erroneous curves were considered appropriate for examination but distinct deviations for the activation energy were expected. The method has been improved by corrections applied by Ozawa (1970) and Anderson *et al.* (1977).

4.2.3 Method of Flynn & Wall (1966)

Flynn & Wall (1966) considered that methods based on the weight loss estimated at different heating rates were subject to experimental scatter of the data which made interpretation difficult. Their method was based on the fractional weight loss at time t where the degree of conversion is α such that $1-\alpha$ is the residual fraction and equals the weight of material lost divided by the total weight loss when t or $T \rightarrow \infty$. The general equation (4.26) was derived,

$$d\alpha/dT = (Z/\phi)f(\alpha)e^{-E/RT} \qquad (4.26)$$

where $f(\alpha)$ is a function of the degree of conversion or weight loss. The method assumes that $Z, f(\alpha)$ and E are independent of T, and Z and E are independent of α. Integrating equation (4.26) and expressing the result in its logarithmic form produces equation (4.27)

$$\log f(\alpha) = \log(ZE/R) - \log \phi + \log P(E/RT) \tag{4.27}$$

For $E/RT > 20$, $\log P(E/RT)$ is closely approximated by equation (4.28)

$$\log P(E/RT_i) \cong -2.315 - 0.457 E/RT_i \tag{4.28}$$

and therefore equation (4.27) can be transformed to equation (4.29)

$$\log f(\alpha) \cong \log ZE/R - \log \phi - 2.315 - 0.457 E/RT \tag{4.29}$$

and differentiating for a constant degree of conversion results in equation (4.30),

$$d \log \phi / d(1/T) \cong (0.457/R)E \tag{4.30}$$

when $R = 1.987$ cal/mole °K equation (4.31) is a used approximation.

$$E \cong -4.35 \, d \log \phi / d(1/T) \tag{4.31}$$

For $20 \leqslant E/RT \leqslant 60$ the approximation of $\log P(E/RT)$ is 0.457, which varies by $\pm 3\%$ (Flynn & Wall, 1966). P is the normal distribution function or probability. Thus the method requires that a plot of $\log \phi$ versus $1/T$ at a constant weight loss will result in straight line whose slope will equal $-0.23E$. Other values of $(1-\alpha)$ may be used to corroborate the determined value of E. Figs. 4.3 and 4.4 give the theoretical use of these values for the degradation of a polymer. Approximate values of E could be

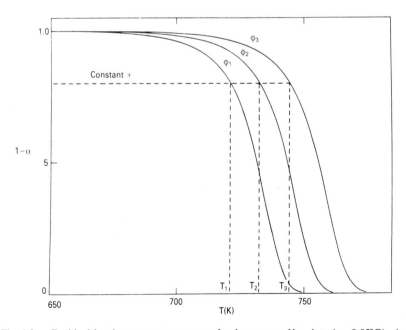

Fig. 4.3 — Residual fraction versus temperature for three rates of heating: $\phi_1 = 0.05$°C/s; $\phi_2 = 0.1$°C/s; $\phi_3 = 0.2$°C/s (reproduced with permission from Flynn, J. H. & Wall, L. A., *Polymer Letters*, **4**, 323–328, copyright (1966) J. Wiley & Sons, Inc).

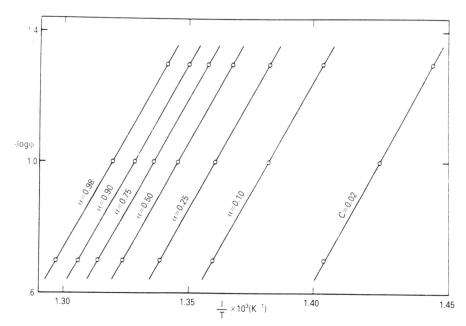

Fig. 4.4 — Logarithm of heating rate versus reciprocal absolute temperature (parameters as Fig. 4.3) (reproduced with permission from Flynn, J. H. & Wall, L. A., *Polymer Letters*, **4**, 323–328, copyright (1966) J. Wiley & Sons, Inc).

sequentially corrected using derived tables if $E/RT<20$ and correction could also make more accurate the values derived from a single approximation. The method anticipates first-order kinetics.

The method examines the weight loss of samples at different heating rates, often between 1 and 20°C/min (DuPont, undated c). A suitable marker of decomposition, e.g. 5% is used and the temperature corresponding to it at each heating rate employed determined. Equation (4.32) (DuPont, undated c) is used to determine the values

$$E = \frac{-R\, d \ln \phi}{b\, d(1/T)}\tag{4.32}$$

In this equation T is the temperature at the selected weight loss and b is an approximation whose value is dependent on the value of E. Zsako & Zsako (1980) developed equation (4.33) (DuPont, undated c), which assumes that a reaction order (n) of unity exists, to determine the pre-exponential factor (Z) of the Arrhenius equation.

$$Z = \frac{-\phi \ln(1-\alpha)}{Rc}\tag{4.33}$$

where c is an approximation whose value depends on E. The function $f(\alpha)$ described by many workers may be written as

$$f(\alpha) = \alpha^a (1 - \alpha)^b \tag{4.34}$$

where a and b are often called the homogeneity factors. Zsako (1968) considered the simple case $a = 0$ and this gives the general rate equation as equation (4.3). The equation may also be rewritten when $b = 0$ to become equation (4.35).

$$d\alpha/dt = k\alpha^a \tag{4.35}$$

Values of a have included 2/3 and 1. A value of $a = 1$ gives the equation (4.36).

$$d\alpha/dt = \alpha(1 - \alpha)^b \tag{4.36}$$

4.3 PHARMACEUTICAL EXAMPLES

Although thermal analysis has been used within the pharmaceutical field to qualitatively assess the stability of pharmaceuticals there are only a few examples where the techniques have been used quantitatively. The remainder of this chapter considers some of these data.

Radecki & Wesolowski (1976) used TG to evaluate the solid-state stability of binary and tertiary systems containing various salicylates and carbonates. Fig. 4.5 is

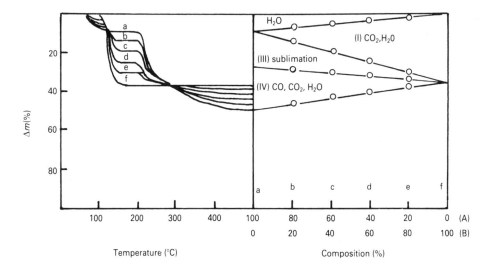

Fig. 4.5 — Thermal decomposition of a system containing disodium salicylate sesquihydrate (A) and sodium hydrogen carbonate (B). Loss in weight as a function of temperature (left) and composition (right). The decomposition of sodium hydrogen carbonate occurs according to scheme I and that of the sesquihydrate according to schemes II, III and IV. (From Radecki, A. & Wesolowski, M., *J. Therm. Anal.*, **10**, 233–245. Reprinted with permission of John Wiley and Sons Ltd. Copyright (1976) J. Wiley & Sons, Inc.)

representative of their data for disodium salicylate sesquihydrate and sodium hydrogen carbonate mixtures. The areas marked on the diagram correspond to the stages of reaction of the mixture. Chemically the reaction breakdown mechanisms corresponding to the transitions are:

Scheme I $2NaHCO_3 \rightarrow Na_2CO_3 + CO_2 + H_2O$

Scheme II $C_6H_4(ONa)COONa.1.5H_2O \rightarrow C_6H_4(ONa)COONa.0.5H_2O + H_2O$

Scheme III $4C_6H_4(ONa)COONa.0.5H_2O \rightarrow 2C_6H_4(ONa)COONa +$
$2Na_2CO_3 + C_6H_4(OH)COOH \uparrow + \text{Volatiles } [CO, CO_2, H_2O]$

Scheme IV $2C_6H_4(ONa)COONa + 2Na_2CO_3 \rightarrow 4Na_2CO_3$
$+ \text{Volatiles } [CO, CO_2, H_2O]$

Their idea of using such diagrams as a means of identification of binary mixtures is cumbersome but the method has potential in the study of the stability of pharmaceutical binary mixtures and in assessing those liable to decompose. It was suggested that the results could be useful for monitoring the course of the commercial-scale manufacture of sodium salicylate and for checking the declared contents of salicylate mixtures. Reaction II is endothermic but a broad split DTA peak at 40–120°C is due to heat absorption. A small endothermic peak at 210–240°C in reaction III was thought to be due to changes within the molecule caused by the hydrolysis of disodium salicylate by its own water of crystallization. This peak overlapped another endotherm due to the sublimation of the liberated salicylic acid. Reaction IV was indicated by a broad DTA endotherm at 240–450°C. The heat loss due to the volatilization of the gases was balanced by the strong exothermic effect of the combustion of the organic moiety and probably the heat of reaction between sodium hydroxide and carbon dioxide over the temperature range examined.

Goldstein & Duca (1982) examined the thermal stability of cyanocobalamin, hydroxycobalamin and cobinamide in the solid state via DTA and TG, using the method of Freeman & Carroll (1958) to determine the stability constants of cyanocobalamin. The TG, derivative TG and DTA curves of cyanocobalamin are indicated in Fig. 4.6. The weight loss up to 140°C and strong DTA endotherm at 135–140°C were due to removal of water. Scanning at 10°C/min showed a plateau at 140–230°C in TG, but at the slower speed of 3°C/min a mass loss of ≈2% was noted at 140–145°C corresponding to removal of the cyanide group. Ammonia evolution was considered responsible for the next weight change. Decomposition progressively occurred above 230°C with extensive structural changes. Four endotherms were noted at 230, 280, 390 and 520°C (the latter at the lower heating rate only) and then a relative stable plateau was noted on the TG curve. This latter corresponded to a compound of phosphorus and cobalt oxides. Hydroxycobalamin lost water up to 140°C and plateaus in the decomposition curve were noted at 140–230 and at 760°C. Cobinamide was stable up to 120°C with only the loss of water apparent. Removal of water occured at 60–120°C, cyan removal occurred at 120°C and the TG curve indicated decompositions from 220 to 600°C which also registered as endotherms at 410, 490, 530 and 610°C.

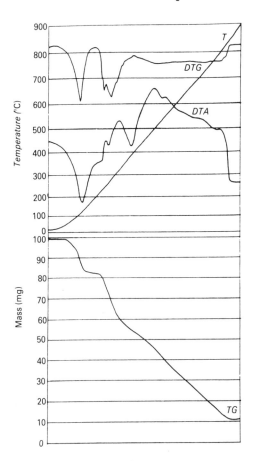

Fig. 4.6 — Thermoanalytical curves for cyanocobalamin (reproduced with permission from
Goldstein & Duca, 1982).

Fig. 4.7 indicates a Freeman & Carroll (1958) treatment of the dehydration data
of Fig. 4.6 in the temperature range 20–140°C. The equation assumed a rate
expression $-dx/dt = kx^n$, where x is the amount of vitamin B_{12} at time t and the rate
constant k is given by the Arrhenius expression. The graph gave a straight line with a
slope equal to an activation energy of 267 ± 1.9 kcal/mole. The zero intercept
indicated a zero-order reaction.

Tomassetti *et al.* (1983) used thermal analysis to assay penicillins and cephalos-
porins. The purity of the antibiotics was determined from the residual sodium or
potassium sulphate residue following analysis through decomposition. Static air or
oxygen were used to produce oxidation but a greater precision of assay was claimed
when either ammonium sulphate or persulphate was added to the compounds prior
to analysis (Tomassetti *et al.*, 1983). The scans of benzylpenicillin potassium differed
only slightly when completed in static air or oxygen. Decomposition commenced
at 240°C and proceeded through two major processes. The first decomposition

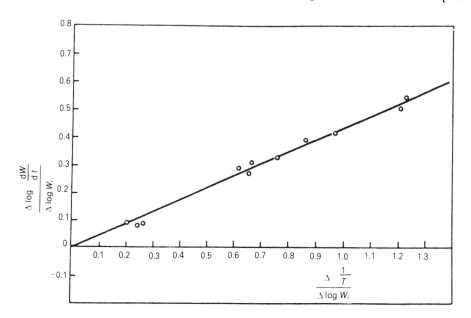

Fig. 4.7 — Freeman & Carroll's mathematical analysis of the TG curve of Fig. 4.6 (reproduced with permission from Goldstein & Duca, 1982).

completed at 420°C produced a residue of potassium sulphate and carbon. The conversion of carbon to carbon dioxide leaving only the sulphate as residue was complete at 720°C. The steps of the process in oxygen were sharper than in the presence of air and the residue of sulphate was formed at a lower temperature of 670°C. In comparison potassium cephalosporin C displayed considerable differences when examined in air or oxygen. The latter atmosphere produced a very fast first process during which violent oxidation of the intact antibiotic was apparent. This was followed by a less sharp process. In air the processes were more gradual, leading to poor differentiation of the two processes (Tomassetti *et al.*, 1983). The interest in this work lies not in the potential ability to assay the antibiotics (the method is insensitive to impurities such as other potassium salts present in potassium benzylpenicillin) but in that it may give a guide to the sensitivity of a drug to oxidation. Another example of non-quantitative analysis was the stability study on the antibiotic cephadrine dihydrate (Jacobson, 1984). This salt was stressed to remove the water of hydration either by fluidized-bed drying or by heating at 80°C before storage at 50°C in an atmosphere of 40% oxygen : 60% hydrogen to create accelerated storage conditions. Both DTA and TG were used. An unstressed sample displayed endotherms at 90 and 102°C and a large exotherm at 202°C. As the severity of ageing was increased, the DTA endotherm at the lowest temperature decreased in intensity and its peak temperature approached 102°C. Severe stress resulted in a single endotherm at 102°C and a splitting of the exotherm to peaks at 192 and 202°C. The former peak remained to the exclusion of the latter as the degree of stress increased and the endotherm at 102°C transferred to higher temperatures. Jacobson (1984) described a stability index (SI) for the drug as equation (4.37)

$$SI = 100\delta + \frac{\text{amplitude exo. } 202°C}{\text{amplitude exo. } 202°C + 192°C} \times 100 -$$

$$- \text{Temp. of major endotherm} \tag{4.37}$$

where δ is the °C between the two endotherms. The greater the value of the SI the greater the stability of the drug. The value for the unstressed dihydrate was +885. The X-ray peak was used to measure undegraded dehydrated material. Accelerated stability ageing further changed the DTA scans giving nondescript poorly defined patterns correlating with the changes in the X-ray diffraction patterns. The theoretical water content of the dihydrate was 9.3%. TG indicated that even the dehydrated sample contained some water (5–6%) which was not bound in the crystalline lattice and was not a measure of the crystalline state.

Thermal analysis has found use in the prediction of the stability of formulated products, e.g., Ager *et al.* (1986) studied aspirin–excipient mixtures. Aspirin displayed a DSC melting endotherm at 138°C and a decomposition endotherm at 160°C. The TG scan (Fig. 4.8) indicated a large weight loss at this latter temperature which was greater than the lost acetic acid and indicates that some salicylic acid may have sublimed. The sharp melting endotherm of aspirin even in blends was indicative of the quality of the aspirin despite the potential to form a eutectic with salicylic acid because it was considered that the interaction would be reduced in the presence of the excipients (Ager *et al.*, 1986). The major phenomenon encountered was peak-broadening. Aspirin was blended with the excipients before storing at ambient temperatures by (a) mortar and pestle, (b) making a tablet or (c) use of chloroform. No significant degradation was noted in blends with potato starch for up to three months' storage. The DSC scans were unchanged and no decomposition was detected by thin-layer chromatography (TLC) or nuclear magnetic resonance (NMR). DSC of blends of aspirin with either silica or alumina indicated that aspirin could be degraded in the solid state by either acid or basic catalysis. Decomposition was inapparent in a 15 : 4 blend (aspirin : silica) after one month but after two days DSC of a 1 : 1 blend lost its definition. Little aspirin was found after two days' storage of a 1 : 10 blend. These results indicated that the excipient must be present at relatively high concentrations for degradation to occur.

Karimain (1981) utilized thermal analysis to study the degradation of the gum tragecanth. The gum was examined by DTA in an atmosphere of either nitrogen or air. In either atmosphere an initial endothermic trend was followed by an exotherm in the range 200–400°C with a maximum at 295°C with a preceeding shoulder. A further exotherm, $3\frac{1}{2}$ times larger than the first exotherm, was observed in the range 400–590°C, peaking at 485°C in the atmosphere of air. These effects were reflected in TG analysis. Weight loss (14%) occurred from ambient to 225°C and was due to the loss of adsorbed water. A second major weight loss (54%) occurred at 225 to 420°C and a third weight loss, apparent in air only, corresponded to 20% weight loss and commenced at 420°C. The weight loss due to moisture at temperatures between 187 and 206°C was examined isothermally and the data expressed as α (fractional decomposition) against time and kinetically analysed using log–log plots and reduced time analysis. The data fitted equation (4.38).

$$(\alpha/[1-\alpha]) = kt^n \tag{4.38}$$

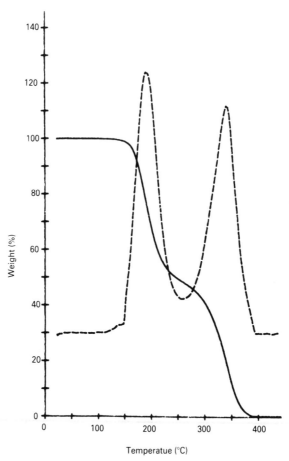

Fig. 4.8 — Thermogravimetric analysis of aspirin. Key: (——) percentage remaining; (– – –)
rate of weight loss (reproduced with permission from Ager *et al.*, 1986).

The log-log plots were linear and established a value of $n=0.75$. The data presented
as Fig. 4.9 was linear for α values 0 to 0.7. The data also fitted equation (4.39)

$$1-(1-\alpha)^{2/3}=kt \tag{4.39}$$

The specific rate constants were used in the Arrhenius equation and the activation
energy so derived was 157.75 and 158.95 kJ/mol using equations (4.38) and (4.39)
respectively. The value was considerably higher than expected and suggested that
the water was strongly bound to the tragecanth. The dependence of the data on the
contracting sphere equation suggested phase-boundary-controlled release of the
water.

Hardy (1982) used the variable heating method of Ozawa (1970) to estimate
stability. Heating rates of 10 or 20°C/min from ambient were employed until
decomposition was complete. Endothermic reactions were considered to be rep-

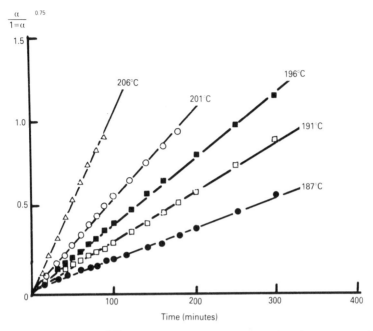

Fig. 4.9 — Plots of $(\alpha/1-\alpha)^{0.75}$ against time for isothermal degradation of gum tragecanth in air (from Karimain, 1981 *Proceedings of the 2nd European Symposium on Thermal Analysis*. Dollimore, D. (ed.). Copyright (1981), Heyden. Reprinted by permission of John Wiley & Sons, Ltd.).

resentative of a thermally stable material. Exothermic reactions were examined by the ASTM E968 method. The specific rate constants at various temperatures were calculated and the time to constant loss, either 10% or 50% (half-life), derived on the basis of first-order kinetics. Hardy (1982) recommended the use of isothermal hold to check the validity of the derived data. Table 4.1 shows the predicted stability for the

Table 4.1 — Stability prediction for ergocalciferol (From Hardy, *Proceedings of the 7th International Conference on Thermal Analysis*. B. Miller (ed.). Copyright (1982). Reprinted by permission of John Wiley and Sons, Ltd.)

Temperature (°C)	Rate constant (min^{-1})	Time to 10% decomposition $t_{0.1}$ (min)	Half-life time $t_{1/2}$ (min)
20	$2.46. \ 10^{-9}$	$4.27. \ 10^{7}$	$2.82. \ 10^{8}$
50	$2.03. \ 10^{-7}$	$5.17. \ 10^{5}$	$3.41. \ 10^{6}$
80	$7.94. \ 10^{-6}$	$1.32. \ 10^{4}$	$8.73. \ 10^{4}$
100	$6.59. \ 10^{-5}$	$1.59. \ 10^{3}$	$1.05. \ 10^{4}$
150	$5.46. \ 10^{-3}$	$1.92. \ 10$	$1.27. \ 10^{2}$

degradation of ergocalciferol. The data suggested that the half-life prediction was 2 hours at 150°C and following isothermal holding at these conditions an area loss of 46.6% was achieved compared to a 48.1% theoretical loss.

4.4 CONCLUSIONS

Perhaps the final words of this chapter should highlight a review article by Li Wan Po (1986) who suggested that corrections should be used for heating rate and time lags. Theoretically DSC and other thermal methods of analysis seem to provide an ideal means of providing information regarding drug stability. Li Wan Po (1986) confirmed that there are very few pharmaceutical data published on such stability data. This is due to the fact that drug instability is rarely a spontaneous problem and that other agents such as moisture or oxygen are required to make the reaction a major problem. These conditions are difficult to mimic under DSC conditions except by using very large sample pans. DSC does however play a major role in examining the interactions between drugs and excipients.

REFERENCES

Ager, D. J., Alexander, K. S., Bhatti, A. S., Blackburn, J. S., Dollimore, D., Koogan, T. S., Mooseman, K. A., Muhvic, G. M., Sims, B. & Webb, V. J. (1986) *J. Pharm. Sci.*, **75**, 97–101.
ASTM E698–79 (Reapproved 1984) *Arrhenius kinetic constants for thermally unstable methods*, 624–631.
Anderson, H., Besch, W. & Hamberland, D. (1977) *J. Therm. Anal.*, **12**, 59–68.
Borchardt, H. J. & Daniels, F. (1957) *J. Am. Chem. Soc.*, **79**, 41–46.
Brown, M. E. & Phillpotts, C. A. R. (1978) *J. Chem. Education*, **55**, 556–560.
Cassel, R. B. (1979) ASTM method of testing for determining the Arrhenius kinetic constants for the screening of potentially hazardous materials, *Perkin-Elmer TAAS 28*. Perkin-Elmer Corp., Norwalk, Connecticut.
Coats, A. W. & Redfern, J. P. (1964) *Nature*, **201**, 68–69.
DuPont (undated a) E-50184. *DSC Kinetics Data Analysis Program, Borchardt & Daniels*.
DuPont (undated b) E-50185. *DSC thermal stability kinetics ASTM E698*.
DuPont (undated c) E-53684. *TGA Decomposition Kinetics Program*.
Duswalt, A. A. (1974) *Thermochim. Acta*, **8**, 57–68.
Flynn, J. H. & Wall, L. A. (1966) *Polymer Letters*, **4**, 323–328.
Freeman, E. S. & Carroll, B. (1958) *J. Phys. Chem.*, **62**, 394–397.
Goldstein, S. & Duca, A. (1982) *Thermochim. Acta*, **59**, 211–220.
Gyulai, G. & Greenhow, E. J. (1974) *J. Therm. Anal.*, **6**, 279–291.
Hardy, M. J. (1982) *Proceedings of the 7th International Conference on Thermal Analysis*. Miller, B. (ed.) John Wiley & Sons, Chichester, 887–892.
Jacobson, H. (1984) *Proceedings of 13th North American Thermal Analysis Society Conference*, Philadelphia, 150–155.
Karimain, R. (1981) Thermal analysis study of the degradation of the gum tragecanth. In *Proceedings of 2nd European Symposium on Thermal Analysis*, Dollimore, D. (ed.) Heyden, London, 211–214.

Kissinger, H. E. (1957) *Anal. Chem.*, **29**, 1702–1706.

Li Wan Po, A. (1986) *Anal. Proc.*, **23**, 391–393.

Ozawa, T. (1970) *J. Therm. Anal.*, **2**, 301–324.

Petty, H. R., Arakawa, E. T. & Baird, J. K. (1977) *J. Therm. Anal.*, **11**, 417–422.

Piloyan, G. O., Ryabchikov, I. D. & Novikova, O. S. (1966) *Nature*, **212**, 1229.

Pope, M. I. (1980) *Anal. Proc.*, **17**, 236–239.

Radecki, A. & Wesolowski, M. (1976) *J. Therm. Anal.*, **10**, 233–245.

Sestak, J. (1980) Kinetic compensation effect: facts and fiction of linear plots using Arrhenius law. In *Thermal analysis, Vol. 1, Theory, instrumentation*. Wiedemann, H. G. (ed.) Birkhauser Verlag, Basel, 29–37.

Tomassetti, M., Campanella, L., Sorrentino, L. & D'Ascenzo, G. (1983) *Thermochim. Acta*, **70**, 303–315.

Torfs, J. C. M., Deij, L., Dorrepaal, A. J. & Heijens, J. C. (1984) *Anal. Chem.*, **56**, 2863–2867.

Zsako, J. (1968) *J. Phys. Chem.*, **72**, 2406–2411.

Zsako, J. & Zsako, J. (1980) *J. Therm. Anal.*, **19**, 333–345.

5

The use of thermal analysis in the purity determination of pharmaceuticals

5.1 INTRODUCTION

The determination of purity is based on the assumption that an impurity will depress the melting point of a pure material whose melting is characterized by a melting point (T_o) and an enthalpy of fusion (ΔH_o). The melting transitions of pure, 100% crystalline materials should be infinitely sharp, but impurities or defects in the crystal structure will broaden the melting range and lower the final melting point to a temperature lower than T_o. Such occurrences are apparent at the extremes of eutectic phase diagrams where the liquidus temperatures decrease below the melting points of the pure materials.

The effect of an impurity on T_o may be predicted from van't Hoff's law, its equation is given as equation (5.1).

$$T_m = T_o - \frac{RT_o^2 \chi_2}{\Delta H_o} \frac{1}{F}$$

(5.1)

where T_m is the sample temperature at equilibrium (K), T_o is the melting point of the pure component (K), R is the gas constant, χ_2 is the concentration of impurity (mole fraction) and F is the fraction molten at T_m. Equation (5.1) predicts a straight-line relationship with an ordinate intersection of T_o when T_m is plotted against $1/F$. The slope is thus equal to $(RT_o^2/\Delta H_o)\chi_2$. When $F = 1$, $T_o - T_m$ corresponds to the melting point depression. Equation (5.1) permits estimation of the amount molten at any T_m and is particularly applicable to DSC melting endotherms, where the area of an endotherm up to a particular temperature can be related to the total area and is equivalent to $1/F$ when expressed as a fraction. Coincidently ΔH_o can be determined and T_o computed.

Bowman & Rogers (1967) used DTA for purity determination. Melting point depressions of impure samples were compared by a differential method that involved running the impure material against a pure sample of the same material as reference. The area under the curve was proportional to the difference in purity between the sample and reference in the impurity range 0.3–2.0 mole% impurity. The method was claimed to be more accurate than melting point depressions. Generally, DTA is used as an estimate of purity only in a qualitative manner and DSC is the method of choice. The derivation of equation (5.1) highlights the theoretical applicability and limitations of the method.

5.2 THERMODYNAMIC CONSIDERATIONS OF MELTING POINT DEPRESSION

The derivation of van't Hoffs law (Marti, 1972; Van Dooren & Muller, 1984) is fundamental to our understanding of its use in purity determination.

The limitations of a thermodynamic approach to purity determination were outlined by Perkin-Elmer (1967a,b) and Marti (1972) and reviewed by van Dooren & Muller (1984). These limitations include:

(a) The melting point must be a triple point, i.e. the liquid, gaseous and solid phases are in equilibrium and have the same vapour pressure at this temperature.

(b) The change in the free energy, ΔG, must be zero to maintain equilibrium throughout the fusion process. In practical terms this means the samples should be small and subjected to low heating rates. Good thermal contacts should be maintained in encapsulation and sample preparation, producing low thermal resistances between the sample, container and the sample heater.

(c) The method is inapplicable to insoluble impurities in the liquid state which do not form ideal solutions. The components must form a eutectic mixture giving ideal solutions in the molten phase of the main component. No chemical interaction, degradation or association should occur in the molten state.

(d) Samples which sublime or volatilize before or during the melting process cannot be examined. Additionally the vapour pressures of impurities should be very small compared to that of the major component.

(e) The molar heat of fusion, (ΔHo), should be constant in the temperature range examined and be independent of impurities.

(f) The activity of any impurity in the solid state should be zero, i.e. the method is inapplicable if solid solutions are formed. However modifications may be made to equation (5.1) to account for this non-ideal behaviour.

(g) The assumption that the activity of the impurity in the liquid state is very low is generally untrue at the eutectic composition when the molar ratios of the major and minor components may approach unity. This implies that only high purity samples may be examined and that low $1/F$ values can only be used.

(h) van Dooren & Muller (1984) emphasized that the activity coefficient of the minor component in the liquid phase should approximate unity. At infinite dilutions where impurity molecules are so diluted that they cannot interact with each other this is valid. This again limits purity determinations to high purity samples and emphasizes that no association between molecules should occur.

(i) The method is valid when T_o equates to T_m, i.e. except when the melting point depression $T_o - T_m$ is great. The latter will occur in very impure samples.

(j) The method is valid only if the initial segment of the liquidus curve of the phase diagram is linear. This may be a poor approximation, especially when the eutectic composition is close to 100% of either component.

(k) The final assumption is that the mole fraction of the impurity must remain constant. As indicated above, purity determination by thermal analysis is invalid if association and evaporation of either component occurs.

Additionally sample decomposition during fusion poses problems. Transitions, such as desolvation, polymorphic charge or sublimation, render the method inappropriate.

Additional to the problems of thermodynamic interpretations and limitations another major problem in purity determination is curve and result interpretation, which when understood, can be partially resolved by correct experimental conditions.

5.3   MELTING CURVE ANALYSIS

5.3.1   Theoretical curves

Perkin-Elmer (1967a) described the effect of purity on peak shape. Under ideal conditions, with no thermal lag and ultrapure mterials, melting should occur over an infinitesimal range. Since power-compensation DSC measures the thermal energy per unit time (dq/dt) transferred to a sample as the sample holder temperature is linearly increased (dT/dt) it follows that with no thermal lags in the system, the instrument read-out is directly proportional to dq/dT_s, the energy change in the sample per degree, the proportionality constant being the scanning rate. Thus equation (5.2) applies

$$\frac{dq}{dt} = \frac{dT_s}{dt} \times \frac{dq}{dT_s} \qquad (5.2)$$

When no transition occurs dq/dT_s is the heat capacity of the sample. When transitions occur dq/dT_s represents the effective heat capacity of the sample due to the energy required to establish the transition because pure materials should melt over a small range (which approximates to zero). dq/dT_s becomes infinite at T_o when impurities are present. However, dq/dT_s is measurable and is a function of sample temperature such that equation (5.3) holds.

$$\frac{dq}{dT_s} = \frac{\Delta q(T_o - T_m)}{(T_o - T_s)^2} \qquad (5.3)$$

and the melting depression due to the impurity is given in equation (5.1) when $F = 1$. These equations produce theoretical DSC peak shapes such as Fig. 5.1. Although the melting depression is only 0.3°C, the melting range exceeds 2°C. Curve analysis is a

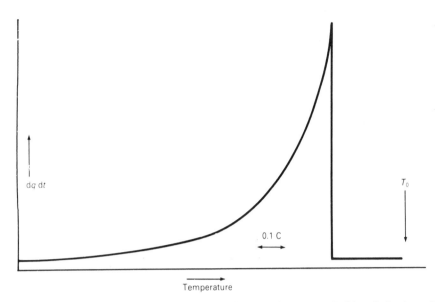

Fig. 5.1 — Theoretical calculated melting peak of an organic compound with typical values of T_o and ΔH_f with 99.5 mole% purity (reproduced with permission from Perkin-Elmer, 1967a).

better indicator of purity then melting point depressions since it covers a wider temperature range. This can be verified by Table 5.1 where calculated values of

Table 5.1 — Calculated melting point depressions (°C), using equation (5.1), corresponding to various fractions melting, for major components of melting points 100 or 250°C and heats of fusion of 7000 or 500 cal/mole

Melting point (°C)	100	100	100	100	100	250
Heat of fusion (cal/mole)	7000	7000	7000	7000	500	7000
Mole fraction impurity	0.005	0.01	0.02	0.05	0.01	0.01
Fraction melted (F)			Melting point depression (°C)			
0.001	197.5	394.9	789.9	1974.6	5529.0	776.4
0.01	19.75	39.49	78.99	197.46	552.90	77.64
0.1	1.975	3.949	7.899	19.746	55.290	7.764
0.2	0.987	1.975	3.949	9.873	27.645	3.882
0.4	0.494	0.987	1.975	4.937	13.822	1.941
0.6	0.329	0.658	1.316	3.291	9.215	1.294
0.8	0.247	0.494	0.987	2.468	6.911	0.971
1.0	0.197	0.395	0.790	1.975	5.529	0.776

melting point depressions at different fractions melted are shown for different impurity levels calculated by equation (5.1). For an enthalpy of fusion of 7000 cal mole^{-1}, 20% of the sample will have melted approximately 1,2,4 and 10°C below the melting point of the pure material at impurity levels of 0.5, 1.0, 2 and 5% respectively. This emphasizes the broadening of DSC curves caused by increasing levels of impurities as evidenced by Fig. 5.2.

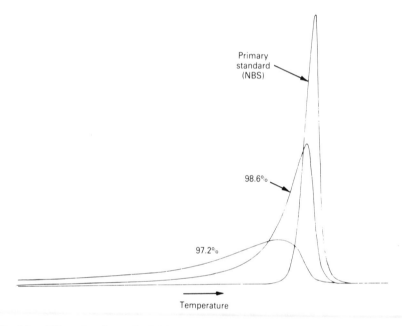

Fig. 5.2 — Effect of purity on the DSC melting peak shapes of benzoic acid (reproduced with permission from Perkin-Elmer, 1976a).

Curve analysis is accomplished using equation (5.4), obtained by integration of equation (5.3) and which may be rearranged into equation (5.5).

$$F = \frac{T_o - T_m}{T_o - T_s} \tag{5.4}$$

$$T_s = T_o - \frac{(T_o - T_m)}{F} \tag{5.5}$$

A plot of sample temperature versus reciprocal F (equation (5.5)) should give a straight line of slope equal to the negative of the melting point depression $(T_o - T_m)$. As equation (5.1) predicts the slope is $RT_o^2 \chi_2 / \Delta H_o$.

5.3.2   Curve structure
In practice curves such as Fig. 5.1 are not obtained. Instrumental lags ensure that the end of the peak is not a vertical drop but is a gradual tailing-off. The tail represents

the energy required for the sample to catch up to the program temperature after melting is completed (Perkin-Elmer (1967a)). This area is included in determining heats of fusion, because the true baseline under the melting peak where the sample temperature is not changing is really lower than the baseline before and after the peak where it is changing. In effect 'heat capacity energy' as it is termed by Perkin-Elmer, is recorded before and after but not during the peak. Such errors do not occur if the total peak area is measured because the areas cut off by baseline interpolation are equivalent to the tail area.

Drawing a continuous straight baseline to join up the pre- and post-transition baselines may not, however, represent the authentic area. Although at the apparent onset of transition the true baseline closely follows the linear projection of the pre-transition baseline (Heuvel & Lind, 1970), the baseline around the endothermic peak displays considerable variation (Fig. 5.3). The deviations are important in the

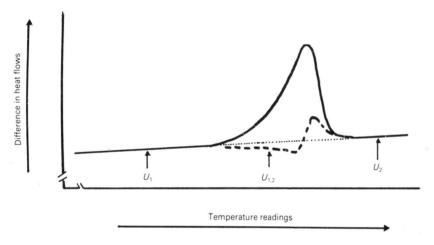

Fig. 5.3 — An appreciable difference in heating rate between sample and reference as encountered in rapid transitions causes a pronounced curvature of the baseline. Reprinted with permission from Heuval, H. M. & Lind, K. J. R., *Anal. Chem.*, **42**, 1044–1048. Copyright (1972) American Chemical Society.

case of rapid transitions, high scanning rates and appreciable capacity shifts (Heuvel & Lind, 1970). In purity determination, with its slow scanning rates and small sample sizes, the emphasis is on equilibrium conditions and consequently such baseline variations may be ignored and the scanning baseline is usually drawn.

Temperature lags can be corrected for by using ultrapure standards such as 99.9999% pure indium which melt over a narrow range. Their leading slope may be used to correct instrument temperature by graphical or computer-assisted extrapolation through any point of the curve to a baseline intercept.

The determination of testosterone purity (Fig. 5.4) is cited from Perkin-Elmer (1967a). Measurements for fraction melted were made preferably on the range 10–50% melted. At point A, for instance, the temperature is read off at point D, the intercept of the corrected lag temperature on the baseline predicted from empty

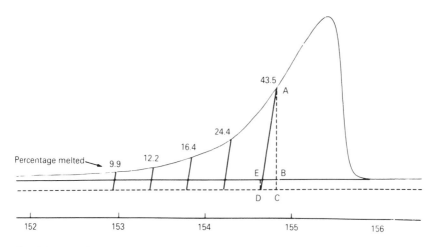

Fig. 5.4 — DSC melting endotherm of testosterone (5.83 mg, 1.25°C/min) showing area calculation tie-lines (reproduced with permission from Perkin-Elmer, 1967a).

sample pans. The area to be measured is up to the line AB and is the whole of the area above the extrapolated baseline joining the pre- and post-transition baseline. The area EBCD must also be included in the measurement of area. This area is equivalent to heat of fusion although appearing to be heat capacity energy. It is equivalent to the tail area following the peak vertex as discussed previously. The area found for any of the arbitrary chosen points on Fig. 5.4 can be expressed as fractions of the total heat of fusion to give values of F. Fig. 5.5 is the plot of the fractions melted data of Fig. 5.4 giving 99.6 mole% as the estimated value of purity.

Variations exist as to precisely what constitutes the measured area. The correction EBCD was used by Decker & Young (1978) but ignored by Davis & Porter (1969), Plato & Glasgow (1969), Marti (1972) and Palmero & Chiu (1976). Driscoll *et al.* (1968) used this correction but the area measured was cut off by the lag temperature projection rather than a vertical drop. The difference between the extrapolated and corrected baselines was 5.5 mcal/s. The published baseline correction may be unnecessary (Grady *et al.* 1973).

5.3.3   Valid area under curve
There is considerable debate as to what represents the most suitable area for plots such as Fig. 5.5, because if the melting process accelerates too quickly, thermal equilibrium as required by equation (5.1) is not maintained. Most authors recommend analysis of the first portion of the melting endotherm only. Perkin-Elmer (1967a) suggested that the area corresponding to 10 to 50% of curve should be used. This was criticised by Driscoll *et al.* (1968) who found that a 10-fold variation in the determined purity occurred depending on where the limits for analysis were drawn. Plato & Glasgow (1969) proposed that the use of the peak should be made at arbitrary intervals up to the point where 50% of the sample is melted, because above this melting is too quick for the maintenance of equilibrium. This corresponds to two-

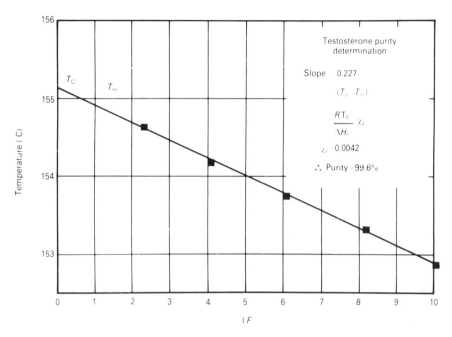

Fig. 5.5 — Testosterone purity determination of data presented in Fig. 5.4 (reproduced with permission from Perkin-Elmer, 1967a).

thirds up the peak. In non-quantitative terms, if 'too much' of the total area is used, the determined purity will be too low (Barrall & Diller, 1970). The 'correct' area was that required by the van't Hoff equation, i.e. the area from onset of melting until the transformation of the last solid to liquid. The total area in integrated fractions, from the onset of melting to the vertex through the peak of the melting endotherm should be used. This vertex is very near the temperature of final melting at low heating rates (below 1.25°C/min) and small samples (less than 3 mg). DeAngelis & Papariello (1968) did not utilize $1/F$ values beyond a certain arbitrary rate of heating, which was the rate at half peak height for a pure standard, beyond this rate errors increased. Barrall & Diller (1970) used the data up to the vertex when the melting of the sample did not exceed this rate. At least six segments and preferably 10 were required to determine correctly $1/F$ values (Barrall & Diller, 1970). The determination of the useable area, however, is only a minor problem when compared to that correection required for undetected melting.

5.4 UNDETECTED MELTING AND ITS CORRECTION

Perkin-Elmer (1967b) drew attention to problems concerned with information lost during DSC. Plots of sample temperature (T_s) against $1/F$ were often not the straight lines predicted but showed upward concavity. The degree of departure from linearity increased at higher impurity levels. This was attributed to either solid solution formation or the probable underestimation of the amount of melting that occurred at lower temperatures. On early instrumentation (e.g. Perkin-Elmer DSC-1B) the peak-to-peak detectable noise level was 0.020 mcal/s. This meant that a significant amount of melting occurred below the threshold of detection. Perkin-Elmer (1967b) estimated that for a 2 mg sample of 99% purity, melting at 400 K with a heat of fusion of 5000 cal/mole, 10% of the sample would have melted below the detectable noise level of the DSC-1B. Therefore the fraction melted, especially at large $T_o - T_s$ values would be seriously underestimated. Various corrections have been applied to determine the extent of the underestimates. Some important examples follow.

5.4.1   Perkin-Elmer method (1967b)

This method attempts to determine the extent of undetected melting. If this is equivalent to x and the apparent measured area equivalent to A, then the real area is $A + x$. The partial areas measured in the determination of fractions melted would be $a_1, a_2, a_3 \ldots a_n$ and following correction for the inapparent area would become $a_1 + x$, $a_2 + x, a_3 + x \ldots a_n + x$. The fraction F_n melted for the nth value is given by equation (5.6).

$$\frac{1}{F_n} = \frac{(A + x)}{(a_n + x)} \tag{5.6}$$

Although x is small compared with A, it is relatively large compared to a_n and causes serious errors, giving incorrectly high values of $1/F_n$ at lower temperatures, i.e. large $T_o - T_s$ values.

The data may be linearized by trial and error. Fig. 5.6 gives an appreciation of gradually increasing the value of x. The exact data which produced the straight-line correction was 6.6% $(100x/A)\%$. Overcorrection (e.g. 10%), produced concave downward curves. Only one value of x will produce a linear profile. Since low values of $1/F$ are affected only slightly by the choice of x, (Perkin-Elmer, 1967b) the estimates are made by extrapolating linearly the initial slope of the curve corresponding to the largest values of $1/F$. This point is then adjusted by choosing x so that it falls on the extrapolated line. All other points could then be similarly adjusted. This technique may also be used to calculate and adjust the heats of fusion for impure materials.

Sanmartin & Regine (1969) developed a less arbitrary method of determining the molar impurity. The value of x may be some 15% of the apparent total area (Plato & Glasgow, 1969). Since no prior knowledge exists as to whether a solid solution is formed or not, the curvature of plots, such as Fig. 5.6, may not entirely be due to instrument sensitivity. Reubke & Mollica (1967) however assumed that instrument instability was the main explanation for deviation from linearity.

The shape of the $1/F$ versus T_s plot may be an 'S' curve (Grady et al., 1973) making correction difficult. Computer analysis may derive the best straight-line fit. The pre-melt correction is varied until the theoretical linear relationship is obtained. Many drugs yielded an 'S'-shaped curve which could not be linearized. Premelt calculation is then guesswork, although Grady et al. (1973) suggested that where the sigmoidal character is strong, impurity values should be calculated, with several x-

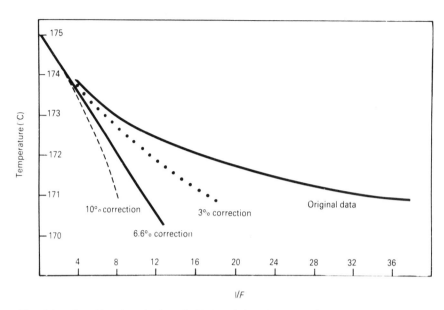

Fig. 5.6 — Graphical linearization of $1/F$ plot of phenobarbitone (99.4 mole% pure) showing effects of nil, under- and over-correction (reproduced with permission from Perkin-Elmer, 1967b).

corrections to assess their confidence. Examples of drugs studied by Grady *et al.* (1973) included glutethimide (0.25 mole% impurity, 7% pre-melt correction) menandione (0.15 mole%, 5%), promazine hydrochloride (0.2 mole%, 3%) tolnaftate (0.72 mole%, 11%) and ethinyloestradiol (1.5 mole%, 13.9%). Grady *et al.* (1973) suggested that the magnitude of the correction correlated roughly with the level of impurity, e.g., 0–0.2% mole% impurities averaged 3.3% pre-melt correction, 0.2–0.5% mole% impurities averaged 6.8% correction and greater than 0.75 mole% averaged 11% correction.

Obviously the calculations are simplified and automated by microprocessors. In describing the use of computers, Driscoll *et al.* (1968) stated that less correction is required to produce linearity when the $1/F$ limit was 4 rather than 2. Additionally limits were used to cut off $1/F$ values above 15, since the line would be undercorrected giving higher calculated impurities, clearly demonstrating that the calculated impurity is a function of the $1/F$ range used. The method of (Driscoll *et al.*, 1968) was most reliable when sliding limits were used based on the fraction melted at the point where the rate of heat input reaches half its maximum value, equivalent to half peak height. This changed the lower limit from 2.5 for impure samples to 8 for pure samples. The upper limit used was $1/F = 50$. Grady *et al.* (1973) confirmed that the choice of area is critical, e.g. in phenacetin analysis pre-melt factors ranging from 1 to 15% produced impurity values ranging from 0.02 to 0.40 mole%. Computer analysis has removed the random nature of the purity determination. Publications using least-square techniques include those of Marti (1972), Barrall & Diller (1970), Driscoll *et al.* (1968) and Grady *et al.* (1973). Two simple, less sophisticated techniques have found usage in the calculation of x. These are:

5.4.2   Sondack's method (1972)

Rather than employ the 'trial and error' method of Perkin-Elmer (1967b), this method determined x based on three points of the fraction-melted curve, using equation (5.7).

$$x = \frac{\dfrac{T_3 - T_2}{T_2 - T_1} A_3 - \dfrac{A_3 - A_2}{A_2 - A_1} A_1}{\dfrac{A_3 - A_2}{A_2 - A_1} - \dfrac{T_3 - T_2}{T_2 - T_1}} \tag{5.7}$$

Where A_1, A_2 and A_3 are the uncorrected areas equivalent to the amount molten at three temperatures T_1, T_2 and T_3 respectively. Sondack (1972) suggested that the points be chosen from the high, middle and low values of uncorrected $1/F$ values. The Sondack method is less accurate than linear least-squares fitting and is less sensitive to variables such as heating rate, sample mass and particle size (van Dooren & Muller, 1983).

5.4.3   Davis & Porter's method (1969)

Davis & Porter criticized the segmental analysis of Perkin-Elmer because very pure samples on correction gave overestimates of the heat of fusion. They developed an alternative method (Fig. 5.7) based on energy transfer concepts. The choice of point B, prior to peak G on the DSC trace, was made on the basis that the peak, even allowing for temperature correction, does not correspond to the temperature at which the major component had melted. The choice of B was not critical because the T_s versus $1/F$ relationship is a straight line with an intercept on the $1/F$ axis equal to 1.0 at the T_s corresponding to point B.

By sectional analysis the uncorrected $1/F$ value for any given temperature (e.g., at point E) is given by equation (5.8)

$$\frac{1}{F} \text{ uncorrected)} = \frac{\text{Area AGC}}{\text{Area AEH}} \tag{5.8}$$

Area ABD corresponds to the energy added by the average power loop during the melting of the sample. Half is due to the sample holder and half to the reference holder. Therefore, energy equivalent to $\frac{1}{2}$ ABD is recorded by the differential loop during the energy transfer represented by area BCD. P can be defined by equation (5.9)

$$P = \text{Area (BCD)} - \tfrac{1}{2} \text{Area (ABD)} \tag{5.9}$$

and represents energy added by the average power loop before the differential power loop began to record. It should be added to each chosen segment such as AEH and implies that power transfer lags power input.

The trace area equivalent to that of a pure sample can be calculated and from this area AGC may be subtracted. This difference can be multiplied by 0.87 and this

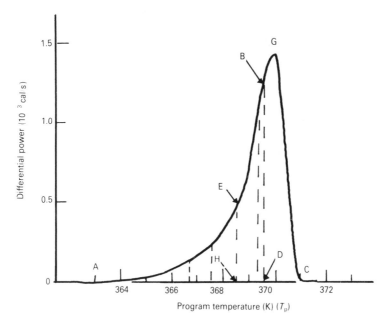

Fig. 5.7 — Cholestryl propionate, crystal to mesophase transition (5.306 mg, 2.5°C/min). (From Davis, G. J. & Porter, R. S. *J. Thermal Anal.*, **1** 449–458. Copyright (1969) John Wiley and Sons Ltd. Reprinted by permission of John Wiley and Sons Ltd.

calculated area corresponds to the energy that was supplied by the average temperature loop and was not recorded. This value may be termed Q. Area AEH is corrected by addition of its share of $\frac{1}{2}$ ABD recorded by the differential temperature loop during the energy transfer represented by area BCD or $\frac{1}{2}$ AEH. The corrected $1/F$ value therefore becomes equation (5.10)

$$\frac{1}{F} \text{(corrected)} = \frac{(\text{Area AGC}) + Q}{(2/3 \text{ Area AEH}) + P + Q} \tag{5.10}$$

If the points of several segments analysed do not pass through $1/F = 1.0$ corresponding to T_s or do not form a straight line, another point similar to B should be chosen. Davis & Porter (1969) claimed that only one or perhaps two attempts needed to be made before favourable results were obtained. Examples were quoted where the intersection of the $1/F$ equal to 1 was within 0.1°C of B. The correction value of 0.87 applied to a rate of 2.5°C/minute and at other rates a similar correction needed calculation. A fraction of 0.87 (87%) was the unrecorded energy of the difference between the energy of the pure standard component (naphthalene) and a 0.05 molar impurity based on a sample size of 3 mg. This estimate of heat loss decreased slightly as the sample weight was increased to 6 mg.

5.4.4   Alternative methods

Plato & Glasgow (1969) considered that small errors in T_o did not affect the accuracy of the melting point depression. For instance a ± 2 K change for T_o around a T_0 of 279.4 K, taken with ΔH_f of 4770 cal mole^{-1} and a ΔT of 0.3 K gave a difference of only 0.03 mole%. Cooksey & Hill (1976) attempted to eliminate the necessity of determining x by measuring peak heights and length in the concave portion of the melting curve. Suitable calculations and standardization with indium allowed adequate determinations of purity of acetanilide and sulphanilamide.

5.5   PHASE EQUILIBRIA CONSIDERATIONS

Ideally knowledge of the phase diagram of the major component with its impurity would be beneficial in purity determination (Grady *et al.*, 1973). However, samples of known purity are required to construct the diagrams and there is no guarantee that an impurity will behave as the samples studied. Additionally impurity identity may not be known.

Eutectic formation has minimal effects on purity determination (Barrall & Diller, 1970) provided that the eutectic temperature is well removed from the melting temperature of the major component. By definition the eutectic melt is insoluble in the solid phase of the major component but is soluble in its melt. Thus eutectic impurities should behave as described by the van't Hoff equation. When the eutectic temperature is near to the melting point of the major component, or possibly at high impurity levels, the melting of the eutectic component will interfere with the melting of the major component and render purity determination impossible. Marti (1972) derived theoretical phase diagrams for several systems, including the phenacetin––benzamide phase diagram. Comparison with experimentally derived diagrams gave a measure of the quality of approximation provided by the chosen theoretical equation for the solubility equilibrium. Theoretical melting curves are useful for proof of the ideality of the experimental curves (Marti, 1972) which is important in assessing the reliability of calculated purity values.

5.5.1   Solid solutions

Since the absence of solid solutions can never be guaranteed, it is reasonable when examining samples to assume that they occur. Driscoll *et al.* (1968) utilized a modified form of the van't Hoff equation (equation (5.1)) as given in equation (5.11) to overcome the problem

$$T_s = T_o - \frac{RT_o^2 X_2}{\Delta H_f} \cdot \frac{1}{\left(\dfrac{K}{1-K} + F\right)} \tag{5.11}$$

$K = k/k^1$, the distribution ratio of the impurity between the solid and liquid phases is difficult to measure. It is zero in the absence of solid solutions Equation (5.11) may linearize some $1/F$ versus T_s curves. However, by using known solid solution formers Driscoll *et al.* (1968) found that DSC could not detect the impurity and that impurities forming true solid solutions altered insignificantly in the heat of fusion. Thus the measured impurity is insensitive to the actual level of impurity. Curvature of plots, due to non-equilibrium solid solution formation, can be corrected using equation (5.11) and by correcting the heats of fusion since for such systems, e.g. toluene in benzene, the calculated impurity was directly related to the measured heat of fusion. A correction for this decrease in the heat of fusion could be made from measurement of that of a pure (99.5% +) standard of the same material. The best estimates of purity (Driscoll, 1968), even allowing for such corrections and equation (5.11), were ± 20% for a single sample and ± 10% for average results with a minimum error of ± 0.02 mole% due to precision. Most errors produced low impurity. No experimental proof of equations (5.11) exists and the validity of the method is doubtful.

The problem caused by solid solutions were highlighted by Marti (1972). During heating, equilibrium concentrations of the component are attained by diffusion of components within each phase. In eutectic systems such diffusion is restricted to the liquid phase only but in solid solutions will occur in both liquid and solid phases. Therefore, equilibrium for solid solutions is not easily maintained. The solid solubility of the impurity is a function of temperature and consequently crystallization frequently occurs. Grady *et al.* (1973) confirmed the inability of DSC to determine solid solution impurities, e.g. 3.6% norethindrone was undetected as an impurity in norethynodrel.

5.5.2 Impurity considerations

DeAngelis & Papariello (1968) considered the influences of impurities structurally either similar or dissimilar to the major component on purity determination. During manufacture analogues will be used which by their very nature will be similar to the final product. Stated examples included isomers, polymorphs or reactants, e.g. *p*-ethoxyacetanilide adulterated with acetanilide or *p*-chloroacetanilde and 7-chloro-1,3-dihydro-3-hydroxy-1-methyl-5-phenyl-2*H*-1,4-benzodiazepin-2-one adulterated with either its own polymorph or 7-chloro-1,3-dihydro-3-hydroxy-5-phenyl-2*H*-1,4-benzodiazepin-2-one. These systems gave excellent agreement between the DSC purity and the actual purity. Rapid re-crystallization of structurally similar materials may lead to solid solution formation and high purity values (DeAngelis & Papariello, 1968). Structurally, dissimilar impurities examined in *dl*-13-ethyl-17α-ethynyl-17-hydroxygon-4-en-3-one (DeAngelis & Papariello, 1968) included salts such as sodium chloride or lactose which were not detected at 5% levels presumably because they did not form a solution with the pure compound in the liquid state, thereby disobeying the van't Hoff equation.

Solid solutions occur in two forms. In substitutional solid solutions the impurities are approximately the same molecular size and shape as the major component and cannot be detected by DSC. Plato & Glasgow (1969) considered *d,l* isomers to be examples of this group. In interstitial solid solutions the impurities fit within the matrix of the major component. These may be alerted by DSC since the melting process involves compounds with low entropies of fusion ($\Delta s = \Delta H_f / T_o$) of less than five. This implies that the molecules are relatively free to move in the solid state and thereby favour solid solution formation. 3-(*p*-Chlorophenyl)-1,1-dimethylurea, aldrin and dieldrin were representative examples of these materials (Plato & Glasgow, 1969).

Elder (1979) considered the influences of two impurities, p-aminobenzoic acid and benzamide, as dopants in phenacetin. In binary systems respective eutectic temperatures of 385 and 370 K were obtained. A eutectic temperature of 353 K was obtained when both impurities were present. These temperatures were determined at a total impurity level of 10 mole%. The fusion enthalpy of the ternary eutectic was less than that of either binary eutectic. At low impurity levels the eutectic endotherms were insignificant in comparison with that describing phenacetin fusion. Table 5.2 gives results for 1.5 total mole% impurities for 3 mg samples. The values of T_o

Table 5.2 — Dynamic DSC purity analyses of doped phenacetin (3 mg samples) (reproduced with permission from Elder, 1979)

Dopant 1 (benzamide) (mole%)	Dopant 2 (p-aminobenzoic acid) (mole%)	Sondack Correction (k_s%)	ΔH_f (kJ/mole)	T_o (K)	ΔT (K)	TOTAL IMPURITY χ_2 (mole%)
0.5	—	4.9	31.83	407.3	0.23	0.53
1.0	—	8.4	30.73	407.3	0.45	0.99
1.5	—	9.3	30.51	407.4	0.68	1.49
—	0.5	3.6	30.50	407.4	0.22	0.48
—	1.0	11.8	31.16	407.4	0.45	1.01
—	1.5	8.9	31.14	407.5	0.67	1.51
0.25	0.25	14.6	29.92	407.4	0.22	0.48
0.50	0.50	13.0	32.00	407.4	0.43	1.00
0.75	0.75	13.2	30.38	407.5	0.67	1.47
0.67	0.33	5.3	31.83	407.5	0.44	1.02
0.33	0.67	9.9	32.34	407.4	0.47	1.09

and the molar enthalpy of fusion (31.09 ± 0.72 KJ mole^{-1}) were remarkably constant and confirmed that at a constant impurity level, purity values were independent of the dopant.

5.5.3 Solidus determination

Kawalec et al. (1982) reviewed equations (5.1) and (5.11) and other data treatments. All used the fraction F and gave problems in estimating initial melting. Problems which contributed to the failure to reach equilibrium were deviations from ideality in the liquid phase and solid solubility in the major component. Although a true equilibrium state was probably not accomplished by dynamic methods Kawalec et al. (1982) considered it was not a source of error, the major problem being the inability of DSC to discern accurately the solidus temperature at low impurity levels. NMR is an appropriate technique to detect accurately this temperature (Kawalec et al., 1982). DSC is unable to detect the start of melting where solid solution occurred (Habash et al., 1982). As the temperature passes through the solidus only a small fraction of the impurity will dissolve since the major portion remains in solid solution. Since any impurity is likely to be present as a result of the manufacturing process and possess a similar structure to the product, solid solution may well occur.

Habash *et al.* (1982) investigated purity in menandione and salol and added further similar data for phenacetin (Kawalec *et al.*, 1982). For doped systems the value of $T_o - T_m$ was determined by computer prediction, manual examination of the trace and NMR analysis, the last technique giving the most accurate estimation of the solidus temperature.

5.5.4 Heats of fusion

The apparent heats of fusion of impure samples are less than those of pure materials. Driscoll *et al.* (1968) proposed that a corresponding correction should be made in the van't Hoff equation. Table 5.3 examines the heats of fusion of benzene in the

Table 5.3 — Heats of fusion (ΔHf, DSC) from DSC melting curves for benzene with various amounts of eutectic impurity compared to literature values (ΔHf, lit) (reproduced with permission from Marti, 1972)

DSC purity (mole%	Heat of fusion, DSC, uncorrected baseline ΔHf, DSC (cal/mole)	(ΔHf, DSC)$-$(ΔHf, lit)(ΔHf, lit)$\times 100$ (%)
99.8	2352	+0.01
99.05	2237	−4.8
99.10	2131	−9.3
97.14	1788	−23.9
91.5	1293	−49.2

presence of added impurity. At a mole% purity of 91.5 the apparent heat of fusion was 49.2% less than the theoretical value. Although this may be explained by instrument insensitivity and pre-melting the heat of fusion needs to be corrected before proper purity determination can be made. Correction will also be made by linearization of the $1/F$ versus Ts diagrams.

5.6 ACCURACY RANGE

Various estimates have been made of the accuracy of DSC purity measurements. DeAngelis & Papariello (1968) concluded that purity values of less than 99 mole% were likely to be erroneous. No systems were encountered in which accurate results were obtained beyond 1.5 mole% impurity. Representative results (Fig. 5.8) revealed that analysis of samples containing 2 mole% impurity yielded values of near 1.5%, and 5 mole% impurity yielded values of only 2.5%. Plato & Glasgow (1969) considered that the method is accurate only for samples which are over 98% pure. Errors in the value of T_o lead only to small changes in the calculated impurity.

Under optimal conditions a precision of ±3% of the concentration of the

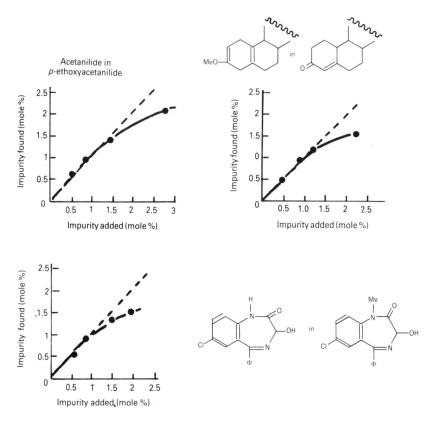

. Fig. 5.8 — Limits of accuracy for three doped systems ———: amount determined by DSC; – – –: actual amount of impurity present (reproduced with permission from DeAngelis & Papariello, 1968).

component of lowest concentrations could be obtained (Barrall & Diller, 1970). Using computer-aided facilities Driscoll *et al.* (1968) claimed that for repeat determinations an accuracy of ±10% around the mole impurity could be obtained with a minimum error of ±0.02 mole% due to precision.

Joy *et al.* (1971) reviewed the accuracy of DSC purity determination and published Fig. 5.9 which indicated that most DSC purities fell into ±20% agreement using either doped samples or samples assayed by other means. Samples suspected of containing solid solutions were not included in this treatment.

Driscoll *et al.* (1968) stated that good correlations for purity were obtained in the range 97.0 to 99.9% purity with 3 mole% as the maximum upper limit of impurity assayable. Joy *et al.* (1971) considered this limit was 2 mole%. Plato (1972) estimated that the standard deviation was less than 4% in the value of the heat of fusion. This % was reflected in the purity value as 4% of the impurity value, e.g. 99.5±0.02 mole%.

An effective upper limit of 99.95% has been placed on the method (Barnard *et al.* 1970; Joy *et al.*, 1971) although values as high as 99.99% have been assigned to ultra-pure zone-refined compounds. Such small values of impurity may be dependent on

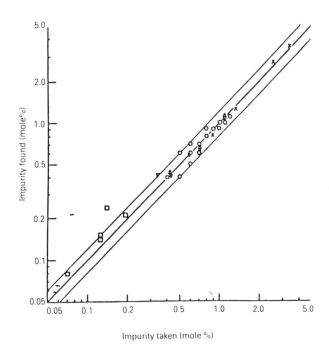

Fig. 5.9 — Correlation of impurity found by DSC measurements and impurity taken for published results. Central line corresponds to exact correlation. Upper and lower lines correspond to ±20% relative departure. Key: ○, DeAngelis & Papariello (1968); □, Driscoll *et al.* (1968); △, Plato & Glasgow (1969); ×, A. P. Gray Microchemical Workshop (1968), Pennsylvania University; ▽, Barrall & Diller, *Abs. Amer. Chem. Soc.*, 159th National Meeting, Houston, Texas, 1970, (reproduced with permission from Joy *et al.*, 1971).

the assumptions made in the calculations, e.g. baseline determination, onset of melting and in the van't Hoff equation.

Raskin (1985) investigated the influence of data treatments on computer-generated purities. The analytical methods of Sondack (1972), Marti (1972), Cooksey & Hill (1976) and Driscoll *et al.* (1968) were compared. The three-point method (Sondack) was satisfactory only if the three values were carefully chosen. The values of $1/F$ required for accuracy were dependent on the impurity, which in real samples would obviously not be known beforehand. The Driscoll *et al.* method was considered better since it utilised a search for the $1/F$ values to be used. The Marti (1972) and Driscoll *et al.* methods involved construction of successive linear approximations of the $1/F$ versus T_s curve although what portion is chosen depends on the level of the impurity. The other methods, however, yielded only one value of purity. Raskin (1985) concluded that under optimum conditions the average uncertainty was about 30% for mole fraction impurities (X_2) in the range 0.0075 to 0.0135 and increased to over 50% in the X_2 ranges 0.0135–0.0185 and 0.005–0.0075. Optimum conditions were defined as a scanning rate of no more than 2°C/min and a sample size of no more than 3 mg, for the better methods of Driscoll *et al.* (1968) and Marti (1972).

5.7   THE STEPWISE METHOD

The problems of non-equilibrium in dynamic DSC methods have presented limitations even at very low heating rates. Staub & Perron (1974) developed an alternative method, a stepwise method based on small incremental increases in temperature followed by isothermal holds to produce melting endotherms. The net effect is illustrated in Fig. 5.10. During the test the temperature is raised in steps to

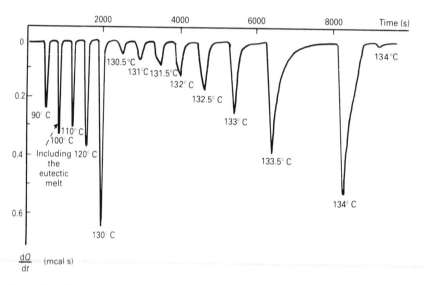

Fig. 5.10 — Purity determination by step heating programming technique. Sample: 10.7 mg phenacetin contaminated with 1.06 mole% benzamide. Reprinted with permission from Staub, H. & Perron, W., *Anal. Chem.*, **46**, 128–130. Copyright (1974), American Chemical Society.

measure the molten fraction F under actual equilibrium conditions. After each rise it is essential to achieve equilibrium. Between the eutectic melt and final melting point uniform temperature steps produce increasingly larger peaks; therefore, as the temperature reaches the final melting point smaller incremental increases are used (Staub & Perron, 1974). The final peak may be smaller due to its closeness to the final melting point. The incremental temperature increase should be small for pure materials, preferably 0.1°C (Palmero & Chiu, 1976). Background readings are taken before eutectic and after final melting to compensate for any differing heat capacities and reference. These areas are subtracted from the areas of the peaks of interest such that each peak area corresponds solely to sample melting. The corrected cumulative peak areas above the eutectic temperature are determined and, as a function of total area, plotted against the corresponding temperatures to produce the usual $1/F$ versus T_s plots. Molar impurities up to 10% could be determined with approximately 5% accuracy (Staub & Perron, 1974).

   Some criticisms and modifications of the method pre-date 1974. The existence of a long equilibrium period following each step is not necessarily due to the slowness o the thermodynamic equilibrium within the sample (Gray, 1972) but may be due to slowness in reaching equilibrium between the furnace and the sample when energy demands of the sample are high and the temperature differences are low. The stepwise method is time-consuming, taking over one hour for a single determination and gives no significantly different results from those obtained by dynamic methods at 0.3125°C/minute (Gray, 1972).

Over the 92–100% purity range the average precision for quantitative analysis was 0.5% and for the heat of fusion was 4% (Zynger, 1975). Stepwise methods detect early melting (Palmero & Chiu, 1976) which improved the linearity of response, e.g. phenacetin doped with 8.82 mole% benzamide gave an onset of melting about 5 K below the peak temperature recorded by dynamic studies whereas the stepwise method detected melting about 25 K below the peak. Correction for undetected melting is required to produce linearity. Palmero & Chiu (1976) determined purity down to 95 mole% without linearization and to 92 mole% with linearization.

Ramsland (1980) described fully automated systems for determining purity by stepwise methods. Ultra-pure samples, because they melt over a very narrow range, cannot be measured by the method. The stepwise technique is more accurate than the dynamic method for samples of low purity due to the difficulty of peak detection in the continuous method (Kiss *et al.*, 1981).

Staub & Perron (1974) suggested that the technique may be simplified by using only the areas of two peaks with Gray (1972) preferring that the second and third last peaks should be used. Gray & Fyans (1972) developed a method based on two consecutive melting peaks.

5.7.1 Stepwise two-peak method
The mathematical concepts of the two-step method of Gray & Fyans (1972) were summarized by Palmero & Chiu (1976). The steps used are as large as possible to check obedience to the van't Hoff equation. Fig. 5.11 shows the derivation of the

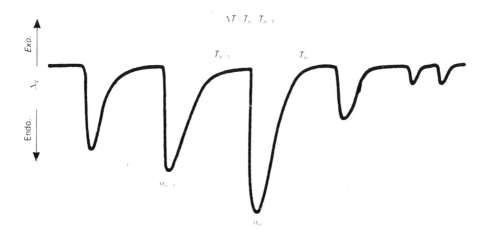

Fig. 5.11 — Representative peaks of stepwise two-peak method (reproduced with permission from Palmero & Chiu, 1976).

method. Remembering the ultimate peak may be smaller than previous peaks, the penultimate peak at T_n with peak area a_n and its immediate predecessor at T_{n-1} with peak area a_{n-1} are considered. The incremental temperature increase (ΔT) is given by equation (5.12).

$$\Delta T = T_n - T_{n-1} \tag{5.12}$$

Following subtraction of the background area from each peak area, T_o may be calculated from equation (5.13)

$$T_o = T_n + \frac{2(\Delta T)(\alpha_{n-1})}{(\alpha_n - \alpha_{n-1})} \tag{5.13}$$

where α_n is the corrected area of the nth stepwise peak and $\alpha_{(n-1)}$ is the corrected areas of the $(n-1)$th stepwise peak. The mole fraction of impurity χ_2, can be calculated from equations (5.14)

$$\chi_2 = \frac{2M\alpha_n \cdot \alpha_{n-1} \cdot \Delta T(\alpha_n + \alpha_{n-1})}{RT_o^2(\alpha_n - \alpha_{n-1})} \tag{5.14}$$

where M is the molecular weight of the pure component. Normally averaged values of T_o and χ_2 from three or four consecutive peak pairs should be used. Palmero & Chiu (1976) claimed to calculate purity down to 90 mole% with an approximate 5% relative error. The method is independent of the heat of fusion.

The time required to determine the penultimate two peaks is 15–20 min (Kiss *et al.*, 1981) depending on the impurity level, but the error in the calculated values may be as high as 100%. The time to measure the penultimate six peaks would be 30–40 min and the error reduced to 20%. Kiss *et al.* (1981) concluded that the time saved in measuring two peaks was outweighed by the increase in error. Obviously greater accuracy would be obtained if all peaks were measured.

The stepwise method has never been fully evaluated and is not much in favour for pharmaceutical purity. It was included in the United States Pharmacopoeia XX (1980).

5.8   PRACTICAL CONSIDERATIONS

For accurate purity determination considerable control is required of variables such as particle size, sample presentation, heating rates and the physico-chemical properties of the samples. Instrument manufacturers have defined methods suitable for their own designs but several general criteria should be followed.

5.8.1   Sample encapsulation
Early studies (e.g. DeAngelis & Papariello, 1968) considered the use of volatile sample pans unnecessary unless the samples were volatile. Since most organic materials have a finite vapour pressure near their melting points, cold-welded

volatile sample sealers should be used at all times (Barrall & Diller, 1970). Sample relocation during the fusion process is a problem, especially with organic materials because the sample formed a point-contact with the pan (Barrall & Diller, 1970). Relocation may alter the appearance of the melting endotherm on a reheating run but may be overcome by using aluminium insert discs on the sample before encapsulation. This sandwich formation favourably maintains contact of the sample with the pan base. The use of volatile sample pans is essential for materials which sublime below their melting points, e.g. anthraquinone (Reubke & Mollica, 1967). Volatile sample pans should also be used for liquids. The samples should be induced to crystallize by the use of coolants, e.g., liquid nitrogen, before the scanning is attempted. Polymorphic modifications may create difficulties by this method. Generally volatile sample pans should be used for all products.

5.8.2 Decomposing samples

Samples which decompose on fusion are often thought unsuitable for purity determination by thermal analysis. Reubke & Mollica (1967), for instance, considered that melting with decomposition eliminated many pharmaceuticals, especially salts. It is possible, however, to examine such materials with computer-facilitated analysis provided the enthalpy of fusion may be determined. Only the first 20% of the melting curve is required to describe purity.

Van Dooren & Muller (1983) examined samples which were thought to decompose on heating including adipic acid, benzoic acid and dydrogesterone. Oxidation did not significantly effect the purity of adipic acid. Benzoic acid sublimed on heating which affected purity determination. Dydrogesterone decomposed slightly on heating and consequently a lower value of T_o was obtained at 0.02 K/s than at 0.04 K/s. Faster heating rates were recommended when examining the purity of decomposing samples and a variation of T_o with heating rate may indicate decomposition. A slower heating rate allows more decomposition and gives a low purity estimate.

Ramsland (1984) developed a technique for unstable materials by analysing the profile of the eutectic peak of two- and three-component systems. Instead of determining the broadening effect on the melting of the major component the technique examined the broadening of the eutectic of two-component systems. Thus the eutectic endotherm of a 95:5 paracetamol:p-aminobenzoic acid system was broadened by the addition of 0.1% phenacetin. Ramsland (1984) claimed that any pure component which formed a eutectic with the unstable compound could be chosen to reduce the eutectic temperature to well below the start of decomposition. Only a minor portion of the decomposing component will melt in the eutectic, further reducing decomposition. Sensitivity was increased by changing the fraction of the eutectic component and impurities in the phenacetin–biotin–paracetamol system were also examined.

Sample loss via vaporization or decomposition when using a pierced sample pan may give purity values in excess of 100 mole%. Decomposition may be reduced by using a nitrogen atmosphere during analysis.

5.8.3 Sample particle size

Few studies have rationalized the influences of particle size on purity. Van Dooren & Muller (1983) found that the influences of particle size varied with the substance and only general trends could be identified. Higher purities were often found with larger particles but lower values were often noted with even larger sizes. Thermal equilibrium is difficult to accomplish in large-size powders because of thermal gradients (Plato & Glasgow, 1969). Extremely small particles and aggregated samples give high purity values (Plato & Glasgow, 1969). Trituration, as directed in the United States Pharmacopoeia XXI (1985) is not therefore always preferable. Decomposition may further modify particle size influences. Some results (Van Dooren & Muller, 1983) best illustrate particle size influences. Adipic acid displayed higher purities for small-size samples and comminuted samples, which corresponded to a sharper melting curve in the $1/F$ range of interest. Smaller particles lead to smaller thermal lags giving sharper peaks and higher purities. Larger size fractions displayed by low T_o values and low enthalpies (248.6 J/g compared with 255 J/g for the other fractions). Amidopyrine showed a low purity for the comminuted fraction and a high T_o for the large particle fraction. The broader peak of the milled samples was attributed to their greater amount of surface defects or to the formation of decomposition products during comminution. Small particles of caffeine aggregated together. Milling enhanced the responsible electrostatic forces giving caffeine aggregates which, in turn, led to higher purities. The influences of sieve fractions on the purity of benzamide is illustrated in Fig. 5.12. The low purity at low particles

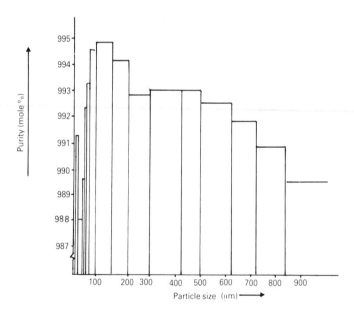

Fig. 5.12 — Purity values obtained for different sieve fractions of benzamide (reproduced with permission from Van Dooren & Muller, 1983).

sizes was due to thermal gradients within the powder bed caused by lower heat transfer properties, broadening the peak. The low purity of fractions of less than 150 μm may have been due to an increase in crystal defects giving a higher energy state thus broadening the peak or to the artifact of the impurity being preferentially distributed onto the crystal surfaces.

5.8.4   Sample size/heating rates

The dual influence of sample size and heating rates cannot be separated. Barrall & Diller (1970) considered that a sample size of 3 mg or less coupled with a heating rate of 1.25°C/min allowed good purity estimation. Higher rates did not allow the maintenance of equilibrium during heating. The rate should not exceed that which adjusts the slope of the endotherm to not exceed the slope of the pure standard at half peak height. Doubling the sample size to 6 mg at 0.625°C/min resulted in an underestimation of impurity of 23%.

Palmero & Chiu (1976) accepted that sample sizes should be kept in the range 2–4 mg and sizes in excess of 5 mg do not allow thermal equilibrium. The maximum heating rate acceptable was 1°C/min, for normal samples. Faster rates, up to 10°C/min could be used for unstable samples or those which gave broad peaks. Van Dooren & Muller (1983) considered that lowering heating rates below 0.04 K/s did not substantially effect the determined purity but higher rates gave low purity values and high T_o estimates, due to thermal gradients within the sample. The values of undetected melting were also greater at the faster heating rates.

5.8.5   Re-examination

Consideration should be given to reheating a sample to obtain a second purity estimate. Reheating following recrystallization may yield information regarding crystal modification (Marti, 1972) and help detect inapparent degradation yielding lower values of purity. Additionally, retreatment will make feasible the purity determination of samples containing mixed polymorphs since only one form should recrystallize on cooling. Marti (1972) suggested isothermally holding unstable samples at 10°C below their melting point for up to 30 min to give an estimate of their stability prior to analysis.

5.8.6   Doping techniques

It is essential that purity determination should be validated. This is usually accomplished by the use of samples containing known quantitites of added impurities. Great care must be taken in preparation of these doped samples. Doping by trituration should be avoided (DeAngelis & Papariello, 1968) because of posible homogeneity when small samples are used. It is vital that both the major component and the impurity are weighed separately into the sample pan prior to analysis. An error of ±0.05 mole% was anticipated in the dopant range of 0–1.5 mole% (Elder, 1979). The use of organic solvents to effect mixing is acceptable only if the solvent can readily be removed and no polymorphic transformation occurs. Rapid solvent removal may produce unstable solid solutions. The scatter of purity results for doped products is often due to mixing and sampling difficulties (Driscoll *et al.*, 1968)). Problems of inhomogeneity are increased by small samples of about 1 mg.

Palmero & Chiu (1976) recommended that solid standards including dopants should be prepared by adding known weights to a sealed test tube, fusing, stirring for 30 min and quenching on dry ice to maintain homogeneity prior to grinding. Phenacetin samples doped with benzamide were stable for over a year. However, Van Dooren & Muller (1983) showed that doping by a resolidified melt technique led to sampling errors and that preweighing of each component in the sample pan was preferable.

5.8.7 Instrument care

For optimum results Plato & Glasgow (1969) recommended that prior to use instruments should be raised to 700 K for a few minutes to remove contaminants and, when not in use, left at 400 K with nitrogen purge on. Contaminated cells in the event of noisy baselines should be boiled in methanol and baked clean at 700 K. These precautions are absolutely necessary since purity determinations require maximum sensitivity.

5.9   PHARMACOPOEIAL TESTS

Purity determination by thermal analysis is included in the United States Pharmacopoeia (USP), XX (1980) and XXI (1985) editions. In the former only the baseline enclosed by the extrapolated baseline is included for fraction melted determination. The method allowed parameters such as the melting range, ΔH_o and calculated eutectic purity to be readily obtained from a single scan. The method does not require multiple, precise, actual temperature measurements. The USP XX (1980) indicated that some chemical similarities are required for solubility in the melt and ionic compounds in neutral organic compounds or thermal decomposition may not be reflected in purity determination.

Both dynamic scanning and stepwise methods were included in the USP XX, (1980) the latter recommending temperature increases at 10°C intervals to within 20°C of melting and gradual stepping through the transition at intervals of 0.2 to 0.5°C, varying according to the purity. Correction methods for fraction of undetected melted were included. More sensibly the USP XXI did not specify a method but stated that procedures and calculations depend on the particular instrument used, thereby recommending recourse to the manufacturers' literature. The limitations due to solid solution formation, insolubility in the melt, polymorphism and decomposition during the analysis were emphasized.

5.10   PURITY ANALYSIS OF PHARMACEUTICALS

The number of compounds whose purities have been checked by DSC is extensive. Plato & Glasgow (1969) examined 95 molecules including organophosphates, amides, substituted ureas, carbamates, heterocyclics, chlorphenoxy compounds, esters and halogenated compounds. Plato (1972) added 64 compounds to this list. Grady *et al.* (1973) examined purity by various methods including thermal analysis, gas–liquid chromatography, thin-layer chromatography and high-pressure liquid chromatography for 115 drugs. Many decomposed on DSC including betamethasone, cyclizine hydrochloride, ethacrynic acid, norethyndrel, spironolactone and tetracaine hydrochloride. Polymorphic modifications were noted for chlorpropamide. Suitable analysis were obtained for many drugs including menadione, norethandrolone, testosterone and bisacodyl. Further examples include studies on testosterone (Perkin-Elmer, 1967a) methylreserpate (Reubke & Mollica, 1967) menandione (Habash *et al.*, 1982), nicotinamide (Marti, 1972), phenacetin (Marti, 1972; Staub & Perron, 1974; Zynger, 1975, Palmero & Chiu, 1976; Kawalec *et al.*, 1982; Habash *et al.*, 1982 and Van Dooren & Muller, 1983). The last reference includes details for adipic acid, amidopyrine, caffeine, dydrogesterone and naphzoline nitrate.

Badwan *et al*. (1986) estimated the purity of triethanolamine alkyl sulphates used as primary detergents in shampoos. Purity levels in excess of 99% were claimed.

Urea provides an example of a problem causer in DSC because it readily converts to biuret in the presence of moisture and heat. Joy *et al*. (1971) obtained values which were lower than the apparent purity of 99.82 ± 0.003 mole% listed by the National Bureau of Standards. Values were between 0.06 and 0.17 mole% less than that stated. This was due either to moisture or the low thermal stability of urea in the presence of even trace levels of water.

5.11 LIMITATIONS

The van't Hoff equation is applicable only to dilute solutions, probably less than 2 mole%. Any impurities which are insoluble in the solid state yet soluble in the melt can be considered. The reason for this is that the impurity must concentrate in the molten phase for the melting point depression to be related linearly to its concentration. Chemicals decomposing near their melting point cannot be examined. Their instability may lead to an increase in impurity within the sample and can give rise to an incorrect ΔH_o. Care should be taken that the sample pans do not induce or accelerate decomposition. The analysis of polymorphic materials is complicated by the need to reduce the material to the same form. This may be accomplished by heating but further emphasizes the usefulness of re-examining curves of materials already subjected to fusion. Materials with high vapour pressures cannot be examined since they may cause the volatile sample pans to rupture. Liquid materials which do not crystallize cannot be examined.

5.12 CONCLUSIONS

Provided its limitations are understood DSC provides a rapid method for purity determination. Decomposing samples can be analysed by more rapid rates than normally considered suitable. Solid solutions create problems. DSC is not the answer to all analytical problems but may support results from other techniques. A low DSC purity value is clear evidence that a compound is not of high purity (Joy *et al*., 1971). High purity obtained by DSC cannot be taken as evidence that the substance is highly pure without supportive evidence from other analytical techniques.

The influences of variables such as scan speed, sample weight and sensitivity have been reduced by the advent of computer facilities to monitor peak changes, enabling more reliance to be placed on DSC purity. Purity estimates should not be based on a single type of data when reporting purity determined by techniques such as thermal analysis, thin-layer chromatography and phase solubility analysis (Grady *et al*. 1973). The method does not yield absolute estimates of purity and requires a knowledge of the history of the sample and the likely impurities. Grady *et al*. (1973) estimated that only about 30% of drug substances yield DSC scans suitable for purity analysis. This is probably a high estimate.

REFERENCES

Badwan, A. A., James, K. C. & Pugh, W. J. (1986) *Anal. Proc.*, **23**, 390–391.
Barnard, A. J., Joy, E. F., Little, K. & Brooks, J. D. (1970) *Talanta*, **17**, 785–799.
Barrall, E. M. & Diller, R. D. (1970) *Thermochim. Acta*, **1**, 509–520.
Bowman, P. B. & Rogers, L. B. (1967) *Talanta*, **14**, 377–383.
Burroughs, P. (1980) *Anal. Proc.*, **17**, 231–234.
Cooksey, B. G. & Hill, R. A. W. (1976) *J. Thermal Anal.*, **10**, 83–88.
Davis, G. J. & Porter, R. S. (1969) *J. Thermal Anal.*, **1**, 449–458.
*DeAngelis, N. J. & Papariello, G. J. (1968) *J. Pharm. Sci.*, **57**, 1868– 1872.
Decker, D. L. & Young, R. W. (1978) *Thermochim. Acta*, **24**, 121–131.
Van Dooren, A. A. & Muller, B. W. (1983) *Thermochim. Acta.*, **66**, 161–186.
+ Van Dooren, A. A. & Muller, B. W. (1984) *Int. J. Pharm.*, **20**, 217–233.
Driscoll, G. L., Duling, I. N. & Magnotta, F. (1968) Purity determination using a differential scanning calorimeter. In *Analytical Calorimetry*, Vol. 1, Porter, R. S. and Johnson, J. F. (eds), Plenum Press, New York, 271–278.
Elder, J. P. (1979) *Thermochim. Acta*, **34**, 11–17.
*Grady, L. T., Hays, S. E., King, R. H., Klein, H. R., Mader, W. J., Wyatt, D. K. & Zimmerer, R. O. (1973) *J. Pharm. Sci.*, **62**, 456–464.
Gray, A. P. (1972). Some comments and calculations on a method of purity determination by stepwise method. *Thermal Analysis Application Study* 3, Perkin-Elmer, Norwalk, Connecticut, U.S.A.
Gray, A. P. & Fyans, R. L. (1972) Methods of analyzing stepwise data for purity determinations by differential scanning calorimetry. *Thermal Analysis Application Study* 10, Perkin-Elmer, Norwalk, Connecticut, U.S.A.
Habash, T. F., Houser, J. J. & Garn, P. D. (1982) *J. Themal Anal.*, **25**, 271–277.
Heuvel, H. M. & Lind, K. C. J. B. (1970) *Anal. Chem.*, **42**, 1044–1048.
Joy, E. F., Bonn, J. D. & Barnard, A. J. (1971) *Thermochim. Acta.*, **2**, 57–68.
Kawalec, B., Houser, J. J. & Garn, P. D. (1982) *J. Thermal. Anal.*, **25**, 259–270.
Kiss, G. V., Seybold, K. & Meisel, T. (1981) *J. Thermal Anal.*, **21**, 57–66.
Marti, E. E. (1972) *Thermochim. Acta*, **5**, 173–220.
Palmero, E. F. & Chiu, J. (1976) *Thermochim. Acta*, **14**, 1–12.
Perkin-Elmer (1967a) Determination of purity by differential scanning calorimetry. *Thermal Analysis Newsletter No.* 5. Perkin-Elmer Corporation, Norwalk, Connecticut, U.S.A.
Perkin-Elmer (1967b) Purity determination by DSC. *Thermal Analysis Newsletter No.* 6. Perkin-Elmer Corporation, Norwalk, Connecticut, U.S.A.
Plato, C. (1972) *Anal. Chem.*, **44**, 1531–1534.
Plato, C. & Glasgow, A. R. (1969) *Anal. Chem.*, **41**, 330–336.
Ramsland, A. C. (1980) *Anal. Chem.*, **52**, 1474–1479.
Ramsland, A. (1984). Proc. 13th North American Thermal Analysis Society Meeting, 138–144.
Raskin, A. A. (1985) *J. Therm. Anal.*, **30**, 901–911.
* Reubke, R. & Mollica, J. A. (1967) *J. Pharm. Sci.*, **56**, 822–825.
Sanmartin, P. & Regine, N. (1969) *J. Thermal Anal.*, **1**, 403–411.
Staub, H. & Perron, W. (1974) *Anal. Chem.*, **46**, 128–130.
Sondack, D. L. (1972) *Anal. Chem.*, **44**, 888, 2089.
United States Phamacopoeia XX, (1980) Thermal Analysis, United States Pharma-

copoeial Convention Inc. Rockville, U.S.A., 984–986.

United States Pharmacopoeia XXI, (1985) Thermal Analysis. United States Pharmacopoeial Convention Inc., Rockville, U.S.A., 1275–1276.

Zynger, J. (1975) *Anal. Chem.*, **47**, 1380–1384.

6

Thermal analysis in the characterization of pharmaceutical solids

6.1 INTRODUCTION

The isolation of a pharmaceutical solid is achieved often by means of a precipitation or crystallization from its solution in a solvent or mixture of solvents. The rate at which the separation of solid material from the solution occurs and the nature of the solvent, the concentration of the drug substance in solution and temperature at which solid separation occurs are all factors which can affect habit, solvate nature and crystalline modification of the separated solid.

In a review, Haleblian (1975) has differentiated habit and crystal chemistry of a drug compound according to the scheme in Fig. 6.1. In addition to the classes included in the scheme of Haleblian (1975) isomers of organic compounds will be discussed and how thermal analysis is useful in the characterization of this type of variation of a chemical substance.

While the definition of isomers may be understood by the majority of readers the definitions of polymorphism, habit and various types of solvent addition solids (solvates, inclusion compounds) may need further clarification.

A drug substance may occur in several different crystal habits, e.g. prisms, plates or needles. Where the habit varies the internal molecular orgnization of the crystal is the same for each habit but different crystal faces have developed to different extents resulting in the different crystal shapes. Unfortunately physico-chemical parameters including thermal analysis do not show differences between different crystal habits. Only their appearance, for example by microscopy, will differentiate them. It is important to identify the different habits as they may have different pharmaceutical properties. For example, Shell (1963) has indicated differing syringabilities of parenteral suspensions prepared from plate habit and needle habit forms of a drug.

In the case of polymorphism, separation of solid material from its solution results in materials of different habits but, in addition, physico-chemical evaluation indicates that the molecular organization within the solids is different. X-ray diffraction

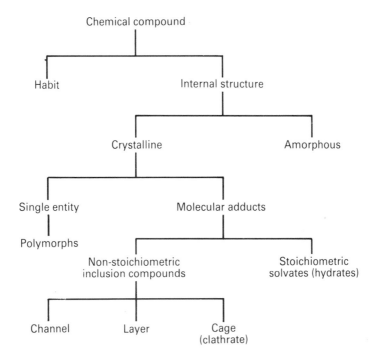

Fig. 6.1 — Outline scheme differentiating habit and crystal chemistry of a chemical compound (reproduced with permission from Haleblian, 1985).

of the physico-chemical methods will definitely confirm the different molecular organization within the solid while thermal analytical methods will provide useful additional information.

Because of the different internal organization within the solid, polymorphs may show different melting points, solubilities, chemical reactivity or stability. These can have an impact on pharmaceutical properties such as dissolution rate and bioavailability. Unfortunately polymorphism is common among pharmaceutical substances and the evaluation of new drug entities for polymorphism is important in early preformulation studies.

Solvates, often termed pseudopolymorphs, are formed when the crystallizing solvent is incorporated into the separating crystals during crystallization.

6.2 THERMAL ANALYSIS IN THE EVALUATION OF ISOMERS

Byrn (1982) has described the case of configurational polymorphism where different configurations such as tautomers or *cis–trans* isomers crystallize in separate crystalline forms. The latter example is well-known within the field of organic chemistry but the ability to crystallize equilibrating isomers in configurational polymorphs is of particular interest as it can allow for the isolation of individual isomers.

The application of thermal analysis in the case of such isomers has been rarely noted in the pharmaceutical literature, perhaps because this phenomenon is much rarer than the less special cases of polymorphism.

Ferrari (1969) has indicated some value of DTA in confirming the nature of two samples that show similar microanalysis and are theoretically (as far as the synthetic chemist is concerned) the same material. He showed that DTA can differentiate two benzyl ketone preparations (Fig. 6.2). Melting points differ by 2°C whereas melting

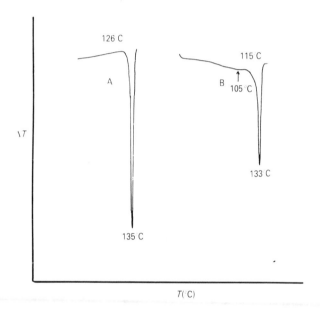

Fig. 6.2 — Comparison of two benzyl ketone preparations by DTA. (A) first preparation; (B) second preparation (reproduced with permission from Ferrari, 1969).

point onsets differ by 11°C. One sample (B) shows an additional shallow endotherm at a temperature below that of the main transition. As microanalysis for the samples was the same the possibility of dealing with a mixture of isomers was suggested.

For the antituberculosis compound ethambutol Ferrari & Grabar (1971) were able to demonstrate the presence of a *meso* isomer in the parent ethambutol by DTA. The mixture of the two-isomers could be analysed by virtue of the formation of a solid solution with a minimum melting point at 175°C. This was a more sensitive approach than attempts at quantifying individual endotherms for the two components themselves, which would be a more routine approach to the problem.

6.3   THERMAL ANALYSIS AND POLYMORPHISM

There is a considerable literature describing the evaluation of polymorphic forms of drug substances by thermal analysis and there are good reasons for the attention pharmaceutical scientists have paid to polymorphism. Varying the polymorphic form for a particular drug substance can alter chemical and physical stability due to altered reactivity and alter the rate of dissolution of the solid substance. As the rate of dissolution can be related to the degree of absorption, particularly of less soluble compounds, polymorphism may have an effect on the therapeutic activity of dose forms whose polymorphic nature was not controlled. Polymorphism is particularly widespread amongst pharmaceutical compounds and the barbiturates, sulphonamides and steroids have been studied in detail both as model compounds for the exhibition of polymorphs and in the evaluation of the effect of polymorphism on pharmaceutical properties. The purpose of this section is to cover the applications of thermal analysis in the study of polymorphs of pharmaceutical compounds.

The characterization of polymorphism within a pharmaceutical compound should be undertaken at the earliest opportunity during development of that compound. Should polymorphism occur this ensures that the most suitable polymorph, such as the one most likely to be routinely produced via scaled-up synthetic routes or the one demonstrating the most desirable stability or biopharmaceutical properties, is identified. This role falls upon the preformulation scientist working in the pharmaceutical and analytical research groups who may screen for polymorphs. This may involve carefully evaluating each new lot emanating from chemical-processes or drug-discovery research groups as the synthesis is scaled up or, perhaps most usually, by carrying out a polymorph screening exercise as soon as adequate material is available. This would involve the crystallization of the drug substance from a variety of solvents where conditions such as the nature of the solvent, temperature of crystallization, rate of cooling, rate of agitation or the nature of precipitation solvent are changed. All may potentially influence the formation of polymorphs.

The end sequence is to carry out a series of physico-chemical examinations on the obtained crystalline materials for evidence of polymorphism. Traditionally the examination methods employed include IR spectroscopy (of Nujol mulls), powder X-ray diffraction and thermal methods, especially DSC.

In the case of DSC, heating rate might be very important in the detection of polymorphism. Tuladhar *et al.* (1983), working with phenylbutazone, indicated that three polymorphs of the compound showed single endotherms on DSC at high heating rates whereas additional peaks were discernible at lower heating rates. This is due to less stable forms melting and recrystallizing as the more stable form prior to that form then melting. The result is a melting endotherm, a recrystallization exotherm and a final melting endotherm (Fig. 6.3). DSC alone is not sufficient evidence for the existence of polymorphism in such studies and it is essential that HSM is applied to visually confirm the events suggested by the DSC curves. It is particularly important to distinguish between endotherms due to melting of a less stable form prior to recrystallization and a second endotherm due to melting of a more stable form and endotherms attributable to desolvation followed by melting of the desolvate form. This can be done by mounting crystals in silicone oil and observing via a hot-stage microscope. An endotherm due to desolvation would correlate with bubbles of vapour appearing in the mountant around crystals at the appropriate temperature. Melting without loss of solvent would show loss of crystalline form without the production of any vapour bubbles.

The thermal stability of a polymorphic substance allows types of polymorphism

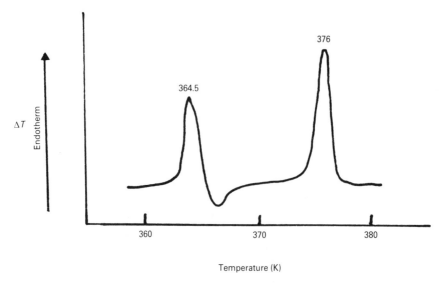

Fig. 6.3 — DSC scan of phenylbutazone polymorph E, obtained at 4°C min^{-1} (reproduced with permission from Tuladhar *et al.*, 1983).

to be defined, namely enantiomorphic and monotropic polymorphism (Kuhnert-Brandstatter, 1971). Enantiomorphic polymorphs can be encountered below the melting point of either polymorph. Hence storage of the less stable polymorph of a compound will result in conversion to a more stable polymorph on storage below the melting point of either. An example of enantiomorphic polymorphism is that of forms I and II and forms I and III of gepirone hydrochloride (Behme *et al.*, 1985). This was discernible by thermal analysis as endotherms on DSC traces which correlated with melting behaviour on HSM. The endotherms are due to solid–solid transitions as the less stable polymorphs convert to the more stable form. Conditioning samples of the lower stable polymorphs at temperatures below their melting points showed the loss of DSC endotherms for the less stable polymorph (Fig. 6.4). Typical DSC traces for the newly crystallized lower stable polymorph show the solid–solid transitions at low scanning rates (2.5–10°C/min) but at higher temperature scanning rates only the melting endotherms of the two polymorphs could be discerned (Fig. 6.5). The high scanning rate traces show exotherms due to recrystallization of the more stable melting form from melts of the less stable form. Such DSC behaviour can be considered typical of enantiotropic polymorphs. The value of HSM in evaluating polymorphism is well indicated by the work of Behme *et al.* (1985). Videorecording of thermal events on the microscope hot stage was useful in interpreting complex melting behaviour prior to which some solid–solid transitions had occurred (Behme *et al.*, 1985).

A further recent sample of enantiotrophy is metoclopramide base (Mitchell, 1985). The DSC trace for metoclopramide base showed two endotherms (at 124–126°C and 146–148°C) but on cooling and reheating (one recrystallization

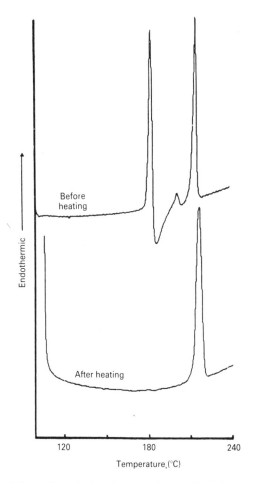

Fig. 6.4 — Differential scanning calorimetric scans of a sample of Gepirone before and after
heat treatment for 3 hours at 150°C which produced form II. The temperature scanning rate was
20°C min^{-1} (reproduced with permission from Behme *et al.*, 1985).

exotherm only was seen on cooling) only one endotherm, the higher temperature
one, was seen. Thermomicroscopy confirmed the lower temperature endotherm as a
solid–solid transition and the higher temperature one as a melting endotherm.
Storing the cooled sample pan for three days at 22°C resulted in the original thermal
behaviour, demonstrating the reversible nature of the enantiotrophy. Monotropic
polymorphism is where one polymorph is stable throughout the temperature range
up to the melting point of the highest melting polymorph. Metoclopramide hydrochloride has been well characterized in exhibiting this behaviour (Mitchell, 1985). By
heating metoclopramide hydrochloride monohydrate under a vacuum in a DSC pan,
dehydration is brought about (73–75°C) and metoclopramide hydrochloride form II
crystallizes from the melt (85–89°C). Form II then melts (154–156°) as the temperature is increased and form I crystallizes from this melt (156–158°C). Further heating

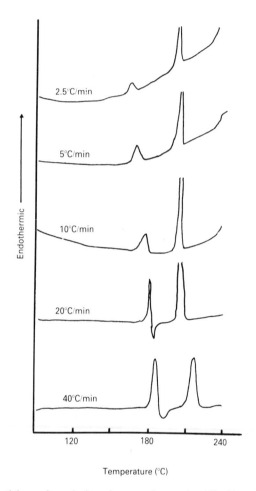

Fig. 6.5 — Differential scanning calorimetric scans of a sample of Gepirone form I recorded at different temperature scanning rates (reproduced with permission from Behme *et al.*, 1985).

results in the melting of form I (186–188°C). Form I cannot be reconverted to the metastable form II easily and spontaneous conversion on storage of the solid does not occur.

A large number of studies on polymorphism, most of which include some form of thermal analysis, have been undertaken. Some are listed in Table 6.1. This list is not exhaustive but includes some of the more useful references, concentrating particularly on significant and marketed compounds.

6.4   THERMAL ANALYSIS AND SOLVATES

During the process of the separation of a solid drug substance from its solution solvent may become entrappped in the crystal. Stoichiometric or non-stoichiometric amounts of solvent may be associated with the crystal.

Table 6.1 — Examples of drugs whose polymorphism has been recently studied by thermal analysis

Compound	Reference[a]
Acetazolamide	Umeda et al. (1985)
Acetohexamide	Graf et al. (1984)
Auranofin	Lindenbaum et al. (1985)
Carbamazepine	Lefebvre et al. (1986)
Chloramphenicol palmitate	Borka (1970)
Chlordiazepoxide hydrochloride	Simmons et al. (1970)
Chlorpropamide	Burger (1975)
	Ford & Rubinstein (1977)
	Simmons et al. (1973)
Chloroquine diphosphate	Van Aerde et al. (1984)
Cimetidine	Shibata et al. (1983)
Clotrimazole	Borka & Valdimarsdotter (1975)
Cortisone acetate	Carless et al. (1966)
Disopyramide	Gunning et al. (1976)
Enalapril maleate	Ip et al. (1986)
Indomethacin	Ford & Rubinstein (1978)
	Kaneniwa et al. (1985)
Lorazepam	Masse et al. (1985)
Maprotiline hydrochloride	Chan & Doelker (1985)
	Kuhnert-Brandstatter et al. (1982)
Meprobamate	Clements & Popli (1973)
Methylpredisolone	Guillory (1967)
Metoclopramide	Mitchell (1985)
Metoclopramide hydrochloride	Mitchell (1985)
Nifedipine	Eckert & Muller (1977)
Oxazepam	Masse et al. (1985)
Phenobarbitone	Mesley et al. (1968)
Phenylbutazone	Tuladhar et al. (1983)
	Muller (1978)
	Matsuda et al. (1984)
Piroxicam	Kuhnert-Brandstatter & Vollenklee (1985)
Propranolol hydrochloride	Kuhnert-Brandstatter & Vollenklee (1985)
Spironolacetone	El-Dalsh et al. (1983)
	Salole & Al-Sarraj (1985)
Sulphabenzamide	Yang & Guillory (1972)
Sulphaethidole	Yang & Guillory (1972)
Sulphaguanidine	Yang & Guillory (1972)
Sulphamethazine	Yang & Guillory (1972)
	Maury et al. (1985)
Sulphamethoxazole	Yang & Guillory (1972)
Sulphamethoxydiazine	Moustafa et al. (1971)
Sulphamethoxypyridazine	Yang & Guillory (1972)
Sulphapyridine	Yang & Guillory (1972)
	Gouda et al. (1977)
Sulphathiazole	Guillory (1967)
	Moustafa & Carless (1969)
	Shami et al. (1972)
	Lagas & Lerk (1981)
Sulphisoxazole	Yang & Guillory (1972)
Tamoxifen citrate	Goldberg & Becker (1987)
Tolbutamide	Leary et al. (1981)
	Ueda et al. (1982)
Trimethoprim	Bettinetti et al. (1976, 1978, 1980)

[a] The references cited contain the most comprehensive data on polymorphism of the compound in question where thermal analysis has been used as one of the characterization methods or is the only literature reference. Where more than one paper is cited this indicates that the papers are complimentary rather than supplementary, e.g. different aspects of polymorphism were investigated and thermal analytical data included. Aspirin has been deliberately excluded from the list because of the controversial nature of its polymorphism.

The general term of crystalline materials containing solvent of crystallization is solvates, where hydrates are a particular case when the solvent entrapped is water. Anhydrates are solvent-free crystalline materials although the term is strictly indicative of a crystal form where water of crystallization is absent. Solvates may be more or less stable, some requiring forcing conditions to produce desolvation, resulting in production of a new crystal form from the anhydrous solid. Other solvates desolvate readily but have an anhydrate form with the same crystal structure as the solvate (as suggested by X-ray diffraction pattern). Such desolvated crystals readily become resolved, e.g. on exposure to an atmosphere of solvent vapour (a high humidity atmosphere for water). Some solids lie between the two extremes, readily attracting the solvent of crystallization but converting to a different crystal form in doing so. Where a different crystal form is produced on desolvation, resolvation has to be accomplished by dissolution and recrystallization.

Thermal analytical methods, DTA, DSC, HSM and TG, are all valuable tools in assisting in the characterization of solvates. An extensive study of drug substances was undertaken by Kuhnert-Brandstatter (1971) employing HSM, showing that a large number of classes of drug substances form hydrates and solvates including β-lactam antibiotics, sulphonamides, corticosteroids and xanthine derivatives. Complex desolvation behaviours for a variety of substances were classified utilizing HSM in conjunction with DSC, evolved gas analysis (EGA) and TG by Kuhnert-Brandstatter & Prohll (1983a, b). Further general examples of the application of DSC, TG and HSM to the analysis of pharmaceutical solvates have been given by Brown & Hardy (1985). These authors used DSC on samples cooled below the freezing point of water in an attempt to detect the melting of ice derived from 'clustered water' in the material. Adsorbed monolayers of water and tightly bound hydrate water will not show in the DSC endotherm for the melting of ice. A solvate may be differentiated from a wet sample because in the former solvent loss occurs over a narrow temperature range(s) whereas in a wet sample the loss slowly occurs over a wide range. Brancone & Ferrari (1966) demonstrated the potential of DTA in analysing for the presence of solvent. In studies on sulphamonomethoxine a shallow endotherm at 120°C was apparent when scanned at atmospheric pressure but under reduced pressure the endotherm decreased to 66°C. Table 6.2 indicates a partial list of important drug substances (or those exhibiting classical behaviour on thermal analysis) that form hydrates or other solvates. Some examples are discussed further below to demonstrate the use of thermal analysis in this field.

Niazi (1978) examined the crystal forms of the antitumour agent mercaptopurine using techniques including DSC. Endotherms for the desolvation and the melting of the monohydrate are seen on the DSC scans for this compound. A heat-treated sample of the monohydrate showed no endotherm for the loss of water of crystallization (Fig. 6.6). DSC was used to calculate the enthalpy of dehydration as 8.27 kcal/mole.

The macrolide antibiotic erythromycin appears to show a number of hydrates that have been evaluated by various physical techniques, notably thermal analysis (Pelizza et al., 1976; Allen et al., 1978). Whilst anhydrous, dihydrate and monohydrate forms are indicated, the role of water in the structure of erythromycin crystals appears to be one of a loose association with the water being easily lost and taken up again (Allen et al., 1978).

Table 6.2 — Examples of hydrates and solvates recently studied using thermal analysis

Compound	Reference
Calcium gluceptate	Suryanarayanan & Mitchell (1986)
Cephalexin	Otsuka & Kaneniwa (1983)
Codeine hydrochloride	Kuhnert-Brandstatter & Proll (1983a)
Erythromycin	Pelizza *et al.* (1976)
	Allen *et al.* (1978)
	Murthy *et al.* (1986)
Ferrous sulphate	Mitchell (1984)
Magnesium stearate	Miller & York (1985)
Magnesium palmitate	Miller & York (1985)
Mercaptopurine	Niazi (1978)
Metronidazole benzoate	Hoelgaard & Møeller (1983)
Piroxicam	Kozjek *et al.* (1985)
Spironolactone	Salole & Al-Sarraj (1985)
Sulphamethoxazole	Abdallah & El-Fattah (1984)
Testosterone	Frøkjaer & Andersen (1982)
Theophylline	Serajuddin (1986)
Trimethoprim	Bettinetti *et al.* (1978)

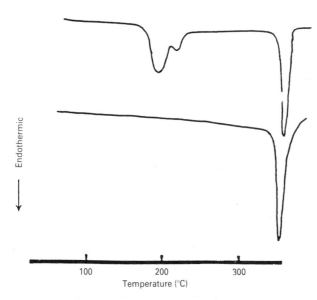

Fig. 6.6 — Differential scanning calorimetry scans of mercaptopurine. Key: top, original drug; bottom, heat-treated sample (held at 200°C for 20 mins) or rerun of sample in the top scan (reproduced with permission from Niazi, 1978).

Cephalexin appears to show similar behaviour (Byrn, 1982) forming a monohydrate and a dihydrate that can be desolvated and resolvated without changes in the crystal lattice. The desolvation in effect produces a sponge-like crystal able to resorb the solvent of crystallization easily.

Otsuka & Kaneniwa (1983, 1984) applied the thermal kinetic methods of Kissinger (1957), Craido et al. (1978) and Barton (1969) (see Chapter 4) to evaluate by DTA the dehydration of various cephalexin hydrates. The activation energy for dehydration of monohydrate was in the range 15–18 kcal/mole and for dihydrate to monohydrate was 15.33 kcal/mol. Typical DTA curves are shown in Fig. 6.7.

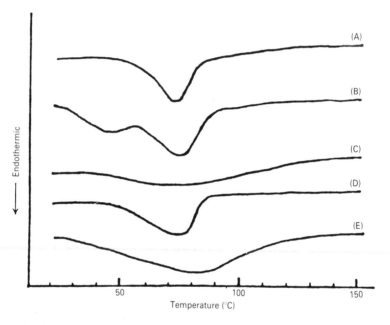

Fig. 6.7 — DTA curves of cephalexin hydrates at heating rates of 10°C/minute. Key: A, cephalexin phase IV monohydrate; B, cephalexin phase II dihydrate; C, cephalexin phase III-hemihydrate; D, cephalexin phase V monohydrate; E, non-crystalline solid cephalexin dihydrate (reproduced from Otsuka & Kaneniwa, 1983 with permission of the publishers).

The use of thermal analysis in the study of pharmaceutical excipients, such as magnesium stearate and lactose, which are also capable of forming solvates, are discussed in greater detail in Chapter 9.

6.5 THERMAL ANALYSIS IN THE EVALUATION OF SUBTLE MODIFICATIONS OF CRYSTAL FORM

To improve certain properties of pharmaceutical solids such as flow, compression and dissolution (and hence possibly bioavailability), the concept of crystal engineering, entailing modification of drug crystals during the crystallization process via a number of procedures including the deliberate addition of a trace impurity has been developed.

Crystals of adipic acid modified in this way (adipic acid being used as a model compound) have been evaluated by complementary physico-chemical techniques including DSC where the heat of fusion was determined (Chow *et al.*, 1984, 1985a). Related fatty acids were used in the case of adipic acid but Chow *et al.* (1985b) utilized a synthetic impurity of acetaminophen (paracetamol), namely *p*-acetoxyacetanilide, to 'poison' crystals of the parent compound and modify their pharmaceutical properties. Again complementary physico-chemical measurements including heat of fusion determined by DSC was utilized to characterize the various products obtained. In further work on adipic acid Chow *et al.* (1986) indicated that DSC was a more sensitive indicator of crystal imperfections than the often applied measurement of density.

Further developing this work York & Grant (1985) proposed the development of a disruption index for such solid solutions whereby heat of fusion data obtained by DSC could be utilized to characterize the disorder created in a crystal structure by such poisoning. The value of such an index remains to be ascertained as work in this field progresses.

REFERENCES

Abdallah, O. & El-Fattah, S. A. (1984) *Pharm. Ind.*, **46**, 970–971.

Allen, P. V., Rahn, P. D., Sarapu, A. C. & Vanderwielen, A. J. (1978) *J. Pharm. Sci.*, **67**, 1087–1093.

Barton, J. M. (1969) *Polymer*, **10**, 151–154.

Behme, R. J., Brooke, D., Farney, R. F. & Kensler, T. T. (1985) *J. Pharm. Sci.*, **74**, 1041–1046.

Bettinetti, G. P., Giordano, F., La Manna, A. & Giuseppetti, G. (1976) *J. Pharm. Pharmac.*, **28**, 87–88.

Bettinetti, G. P., Giordano, F., La Manna, A. & Giuseppetti, G. (1978) *Boll. Chim. Farm.*, **117**, 522–529.

Bettinetti, G. P., Giordano, F. & Ferloni, P. (1980) *Il Farmaco Ed. Sci.*, **35**, 706–714.

Borka, L. (1970) *Acta Pharm. Suec.*, **7**, 1–6.

Borka, L. & Valdimarsotter, S. (1975) *Acta Pharm. Suec.*, **12**, 479–484.

Brancone, L. M. & Ferrari, H. J. (1966) *Microchem. J.*, **10**, 370–392.

Brown, D. E. & Hardy, M. J. (1985) *Thermochim. Acta*, **90**, 149–156.

Burger, A. (1975) *Sci. Pharm.*, **43**, 152–161.

Byrn, S. (1982) *Solid state stability of drugs*, Academic Press, New York.

Carless, J. E., Moustafa, M. A. & Rapson, H. D. C. (1966). *J. Pharm. Pharmac.*, **18**, 190S–197S.

Chan, H. K. & Doelker, E. (1985) *Drug Dev. Ind. Pharm.*, **11**, 315–332.

Chow, K. Y., Go, J., Mehdizadeh, M., & Grant, D. J. W. (1984) *Int. J. Pharm.*, **20**, 3–24.

Chow, A., K. Y., Go, J., Zhongshan, W., Mehdizadeh, M., & Grant, D. J. W. (1985a) *Int. J. Pharm.*, **25**, 41–55.

Chow, A. H.-L., Chow, P. K. K., Zhongshan, W. & Grant, D. J. W. (1985b) *Int. J. Pharm.*, **24**, 239–258.

Chow, K. Y., Go, J., & Grant, D. J. W. (1986) *Drug Dev. Ind. Pharm.*, **12**, 247–264.

Clements, J. A. & Popli, S. D. (1973) *Can. J. Pharm. Sci.*, **8**, 88–92.

Craido, J. M., Morales, J. & Rives, V. (1978) *J. Therm. Anal.*, **14**, 221–228.

Eckert, T. & Muller, J. (1977) *Arch. Pharm.*, **310**, 116–118.

El-Dalsh, S. S., El-Sayed, A. A., Badawi, A. A., Khattab, F. I. & Fouli, A. (1983) *Drug Dev. Ind. Pharm.*, **9**, 877–894.

Ferrari, H. (1969) DTA in pharmaceutical sciences, in *Thermal analysis, vol 1: instrumentation, organic materials and polymers'* Academic Press, New York, 41–64.

Ferrari, H. & Graber, D. G. (1971) *Microchemical J.*, **16**, 5–13.

Ford, J. L. & Rubinstein, M. H. (1977) *J. Pharm. Pharmac.*, **29**, 209–211.

Ford, J. L. & Rubinstein, M. H. (1978) *Pharm. Acta Helv.*, **53**, 327–332.

Frøkjaer, S. & Andersen, V. S. (1974) *Arch. Pharm. Chemi. Sci. Ed.*, **2**, 50–59.

Goldberg, I. & Becker, Y. (1987) *J. Pharm. Sci.*, **76**, 259–264.

Gouda, M. W., Ebian, A. R., Moustafa, M. A. & Khalil, S. A. (1977) *Drug Dev. Ind. Pharm.*, **3**, 273–290.

Graf, E., Beyer, C. & Abdallah, O. (1984) *Pharm. Ind.*, **46**, 955–959.

Guillory, J. K. (1967) *J. Pharm. Sci.*, **56**, 72–76.

Gunning, S. R., Freeman, M. & Stead, J. A. (1976) *J. Pharm. Pharmac.*, **28**, 758–761.

Haleblian, J. K. (1975) *J. Pharm. Sci.*, **64**, 1269–1288.

Hoelgaard, A. & Møeller, N. (1983) *Int. J. Pharm. Sci.*, **15**, 213–221.

Ip, D. P., Brenner, G. S., Stevenson, J. M., Lindenbaum, S., Douglas, A. W., Klein, S. D. & McCauley, J. A. (1986) *Int. J. Pharm.*, **28**, 183–191.

Kaneniwa, N., Otsuka, M. & Hayashi, T. (1985) *Chem. Pharm. Bull.*, **33**, 3477–3455.

Kissinger, H. E. (1957) *Anal. Chem.*, **29**, 1702–1706.

Kozjek, F., Golic, L., Zupet, P., Palka, E., Vodopivec, P. and Japelj, M. (1985) *Acta Pharm. Jugosl.*, **35**, 275–281.

Kuhnert-Brandstatter, M. (1971) *Thermomicroscopy in the analysis of pharmaceuticals*, Pergamon Press, New York, 34–42.

Kuhnert-Brandstatter, M. & Proll, F. (1983a) *Mikrochim. Acta*, **2**, 463–476.

Kuhnert-Brandstatter, M. & Proll, F. (1983b) *Mikrochim. Acta*, **3**, 287–300.

Kuhnert-Brandstatter, M. & Vollenklee, R. (1985) *Fresenius Z. Anal. Chem.*, **322**, 164–169

Kuhnert-Brandstatter, M., Wurian, I. & Geiler, M. (1982) *Sci. Pharm.*, **50**, 208–216.

Lagas, M. & Lerk, C. F. (1981) *Int. J. Pharm.*, **8**, 11–24.

Leary, J. R., Ross, S. D. & Thomas, M. J. K. (1981) *Pharm. Weekblad. Sci. Ed.*, **3**, 62–66.

Lefebvre, C., Guyot-Hermann, A. M., Draguet-Brughmans, M., Bouche, R. & Guyot, J. C. (1986) *Drug Dev. Ind. Pharm.*, **12**, 1913–1927.

Lindenbaum, S., Rattie, E. S., Zuber, G. E., Miller, M. E. & Ravin, L. J. (1985) *Int. J. Pharm.*, **26**, 123–132.

Masse, J., Chauvet, A., DeMaury, G. & Terol, A. (1985) *Thermochim. Acta.*, **96**, 189–206.

Matsuda, Y., Tatsumi, E., Chiba, E. & Miwa, Y. (1984) *J. Pharm. Sci.*, **73**, 1453–1460.

Maury, L., Rambaud, J., Pauvert, B., Lasserre, Y., Berge, G. & Audran, M. (1985) *J. Pharm. Sci.*, **74**, 422–426.

Mesley, R. J., Clements, R. L., Flaherty, B. & Goodhead, K. (1968) *J. Pharm. Pharmac.*, **20**, 329–340.

Miller, T. A. & York, P. (1985) *Int. J. Pharm.*, **23**, 55–67.

Mitchell, A. G. (1984) *J. Pharm. Pharmac.*, **36**, 506–510.

Mitchell, A. G. (1985) *J. Pharm. Pharmac.*, **37**, 601–604.

Moustafa, M. A. & Carless, J. E. (1969) *J. Pharm. Pharmac.*, **21**, 359–365.

Moustafa, M. A., Ebian, A. R., Khalil, S. A. & Motawi, M. M. (1971) *J. Pharm. Pharmac.*, **23**, 868–874.

Muller, B. W. (1978) *Pharm. Acta Helv.*, **53**, 333–340.

Murthy, K. S., Turner, N. A., Nesbitt, R. U. & Fawzi, M. B. (1986) *Drug Dev. Ind. Pharm.*, **12**, 665–690.

Niazi, S. (1978) *J. Pharm. Sci.*, **67**, 488–491.

Otsuka, M. & Kaneniwa, N. (1983) *Chem. Pharm. Bull.*, **31**, 1021–1029.

Otsuka, M. & Kaneniwa, N. (1984) *Chem. Pharm. Bull.*, **32**, 1071–1079.

Pelizza, G., Nebuloni, M. & Gallo, G. G. (1976) *Il Pharmaco Ed. Sci.*, **31**, 254–263.

Salole, E. G. & Al-Sarraj, F. A. (1985) *Drug Dev. Ind. Pharm.*, **11**, 855–864.

Serajuddin, A. T. M. (1986) *J. Pharm. Pharmac.*, **38**, 93–96.

Shami, E. G., Bernardo, P. D., Rattie, E. S. & Ravin, L. J. (1972) *J. Pharm. Sci.*, **61**, 1318–1320.

Shell, J. (1963) *J. Pharm. Sci.*, **52**, 100–101.

Shibata, M., Kokubo, H., Morimoto, K., Morisaka, K., Ishida, T. & Inque, M. (1983) *J. Pharm. Sci.*, **72**, 1436–1442.

Simmons, D. L., Ranz, R. J., Picotte, P. & Szabolcs (1970) *Can. J. Pharm. Sci.*, **5**, 49–51.

Simmons, D. L., Ranz, R. J. & Gyanchandani, N. D. (1973) *Can. J. Pharm. Sci.*, **8**, 125–127.

Suryanarayanan, R. & Mitchell, A. G. (1986) *Int. J. Pharm.*, **32**, 213–221.

Tuladhar, M. D., Carless, J. E. & Summers, M. P. (1983) *J. Pharm. Pharmacol.*, **35**, 208–214.

Ueda, H., Nambu, N. & Nagai, T. (1982) *Chem. Pharm. Bull.*, **30**, 2618–2620.

Umeda, T., Ohnishi, N., Yokoyama, T., Kuroda, T., Kita, Y., Kuroda, K., Tatsumi, E. & Matsuda, Y. (1985) *Chem. Pharm. Bull.*, **33**, 3422–3428.

Van Aerde, P., Remon, J. P., Van Severen, R. & Braeckman, P. (1984) *J. Pharm. Pharmac.*, **36**, 190–191.

Yang, S. S. & Guillory, J. K. (1972) *J. Pharm. Sci.*, **61**, 26–40.

York, P. & Grant, D. J. W. (1985) *Int. J. Pharm.*, **25**, 57–72.

7

The use of thermal analysis in the study of solid dispersions

The concept of solid dispersions dates back to 1961 when Sekiguchi & Obi found that administration of the fused mixture of the poorly water-soluble drug sulphathiazole and the water-soluble carrier urea produced an enhanced absorption of the drug in rabbits. Subsequently over 400 papers have appeared detailing the physico-chemical properties, phase equilibria, pharmacological activity and pharmacodynamics of many drug–carrier systems. Thermal analysis has played an important role in our understanding of the first two phenomena.

7.1 INTRODUCTION

A solid dispersion was defined by Chiou & Riegelman (1971) as 'a dispersion of one or more active ingredients in an inert carrier or matrix at the solid state prepared by a melting (fusion), solvent or melting–solvent method'. Fusion generally implies subjecting both drug(s) and carrier(s) to a temperature above their eutectic point followed by rapid cooling in an attempt to supersaturate the drug in the carrier in the solid state. Spray congealing and liquid filling of gelatine capsules with the melt prior to solidification have been considered as means of producing the dispersion by fusion methods (Ford, 1986) but problems such as incompatabilities and decomposition have lead to the development of solvent processes. The simplest approach involves dissolving the drug and carrier in a common solvent to effect solution followed by evaporation to leave an intimate blend of drug and carrier. Alternatively the addition of another solvent or solute may mutually precipitate the drug and carrier to form a co-precipitate. Spray-drying or freeze-drying are processes which can be adapted for this method of preparation. When the carrier has a low melting point, a process where the drug dissolved in a suitable solvent is added to the molten carrier prior to solvent removal may be used.

The structure of the resultant dispersion may be represented by one or more of the following phase interactions (Chiou & Riegelman, 1971): simple eutectic mixtures, solid solutions, glass solutions or suspensions, amorphous precipitations in a crystalline carrier or compound or complex formation.

When solid dispersions are used to accelerate the dissolution rate of poorly soluble drugs the chosen carrier will be freely water soluble. Alternatively solid dispersions intended to retard dissolution rate use carriers of low aqueous solubility.

It is the general aim to prepare dispersions in which the drug is dispersed in as near a molecular state as possible to provide a thermo-energetic state of the drug of high aqueous solubility once the carrier has dissolved. Thermal analysis, whether DTA or DSC, has proved a powerful tool in evaluating the drug–carrier interactions. However, since many dispersions contain amorphous or molecularly dispersed drugs they are susceptible to changes during storage which may modify their dissolution rates. Ageing characteristics and stability problems may also be predicted from thermal analysis.

It is the aim of this chapter to review the use of thermal analysis in characterizing solid dispersions. Certain principles, outlined in Chapter 2, are discussed in greater detail with particular reference to solid dispersions.

7.2   THERMAL ANALYSIS OF COMMONLY USED WATER-SOLUBLE CARRIERS

Water-soluble carriers create problems in selecting a solvent in which they and the drug are mutually freely soluble. Therefore fusion is the method of choice in preparing dispersions despite the potential problem of chemical instability. The viscous nature of the melt will tend to entrap the drug in a near molecular state. The polymeric carriers polyethylene glycol and polyvinylpyrrolidone are soluble in organic solvents and consequently may be used in solvent processes. This is necessary for the latter, which melts with thermal decomposition.

7.2.1   Citric acid

Citric acid is a material which, on cooling through its recrystallization temperature, forms a hard vitreous glass. Its glass transition temperature (T_g) has been the subject of considerable debate. Summers & Enever (1980) stated that T_g cannot be given an absolute temperature other than that extrapolated to a zero heating rate. This is because the T_g is dependent on the thermal history of the glass and the heating rate employed in its estimation. Discrepancies in the literature exist. Summers & Enever (1977) found that the T_g of citric acid was $-23°C$ at a scanning rate of 2°C/min yet Timko & Lordi (1979) reported it to be 13.5 and 10.2°C for bulk and *in situ* prepared glass at 10°C/min. The discrepancy is further emphasized by entrapped moisture. Summers & Enever (1980) showed that the T_g ranges were -13 to -8, -18 to -13 and -26 to $-23°C$ for moisture contents of 3, 4, and 5% w/w respectively. The T_g of citric acid glass decreased with an increase in either the exposure time or temperature above its melting point (Timko & Lordi, 1982) owing to an accumulation of its breakdown product, aconitic acid. The problem in the discrepancies may be related to Summers & Enever (1977, 1980) using the monohydrate, since water is difficult to remove from its melt at 150°C, whereas Timko & Lordi (1979) used the anhydrous product. These differences highlight the requirement that pre-history and experimental conditions should be specified when examining glass transition temperatures. Anhydrous citric acid displayed an endotherm at 152–156°C (Timko & Lordi, 1979) which is somewhat higher than the melting point of the hydrate.

7.2.2 Sugars

Not all sugars are suitable as carriers. Most are unstable at their melting points whilst many are insufficiently viscous in solution in organic solvents to prevent or retard drug crystallization when used in solvent processes. Timko & Lordi (1984) determined T_g values for dextrose and sorbitol at 37.4° and −2°C respectively. Crystalline sorbitol displayed endothermic peaks at 85 and 94°C corresponding to the melting of two polymorphic modifications.

7.2.3 Urea

Urea melts at about 132°C giving a single endotherm (Ford & Rubinstein, 1977). Although trace quantities of water may alter purity values by promoting the conversion, during thermal analysis, to biuret, the main interest has centred around its phase diagrams with drugs. Urea is used for melt-prepared dispersions and consequently thermal analysis is used to examine the resultant phase equilibria.

7.2.4 Cyclodextrins

Cyclodextrins have been increasingly investigated because of their ability to complex with, mask the taste of, stabilize, solubilize and increase the bioavailability of drugs (Jones *et al.*, 1984). They are produced by the hydrolysis and cyclization of starch and may produce α, β or γ cyclodextrins. These molecules may form inclusion complexes because their molecular ring encloses cavities of 0.6, 0.8 and 1.0 nm in diameter respectively. Freeze-drying and co-precipitation are usually used to form the solid dispersions. Methods for analysing the formed inclusion complexes include NMR, IR, circular dichroism, fluorescence spectroscopy, chromatography and X-ray diffraction coupled with DSC, DTA, EGA or TG. The last is popular because cyclodextrins are used to stabilize volatile drugs.

   Sztatisz *et al.* (1980) reported a complete thermal analysis of β-cyclodextrin using DSC, TG and EGA (Fig. 7.1). The TG curve showed weight loss at about 100°C corresponding to ≈14% moisture loss. The DSC scan showed a minor endotherm at 220°C corresponding to a reversible transformation within the molecule. Decomposition commenced at above 250°C and was accompanied by melting at 300°C prior to ignition of the sample.

7.2.5 Polyvinylpyrrolidone

Polyvinylpyrrolidone (PVP) is unstable at its melting point (above 270°C). The T_g values of PVP reported in the literature vary from 54 to 175°C. Tan & Challa (1976) attributed this to varying levels of moisture and by using PVP 750 000 containing known levels of water estimated the T_g of anhydrous PVP as 175±1°C from the intercept of the plot of T_g versus composition (Fig. 7.2) obtained from DSC determination. The data plotted as reciprocal T_g against water content (see equation (7.2) yielded a straight-line plot and a T_g for water of −145°C. Water can readily plasticize PVP. PVP samples should be carefully dried before thermal analysis when determining T_g values (Tan & Challa, 1976).

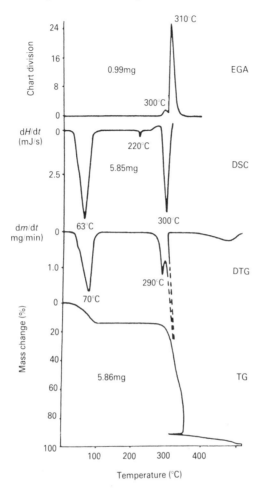

Fig. 7.1 — Thermoanalytical curves of β-cyclodextrin (reproduced with permission from Sztatisz *et al.*, 1980).

The presence of water may also manifest itself as a broad endotherm during scanning. For PVP molecular weights of 10 000, 40 000 and 360 000, Turner & Schwartz (1985) found that repeat DSC analysis of the same sample gave different scans. First runs gave broad endotherms corresponding to water removal at 110–120°C which were lost on repeat scanning. Storage of the sample in atmospheric conditions prior to the re-run allowed this transition to redevelop. A minor endotherm was also apparent in the first run at 62°C and was thought to be due to formation of internal stress within the sample. It was not reproducible. Several runs on the same sample were considered necessary before reliable consistent T_g values were obtained. The T_g increased with increasing molecular weight until a limiting value of 180°C was obtained. In addition to variations attributed to moisture content Turner & Schwartz (1985) considered that residual monomers may reduce the T_g.

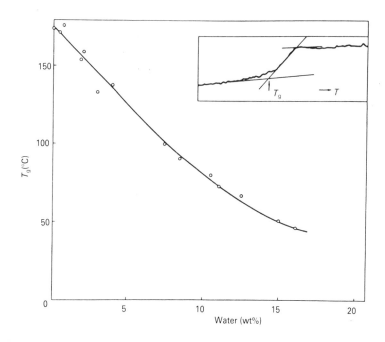

Fig. 7.2 — The lowering effect of water content on the glass transition temperature of PVP. Inset: a typical DSC scan; scan speed 8°C/min; sensitivity range 1 (reproduced from Tan & Challa, Fig. 1, page 739, *Polymer*, 1976 by permission of the publishers, Butterworth & Co. (Publishers) Ltd ©.

7.2.6 Polyethylene glycol

The molecular weight of polyethylene glycol (PEG) varies from 300, when it is a liquid, to over 4.10^6, when it is a crystalline solid and often termed polyethylene oxide. The molecular weights used for solid dispersions are in the range 1000–20 000. Beaumont *et al.* (1966) found that the enthalpies of fusion for PEG 750, PEG 4000 and PEG 20 000 were 39.5, 46.1 and 44.5 cal/g with melting points 30, 59.6 and 63.1°C. These indicated that PEG 4000 was the most crystalline fraction and NMR and X-ray diffraction studies revealed that PEG 4000 was 95% crystalline. They calculated that the enthalpy of fusion of a 100% crystalline material was 51.5 cal/g with a melting point of 336 K. These values have been used as standards by many later workers to estimate the degree of PEG crystallinity.

The PEG weight fraction in the range 4000–6000 is particularly sensitive to changes in crystalline structure. Beech *et al.* (1972), using dilatometry, showed that PEG with a number molecular weight (M_n) of 4000 or below gave one melting transition whereas an M_n of 6000 or above gave two transitions. PEG of $M_n \simeq 6000$ showed one transition when crystallized at above 55°C but two transitions when crystallized at below 55°C. X-ray diffraction indicated that the lower transition corresponded to the melting of once-folded-chain crystals whereas the higher transition corresponded to the melting of the extended-chain crystals. This PEG of $M_n \leqslant 4000$ crystallized as extended-chain crystals, of $M_n = 6000$ as extended- and folded-chain crystals and of $M_n > 6000$ increasingly as folded-chain crystals only.

Chatham *et al.* (1986) confirmed that for PEG 4000 extended- and once-folded chain forms may be formed but twice-folded forms were unstable. Untreated PEG 4000 displayed a single endotherm at 57°C and a crystallinity of 83%. Following fusion and cooling to room temperature at 8°C/min two endotherms were apparent at 52 and 57°C, corresponding to fusion of the once-folded and extended-chain form crystals respectively. The predominant stable crystalline form varies both with the molecular weight of the polymer and the heating rate used in the studies. Lang *et al.* (1977) used a molecular weight of 9000–10 000 and therefore the extended-chain form crystals were the less stable form.

Kambe (1980) further confirmed the dependence of DSC traces of PEG on the previous history. For PEG of $M_n = 4000$ and a crystallization temperature of 303 K the major melting endotherm was preceded by a smaller endotherm and a subsequent exotherm. The area of both these transitions decreased with increased crystallization time. The endotherm corresponded to melting of the thermodynamically unstable modification i.e. the folded crystals and the exotherm to recrystallization of the favoured form, i.e. the extended crystal form.

Transitions in PEG below the apparent melting point also exist. Lang *et al.* (1977), using a heating rate of 20°C/min, samples quench-cooled to −170°C and derivative DSC, found several second-order transitions below the first-order melting transition. These transitions included a T_γ transition at 130–140 K corresponding to the vibration of submolecular units of 3–5 methylene units and glass transition temperatures at 190–240 K. Two transitions in this later range were apparent (Fig. 7.3), one at 190–200 K ($T_{g(L)}$) and one at 230–240 K ($T_{g(U)}$), equivalent respectively to a glass transition in an amorphous phase not under restraint from crystalline areas and to a glass transition in an amorphous phase under restraint from crystalline regions. Further transitions at 263–313 K were T_α transitions (Lang *et al.*, 1977) and due to rotational or oscillatory movements of the polymer molecule. Several endotherms were associated with these transitions and their relative intensities were dependent on the previous history of the sample. Additionally the melting transition may consist of three separate transitions (Lang *et al.*, 1977) corresponding to the melting of less stereo-regular fractions of the polymer, the melting of crystals associated with the primary crystallization process and to the melting of folded-chain crystals.

Ford (1987) characterized the thermal properties of a PEG 6000 sample prior to its incorporation into melt-prepared dispersions. Untreated PEG 6000 gave a weak endothermic change in baseline at ≃−50°C and another baseline change at 29 to 40°C before fusion at 63–64°C. The trace was identical whether obtained at either 5 or 10°C/min. Fusion of PEG 6000 altered the appearance of the scan. The second-order transition in the range −50 to −28°C apparently increased in strength, implying that fusion reduced the crystallinity of the polymer, rapid recrystallization prior to the analysis probably creating amorphous, non-crystalline PEG. A lower heating rate of 5°C/min resulted in a second endotherm at 57°C. This probably corresponded to the melting of fully extended chain crystals. Interestingly Ford (1987) reported the

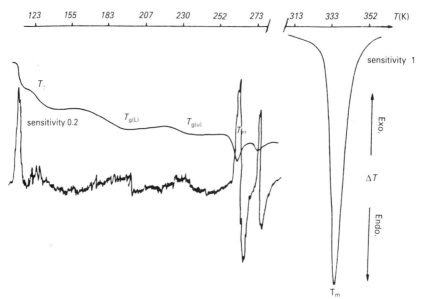

Fig. 7.3 — DSC scan and its derivative curve for an unannealed PEG 9000–10 000 sample (reproduced with permission from Lang *et al.*, 1977).

thermal analysis of this batch of PEG solid dispersed with seven drugs and no double endotherms were apparent for any of the systems. Corrigan (1986) reported that the DSC scans of four batches of PEG 6000 were different and that as many as three peaks were apparent in some samples.

7.3   TYPES OF SOLID DISPERSIONS

Wherever possible thermal analysis of solid dispersions should be performed on the final dispersion rather than on physical mixes or else the true nature of the dispersion may not be revealed. This section highlights some of the phase equilibria that have been determined by thermal analysis. It is not intended to be a thorough review of the subject.

7.3.1   Eutectics, monotectics and solid solutions

Although many solid dispersions form metastable structures on preparation, these may gradually age to form more stable solid solutions or eutectics. Systems which are metastable will, more often than not, display crystallization phenomena on thermal analysis. The simplest phase diagram is one revealing simple eutectic formation without solid solutions. These are characterized by a eutectic horizontal continuing across the phase diagram and rapid dissolution rates are due simply to decreased drug particle size. For instance, DTA revealed an endotherm corresponding to eutectic melting at 104°C in the chloramphenicol–urea dispersion even for dispersions containing only 2% chloramphenicol (Chiou, 1971). The level of solid solubility for chloramphenicol at the eutectic temperature was thus less than 2% w/w chloramphenicol. This was supported by X-ray diffraction data and illustrates the principle that at low levels of carrier or drug it is essential that thermal analysis be supported by results obtained from other analytical methods.

The dissolution rates of sulphathiazole from its dispersions in urea were over 700 times higher than from the drug alone (Chiou & Niazi, 1971). DTA of physical mixes revealed that the eutectic contained 52% sulphathiazole. However, their eutectic temperature obtained varied with the polymorph of sulphathiazole used to prepare the mix. Eutectic temperatures were 118° and 121°C for the form I and form II respectively (Fig. 7.4). No eutectic endotherms were noted for melts containing 5%

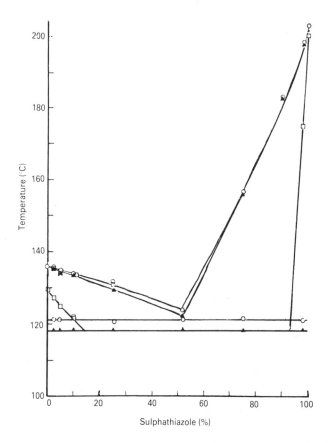

Fig. 7.4 — Phase diagrams of the sulphathiazole–urea binary system determined from different mixtures. Key: ▲, sulphathiazole form I–urea physical mixture; ○, sulphathiazole form II–urea physical mixture, and □, solid dispersion (reproduced with permission from Chiou & Niazi, 1971).

sulphathiazole. The solubility of the drug in urea was 10% at 121°C when form II would precipitate and 15% at 118°C when for I would precipitate (Chiou & Niazi, 1971). X-ray diffraction supported the claim for the presence of solid solutions.

Where the carrier does not melt readily on scanning, DSC or quantitative DTA may be used to estimate the limit of solid solubility (Shefter & Cheng, 1980; Theeuwes *et al.*, 1974). Solid dispersions containing incremental increases of drugs in PVP 40 000 were prepared by the solvent method (Shefter & Cheng, 1980). Following DSC analysis, plots of the calorific values of the endotherms corresponding to drug melting versus drug content produced straight-line plots with intercepts on the content axis equivalent to the solid solubility. Solid solubilities were determined for cholesterol (13%), methyl-π-hydroxybenzoate (54%), methyl-*p*-aminobenzoate (37%), ethyl-*p*-aminobenzoate (42%), 4-amino-antipyrine (43%) and griseofulvin (12%), each in PVP 40 000. The technique is described in more detail in Chapter 8.

Numerous publications deal with the phase interactions between drugs and PEG of varying molecular weights. Hoelgaard & Møller (1975) detailed the phase equilibria using DSC of PEG 6000 or PEG 20 000 with testosterone. Each molecular weight resulted in a simple eutectic at 2.5% testosterone with no evidence for a solid solution of the drug in either PEG. The eutectic temperatures were 56° and 48°C for PEG 6000 and PEG 20 000 respectively. Their method of determining transition temperatures has been adapted by many subsequent researchers and is illustrated in Fig. 7.5. The onset of PEG fusion corresponded to the solidus line and at certain drug

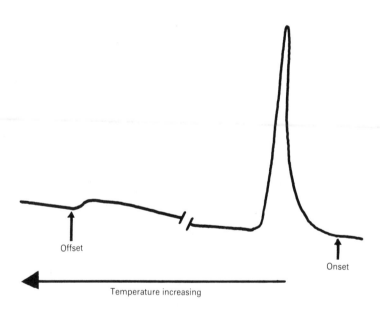

Fig. 7.5 — DSC curve showing derivation of onset and offset temperatures (redrawn with permission after Hoelgaard & Møller, 1975).

levels the offset temperature corresponded to the liquidus line of the phase diagram. Ford & Rubinstein (1978) used a similar approach in determining the phase equilibria of indomethacin–PEG 6000. The phase diagram is more complicated than usual phase diagrams in displaying a eutectic containing 13% drug, two polymorphic forms of indomethacin, a solid solution of indomethacin in PEG 6000 and glassy dispersions in melts containing in excess of 80% indomethacin (Fig. 7.6). The phase

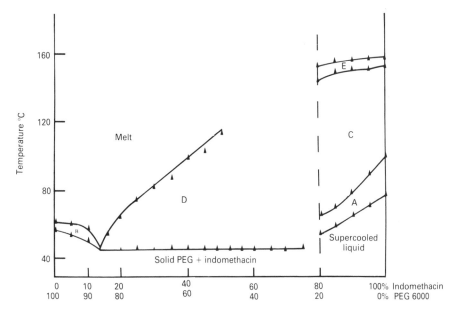

Fig. 7.6 — Phase diagram of the indomethacin–PEG 6000 system constructed from melts. Key: (A) zone of recrystallization; (B) PEG 6000 and liquid; (C) indomethacin forms I and II and liquid; (D) indomethacin form II and liquid; (E) indomethacin form I and liquid (reproduced from Ford & Rubinstein, 1978, with permission of Schweizerischer Apothekerverein).

diagram was used to explain that rapid dissolution rates of the dispersion were due mainly to the molecular dispersion of indomethacin in PEG. The DSC data (Fig. 7.7) used to construct Fig. 7.6 revealed that indomethacin solidified from its melt as a glassy solid and examination by DSC produced traces in which recrystallization was evident in the range 74.5–100.5°C as a broad exotherm and endotherms at 151.5 and 157°C reflected that two polymorphs were formed with some interconversion at about 155°C of the lower melting metastable form II and the stable form I. The DSC scans indicated that incorporation of PEG 6000 into the melt gradually favoured form II, or that the interconversion did not as readily occur. No transitions corresponding to indomethacin fusion were noted in the range 50–80% drug. Ford & Rubinstein (1978) assumed that the drug was present in the melt as the molecular state. HSM was used as an adjunctive process to confirm the polymorphic modifications since the indomethacin forms crystallized as two different structures.

Similar information was provided by the phenytoin–PEG 6000 melt (Yakou *et al.*, 1984). The drug had a sharp endothermic peak at 299°C. The dispersion was prepared by fusion at 70°C to melt the PEG and then at 250°C to disperse the drug. The melt containing 10% phenytoin displayed only an endotherm equivalent to the melting of the PEG and as the characteristic peaks of the drug were inapparent in X-

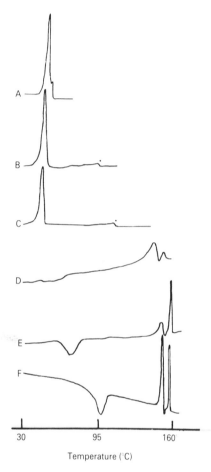

Temperature (°C)

Fig. 7.7 — DSC scans of (A) 10% indomethacin: 90% PEG 6000, (B) 40% indomethacin: 60% PEG 6000, (C) 50% indomethacin: 50% PEG 6000, (D) 80% indomethacin: 20% PEG 6000, (E) 90% Indomethacin: 10% PEG 6000 and (F) glassy indomethacin (reproduced from Ford & Rubinstein, 1978, with permission of Schweizerischer Apothekerverein).

ray diffraction spectra and combined evidence supported the view that the drug was dispersed in the molecular or amorphous state.

DSC or quantitative DTA may be used to quantifty the crystallinity of solid dispersions. Ford *et al.* (1986) examined the crystallinity of solid dispersions containing indomethacin or phenylbutazone melt dispersed in PEG 1500, PEG 4000, PEG 6000 or PEG 20000. Dissolution rates from phenylbutazone dispersions were considerably less than anticipated and the contents giving maximum dissolution rates were ≈2% for the phenylbutazone systems but ≈12.5% for the indomethacin dispersions. The heats of fusion were determined for untreated polymers and following their fusion and subsequent recrystallization. The results (Table 7.1) support previous findings that maximum crystallinity for PEG occurs around a molecular weight of 6000. The relevance of Table 7.1 is that the heats of fusion,

Table 7.1 — Heats of fusion of untreated or recrystallized polyethylene glycol and its solid dispersions with indomethacin or phenylbutazone (reproduced with permission from Ford *et al.*, 1986)

	Heats of fusion (cal/g)[a]			
	PEG 1500	PEG 4000	PEG 6000	PEG 20000
Untreated	44.7±1.5 (4)	46.4±1.8 (4)	47.2±1.0 (5)	39.5±0.7 (4)
Recrystallized	42.5±1.1 (4)	43.4±2.2 (5)	44.9±0.6 (4)	41.6±2.8 (4)
2% indomethacin	36.8±1.9 (4)	43.7±0.5 (3)	51.3±2.9 (7)	42.2±2.9 (4)
5% indomethacin	35.8±3.3 (4)	41.6±3.6 (4)	42.6±2.3 (7)	42.2±4.0 (4)
10% indomethacin	30.6±1.8 (2)	34.6±3.3 (2)	38.9±3.9 (3)	44.8±2.6 (4)
2% phenylbutazone	39.2±3.1 (3)	43.9±1.2 (4)	47.9±3.7 (6)	41.9±6.6 (4)
5% phenylbutazone	40.5±4.3 (7)	42.4±3.0 (4)	47.8±2.8 (5)	40.3±3.9 (4)
10% phenylbutazone	37.9±3.4 (4)	39.8±1.8 (4)	42.9±0.9 (4)	42.7±1.6 (4)

[a] Mean ± standard deviation (number of replicates).

which correspond to the PEG/eutectic moiety of the dispersions, generally decreased more prominently for the indomethacin dispersions than for the phenylbutazone dispersions, thus indicating that the indomethacin dispersions were less crystalline (more amorphous) than the phenylbutazone dispersions of comparative composition. Disruption and loss of the crysallinity of the carrier are therefore prerequisites for increased release rates from solid dispersions containing PEG. The change in crystallinity induced by the two drugs was confirmed using FTIR.

Few comparisons of the properties of dispersions containing the same drug dispersed in different molecular weights of PEG have been made. Ford (1984) examined the properties of glutethimide melt dispersions with PEG 1500, PEG 6000, PEG 14000 and PEG 20000. The respective DSC melting ranges of the PEGs were 42–45, 54–61, 58–61 and 59–62°C. The sample of PEG 6000 displayed a double melting endotherm and DSC of the 24-hour-old melts indicated that the dual melting point was apparent in dispersions to the drug-rich side of the eutectic of 32% glutethimide and in PEG 6000 itself. This was probably due to the presence of both extended- and folded-chain-crystal forms in the melt. The dispersions containing PEG 1500 at high glutethimide content showed a propensity to supercool to a vitreous glassy state which was reflected by crystallization exotherms in the 24-hour-old melts containing more than 96% glutethimide. Storage for seven days resulted in devitrification and loss of this transition. Eutectic compositions (percentage glutethimide) and temperatures were 24% at 35°C, 32% at 45.5°C and 32% at 48°C for PEG 1500, PEG 6000 and PEG 14000 respectively. This knowledge was essential in explaining the release rates from the dispersions. The PEG 1500 dispersions, containing 2.5–10% drug, gave three- and fourfold higher dissolution rates than the PEG 6000 and PEG 14000 dispersions respectively of similar composition, because dissolution rates were determined at 37°C where there was a tremendous contribution from the PEG 1500 dispersions melting. Dissolution rates were not related to the eutectic composition. PEG 20 000 dispersions containing more than 30% glutethimide showed three endotherms corresponding to the eutectic, excess glutethimide and a supposed degradation product formed on production of the melt. Decomposition was confirmed by hot-stage microscopy which revealed that bubble formation occurred at 60–100°C. This apparent decomposition was assumed to be due to the presence of peroxides, a by-product often associated with PEG. Polyoxyethylene-40-stearate (P40S), with a melting range of 45–50°C, appears to be a suitable alternative to PEG as a carrier (Kaur *et al.*, 1980a, b). DTA was used to evaluate the phase interactions of griseofulvin and tolbutamide with P40S in physical mixes, co-precipitates and melts. The griseofulvin dispersions displayed a monotectic phase diagram with negligible solubility of the drug in the carrier at its melting point. The tolbutamide systems however showed phase diagrams (e.g. Fig. 7.8) in which the

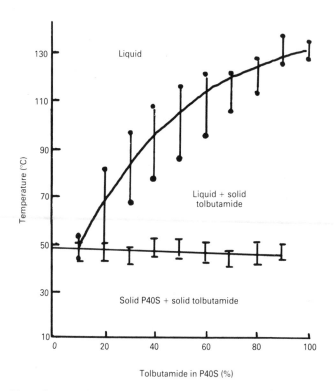

Fig. 7.8 — Phase diagram of coprecipitated mixtures of tolbutamide and poloxyethylene 40 stearate (P40S) determined using DTA and HSM (reproduced with permission from Kaur *et al.*, 1980a).

left-hand liquidus tie lines were absent. The intercept of the right-hand solidus with the eutectic horizontal represents the solubility of tolbutamide in the 'just-melted' P40S. This value varied according to the method of preparation, the co-precipitate giving a value of 10% and the physical mixes and melts giving apparent solubility values of 20%. Kaur *et al.* (1980b) indicated that P40S was a more effective carrier than PEG 2000 (similar molecular weight) in increasing the dissolution rates of both drugs because P40S is a surfactant and capable, unlike PEG 2000, of solubilizing the drugs. PEG 2000 has a melting range of 48–58°C and its phase diagrams with either tolubutamide or griseofulvin were revealed by HSM (Kaur *et al.*, 1980a) to be monotectics irrespective of whether the dispersions were prepared by physical trituration, fusion or co-precipitation. DTA revealed the existence of two transitions irrespective of drug concentration although unable to detect an endotherm corresponding to drug fusion at less than 50% drug content (Kaur *et al.*, 1980a).

7.3.2 Complexes

Complex formation is a relatively common occurrence following solid dispersion preparation. It is generally recognized on DSC or DTA by the appearance of additional endotherms or exotherms. For instance, DSC analysis of the sulphamethozaxole-polyether 18-*crown*-6 dispersion indicated that its complex displayed a single endotherm near 130°C (Takayama *et al.*, 1977). The physical mix showed a more complicated trace with endotherms at 100° and 122°C corresponding to the melting of the eutectic and the complex respectively and an exotherm at 105°C corresponding to complex formation and its solidification. The suitability of a complex as a solid dispersion depends on its solubility and ability to dissociate in aqueous solution. However the techniques of examining complexes by thermal analysis is independent of these properties.

Urea is capable of forming complexes with many drugs. Examples which have been studied by thermal analysis include DSC of its salicylic acid complex (Collett *et al.*, 1976) and DTA of its phenobarbitone complex (Winfield & Al-Saidan, 1981). The cooling rate used to prepare the dispersions is very important and crystal size may be reflected in the DSC traces, for example the 95% salicylic acid : 5% urea dispersion prepared by slow cooling displayed a thaw point at 144°C and a melting point at 151°C, whereas the values were 141 and 152°C for a rapidly cooled dispersion. The 30% salicylic acid : 70% urea dispersion showed less defined peaks with endotherms in the range 47–67°C and 95–117°C whether rapidly or slowly cooled (Collett *et al.*, 1976). Additionally the slowly cooled 30% dispersion showed another endotherm in the range 132–187°C. X-ray diffraction confirmed that complex formation occurred. Rapid cooling produced smaller crystallites due to the formation of many nucleation sites whereas slow cooling produced fewer nucleation sites and larger crystallites.

Winfield & Al-Saidan (1981) examined physical mixtures of phenobarbitone and urea. Mixtures containing in excess of 20% phenobarbitone displayed at least three endotherms. The lowest melting endotherm gave an onset temperature of 105°C, equivalent to a eutectic temperature. The highest temperature corresponded to the liquidus temperature. Only one endotherm was apparent in the range 55–75% phenobarbitone. The phase diagram constructed from DTA scans suggested formation of a stoichiometric 1 : 2 phenobarbitone : urea complex. Scans of mixtures of 87–95% showed a fourth endotherm peak at 152.5°C possibly indicative of polymorphic changes. Differences with pre-melted samples were noted in the value of the apparent eutectic composition but both physical mixes and melts gave a eutectic peak at 111°C. Solid solution formation was indicated at compositions in excess of 90% phenobarbitone since the solidus temperature increased. Fused systems displayed complex formation in the ranges 40–55% and 70–85% but in the intermediate range only one peak was apparent. Thus an unstable peritectic was predicted (Winfield & Al-Saidan, 1981).

Several barbiturates including phenobarbitone, cyclobarbitone, barbitone and pentobarbitone form complexes with PEG 6000 (Nakai *et al.*, 1981). X-ray diffraction and DSC showed that complexes were formed by co-precipitation and trituration. One drug molecule interacted with two repeating units of –(–CH$_2$–CH$_2$–O–)– in the PEG molecule. Corrigan (1986) further examined the properties of phenobarbitone dispersed in PEG 4000 by physical mixing. As the PEG content increased the endotherm corresponding to its melting increased whereas that corresponding to phenobarbitone at 448 K decreased and was inapparent at ≈30–40% PEG. In this range another endotherm at 422 K developed which X-ray diffraction confirmed was due to complex formation. This endotherm was lost at PEG concentrations greater than 80%. The endotherms corresponding to the complex and PEG were followed by small exotherms. This complex retarded dissolution rates because it was poorly water-soluble. DTA or DSC are particularly useful techniques for drugs that form inclusion complexes with cyclodextrins, usually evidenced by loss of the drug melting endotherm. Corrigan & Stanley (1982) examined freeze-dried β-cyclodextrin dispersions of thiazide diuretics. β-Cyclodextrin itself showed endotherms at 375 and 580 K corresponding to dehydration and decomposition respectively. Bendrofluazide-β-cyclodextrin physical mixes showed an endotherm at 500 K corresponding to drug fusion but the peak was moved to 525 K in the freeze-dried product. This was attributed to a solid-state complex which did not occur in freeze-dried β-cyclodextrin mixtures with hydrochlorothiazide, hydroflumethiazide or chlorthiazide, for which the DSC scans of the freeze-dried product resembled those of the physical mixes.

Uekama *et al.* (1983a) examined diazepam–γ-cyclodextrin systems. Physical mixes displayed an endotherm at 130°C corresponding to diazepam fusion and a broad endotherm at 290°C due to cyclodextrin decomposition. The diazepam melting endotherm in the 2:3 diazepam:γ-cyclodextrin complex was lost. Circular dichroism and X-ray diffraction confirmed the existence of the complex. The loss of drug fusion endotherms has similarly occurred for flurbiprofen inclusion complexes with β- and γ-cyclodextrins when increased crystallinity was apparent using X-ray diffraction (Otagiri *et al.*, 1983) and with complexes of femoxetine-β-cyclodextrin (Andersen *et al.*, 1984), prostaglandin F$_{2\alpha}$–γ-cyclodextrin (Uekama *et al.*, 1984), prednisolone–β- or γ-cyclodextrin (Uekama *et al.*, 1985), ibuprofen–β-cyclodextrin (Chow & Karara, 1986) and the 1:1 aspirin:di-*O*-methyl-β-cyclodextrin complex (El-Gendy *et al.*, 1986). In the later system the 1:1 physical mix showed endotherms at 413 and at 460 K equivalent to aspirin fusion and fusion of a complex formed on heating. Reheated samples displayed only the later endotherm. Vigorous grinding via vibrational mixing produced an exotherm which X-ray diffraction confirmed was due to crystallization of the complex. The endotherm at 413 K concomitantly decreased as the exotherm increased with increased duration of grinding.

When the drug is volatile similar observations have been made with the loss of the boiling endotherm. Additionally TG has proved a useful tool in the analysis of cyclodextrin complexes with volatile drugs and has shown that the volatility of cinnamic acid derivatives from their cyclodextrin complexes varied according to their substitution type (Uekama *et al.*, 1979). Methyl-cinnamate, ethyl-cinnamate and cinnamaldehyde were stabilized by both α- and β-cyclodextrin complexes. Uekama *et al.* (1983b) examined clofibrate–cyclodextrins formed by precipitation from water. α-Cyclodextrin formed only a water-soluble complex by γ- and β-cyclodextrins formed 1 : 1 complexes. DTA showed that an endotherm, apparent at 150°C in physical mixes due to boiling, was lost in the inclusion complex. TG confirmed that the volatility was reduced (Fig. 7.9) by complex formation which in turn reduced

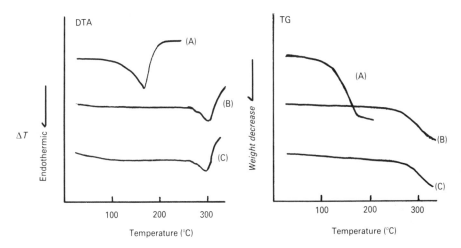

Fig. 7.9 — DTA and TG scans of clofibrate–cyclodextrin systems. (A) clofibrate, (B) clofibrate–β-cyclodextrin complex, and (C) clofibrate–γ-cyclodextrin complex (reproduced from Uekama *et al.*, 1983b, with permission from Schweizerischer Apothekerverein).

photodegradation of the drug and increased the dissolution rate. Similar decreased volatility and loss of a boiling endotherm was noted (Uekama *et al.*, 1983c) in benzaldehyde complexes with α-, β- and γ-cyclodextrins. Broad endotherms at 100–150°C during DTA analysis were attributed to water molecules trapped within the complex. X-ray diffraction revealed that the complexes involving the β- and γ-cyclodextrins were less crystalline than those involving the α-derivative. DTA did not pick out these differences (Uekama *et al.*, 1983c).

7.3.3 Glassy dispersions
When either the drug or carrier is capable of glass formation various possibilities exist as to the structure of the glassy dispersion. Thus both materials may form glasses or either or neither material may form a glass. Additionally total miscibility, total immiscibility or partial miscibility may be found.

The relationship between T_g and composition is difficult to predict and is not a straight-line relationship. Fig. 7.10 illustrates such a relationship for the primidone–

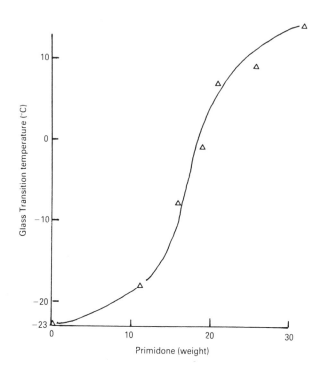

Fig. 7.10 — Variation of the glass transition temperature of citric acid–primidone melts with primidone concentration (reproduced with permission from Summers & Enever, 1977).

citric acid system (Summers & Enever, 1977). For a given temperature increment above the T_g the system displaying the highest T_g will be the least viscous. This principle explains why the systems containing low levels of primidone were the most rapid to devitrify. Dispersions containing up to 32% primidone (Summers & Enever, 1976) could be prepared before drug decomposition became excessive. Use of citric acid monohydrate protected against primidone degradation. The glasses were unstable, those containing 1% drug started to devitrify within 15 minutes, whereas those containing 30% primidone did not start to crystallize at room temperature within six months. The phase diagram of the system devitrified by storage at 60°C for three days showed a eutectic containing 20% primidone at 141°C. Although DSC suggested the possibility of solid solution formation HSM and X-ray diffraction studies confirmed that there was no change in the crystal lattice of citric acid and therefore no solid solution formation. Equation (7.1) (Gordon & Taylor, 1952) was used by Summers & Enever, (1977) to explain the variation of T_g with composition.

$$T_g = -k(T_g - T_2) W_2/W_1 + T_1 \qquad (7.1)$$

T_1 and T_2 are the glass transition temperatures of pure components 1 and 2, and W_1 and W_2 are the weight fractions of the components. Coefficient k is related to the expansion in the vitreous and liquid states. Equation (7.1) is valid if the bonding energies between the molecules in a mixture are equal to the mean of the bonding energies between molecules of the individual components. Since Fig. 7.11 which

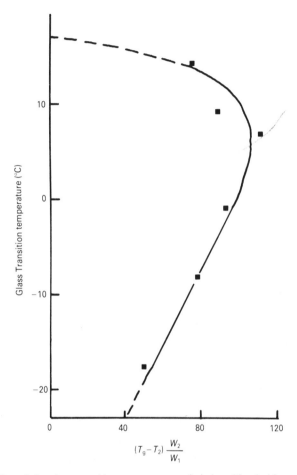

Fig. 7.11 — Plot of the glass transition temperature of citric acid–primidone melts against $(T_g - T_2)W_2/W_1$. Data from Fig. 7.10 (reproduced with permission from Summers & Enever, 1977).

shows the data of Fig. 7.10 plotted according to equation (7.1) is not a straight line it suggests that the concentration-dependent devitrification may be due to the formation of strong hydrogen bonds between the component molecules in the vitreous state.

Phenobarbitone formed a glass on rapid cooling with a T_g of 41.9°C and an exotherm around 90°C corresponding to recrystallization (Timko & Lordi, 1984) and recrystallized to one of two polymorphs (form I or IV) depending on the technique used to prepare the glass. Equation (7.2) (Nielson, 1974) was used to estimate the T_g of two miscible glasses.

$$\frac{1}{T_g} = \frac{W_a}{T_{g(a)}} + \frac{W_b}{T_{g(b)}} \qquad (7.2)$$

where $T_{g(a)}$ and $T_{g(b)}$ are the glass transitions of the two components and W_a and W_b are their respective weight fractions. Calculated values for the mixtures of citric acid and phenobarbitone were higher than the determined values (Timko & Lordi, 1984) indicating that less bonding occurred between the citric acid and phenobarbitone molecules than between either groups of molecules alone, Anomalous deviations from this simple equation in sulphonamide–dextrose mixtures indicated either decomposition or an interaction with sulphonamides and compounds containing carbonyl groups (Timko & Lordi, 1984).

The hexobarbitone–dextrose dispersion provides an example of immiscible glasses (Timko & Lordi, 1984). Each T_g, 13°C for hexobaritone and 37°C for dextrose, was apparent in the prepared glasses and were invariant with composition. Dextrose–paracetamol (Timko & Lordi, 1984) was an example where the glasses were partially miscible. Below the limits of miscibility, a single glass transition was apparent whereas where the components were immiscible two transitions were apparent. It is not uncommon for phase diagrams to display glass formation only in the excess of one ingredient, e.g. chlorpropamide–urea dispersions (Ford & Rubinstein, 1977). DSC of physical mixtures predicted the eutectic to contain 89% chlorpropamide. Chlorpropamide undercooled to a glass solid, a metastable state that showed no transitions when scanned at 4°C/minute. Peaks corresponding to the liquidus curve in the cooled melts were found only in mixtures corresponding to 0–40% and 94–100% chlorpropamide. Only small endothermic transitions were noted in the range 50–92% chlorpropamide in freshly prepared melts and these were attributed to possible glass transactions. Ford & Rubinstein (1977) claimed that the eutectic composition may exist as a glass.

7.4 ASSESSMENT OF SOLID DISPERSION POTENTIAL

Ford (1987) used DTA to assess the potential of several PEG 6000 dispersions. Since these rely on a molecular distribution of the drug throughout the carrier to produce high dissolution rates, the ability of drugs to form stable glasses was assessed. Melts of chloramphenicol, glutethimide, griseofulvin, indomethacin and paracetamol solidified to form glasses. The values of T_gs, recrystallization temperatures and melting-points were determined (Table 7.2). The appearance of polymorphs was confirmed by HSM. Phenylbutazone and phenacetin did not undercool to form stable glasses. The phase diagrams of their dispersions with PEG 6000 were determined from melts which, following fusion, were rapidly chilled either to 4°C or using liquid nitrogen to -120°C. Consequently the phase diagrams showed glass transitions and melting and recrystallization zones. An example is provided by Figs. 7.12 and 7.13. Indomethacin displayed a T_g at 41°C and thereafter no transitions when its melt was examined at 10°C/min. This is in contrast to Fig. 7.6 which was obtained at 4°C/min. At 10°C/min T_g values were obtained in the range 40–100% indomethacin. Some PEG recrystallization occurred in melts containing 50–60% indomethacin and PEG endotherms occurred in melts containing up to 60%

Table 7.2 — Glass transition temperatures (T_g, °C), recrystallization temperatures (T_c, °C) and fusion temperatures (T_m, °C) of
seven drugs obtained by DTA of their melts following rapid cooling
to and scanning at 10°C/min (data abstracted from Ford, 1987)

Drug	T_g	T_c	T_m
Chloramphenicol	28	76	152
Glutethimide	0	n/o	n/o
Griseofulvin	89	149	222
Indomethacin	41	n/o	n/o
Paracetamol	24	74	171
Phenacetin	n/o	n/o	138
Phenylbutazone	n/o	n/o	98 and 108

n/o, not obtainable.

indomethacin. Melts containing up to 40% indomethacin recrystallized when cooled to 4°C, but in melts rapidly cooled to −120°C T_g values were detected in melts containing 20% or more indomethacin. T_α transitions arising from the PEG moiety were apparent in melts containing 0–30% indomethacin and cooled to 4°C. PEG inhibited indomethacin recrystallization. DTA studies at 5°C/min gave slightly different values of the transition temperatures. Fig. 7.13 displays the experimentally determined T_g values and those calculated from equation (7.2); such diagrams were used to assess the potential of the dispersions. The factors affecting rapid dissolution rates included (a) the composition range over which PEG endotherms were apparent with or without being superseded by PEG recrystallization, (b) the range over which drug melting endotherms were apparent with or without prior drug recrystallization exotherms and (c) the position and extent over which the T_gs were apparent. Since fast release should occur in dispersions in which drugs are molecularly entrapped, range (b) should be minimal and range (c) should extend through the phase diagram to low PEG levels. Range (c) should be as small as possible. It was predicted that systems which displayed PEG melting endotherms at drug contents from 0 to 70% or above and drug melting endotherms at contents in excess of 50% drug made unsuitable dispersions because increases in dissolution rate occurred over a limited range at low drug content. Phenacetin and phenylbutazone were unsuitable drug candidates and displayed maximum dissolution rates at low drug contents.

7.5 STABILITY OF SOLID DISPERSIONS

Spurious peaks or peak broadening on DSC or DTA are indicative of decomposition. Succinic acid provides an interesting alternative to citric acid as a carrier because it does not readily form a glass. Endotherms corresponding to fusion of the eutectic component were apparent throughout the range 1–95% drug during DTA of its physical mixes with griseofulvin, indicating a lack of solid solution. Their thaw

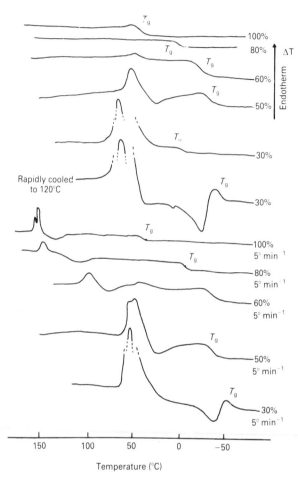

Fig. 7.12 — DTA scans of indomethacin–PEG 6000 melts rapidly cooled at 4°C or, where indicated, to −120°C, and scanned at 10°C/min except where stated (percentages given are percentages of drug). (Reprinted from Ford (1987) p. 1758, by courtesy of Marcel Dekker, Inc.)

points, equivalent to the eutectic temperature, were around 170°C (Chiou & Niazi, 1973). Succinic acid displayed one endotherm at 185–187°C. However previously fused samples showed an extra endotherm commencing at 118°C with concomitant broadening of the fusion endotherm which was attributed to succinic anhydride formation (confirmed by X-ray diffraction). Griseofulvin, following fusion and resolidification, gave DTA scans identical to the starting product and was thought to be stable (Chiou & Niazi, 1973). The endotherm corresponding to the eutectic in fused samples containing 10% griseofulvin was lost during DTA but a smaller endotherm starting at 109°C, attributed to succinic anhydride, was noted. It was estimated that as much as 7% anhydride was present with reference to samples of known adulteration. Storage at 80°C removed the earlier peak due to sublimation of succinic anhydride.

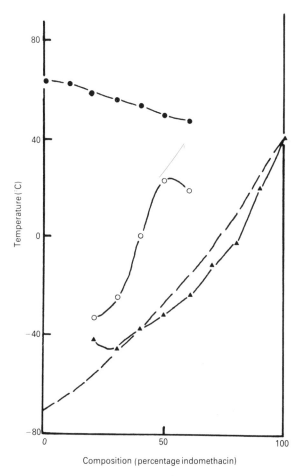

Fig. 7.13 — Phase diagram of indomethacin–PEG 6000, constructed from data as Fig. 7.12. Key: ●, PEG fusion endotherm; ○, PEG recrystallization exotherm; ■, drug fusion endotherm; □, drug recrystallization exotherm; ▲, T_g values; broken line, predicted T_g values. (Reprinted from Ford (1987) p. 1759, by courtesy of Marcel Dekker, Inc.)

EGA is a suitable method for evaluating the loss of volatile components, especially from cyclodextrin complexes. Sztatisz *et al*. (1980) showed that an allyl mustard oil volatilized at 35°C but when complexed with β-cyclodextrin did not volatilize until above 50°C, indicating greater stability for the complex. Similar increases in stability for β-cyclodextrin complexes with vitamin D_3, vitamin E acetate and paracetamol were claimed by Sztatisz *et al*. (1980). EGA was similarly used in the evaluation of the stability of β-cyclodextrin complexes with cholecalciferol (Szejtli *et al*., 1980) in combination with DSC and DTG. The DSC scan of β-cyclodextrin showed an endotherm at below 100°C corresponding to moisture loss and a decomposition/melting endotherm at above 290°C, EGA confirming the decomposition temperature. Cholecalciferol melted at 85°C and oxidation commenced, giving an apparent weight gain on the TG curve (Fig. 7.14) before weight loss at above 205°C confirmed the decomposition of the drug. The DSC scan of the physical mix of equivalent weight fraction to the complex was similar to that of β-cyclodextrin, with a slight weight loss at 150°C probably characteristic of drug decomposition. The EGA curve showed characteristics of the individual components. Thermal analysis of the complex gave a TG peak at 255°C and a corresponding DSC endotherm. A DSC endotherm at 290°C and gas evolution at 310°C confirmed the decomposition of the sample at high temperatures. The absence of any evolved gas prior to this decomposition indicated the lack of any solvent residue from the preparation within the sample. Similar findings were reported for a prostacyclin complex with β-cyclodextrin (Szeman et al., 1987).

A further example of the use of thermal analysis in instability studies involves the chlorthalidone–urea system (Bloch et al., 1982). The drug displayed only one endotherm but urea, following a fusion endotherm at 136°C, displayed several endotherms at around 200°C characteristic of decomposition. In physical mixes later endotherms marked decomposition. The system showed a eutectic at 35% chlorthalidone at a temperature of 127°C. Decomposition of mixes occurred at above 190°C. Melts containing up to 30% drug were hard and crystalline whereas melts of more than 40% were soft and glassy. Pre-molten samples containing up to 40% drug showed only one endotherm corresponding to the eutectic. Higher concentrations displayed only the endotherms corresponding to decomposition with no endotherms in the range 47–227°C (Bloch et al., 1982).

7.6 MISCELLANEOUS CARRIERS

Several other carriers not previously mentioned in this chapter have been evaluated by DSC or DTA. Kim et al. (1985) evaluated the properties of a novel drug, HP573, dispersed onto colloidal silicon dioxide by fusion. The drug had a melting point of 53°C and silica-74 (colloidal silicon dioxide) showed no DSC transitions in the range 20–300°C. Their physical mix gave a DSC scan typical of the drug only, but on storage the enthalpy of fusion decreased indicative of transformation of the drug to an amorphous form. X-ray diffraction confirmed this transformation. After storage the dispersion containing 50% drug displayed no endothermic transitions and that containing 66% of the drug showed a much reduced endotherm, peaking at 41°C. The ratio of the drug to silicon dioxide was therefore of immense importance in determining the amount of crystallinity of the drug. Additionally silica-74, described as mesoporous and having a pore size of 15 nm, behaved similarly to silica-244, which has a pore size of 20 nm. The solid dispersion containing microporous silica-63 (pore size 2 nm) was crystalline at a 1:1 ratio. Subsequent dissolution data for the 1:1 systems indicated that the mesoporous silica dispersions gave greater dissolution rate increases than the microporous systems. The transformation kinetics were followed by using the ratio of the enthalpy of fusion of the stored systems to that of the initial unaged system (Kim et al., 1985). The ratio, plotted as a first-order plot revealed that transformation occurred in a biphasic manner for the silica-244 and silica-74 system. The size of the pore openings was an important factor in determining the rate of conversion, the non-porous silica and silica-63 displaying no transformation. It appears therefore that the model drug was hydrogen-bonded to the silica surface via

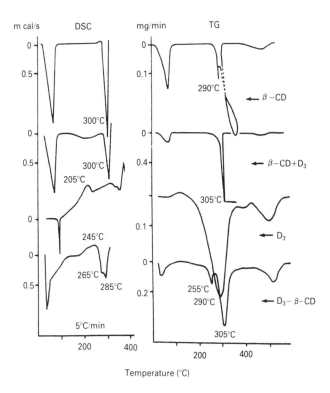

Fig. 7.14 — Thermoanalytical curves β-cyclodextrin, cholecalciferol, their mechanical mixture and their inclusion complex (reproduced with permission from Szejtli *et al.*, 1980).

the butanol moiety. The DSC scan symmetry was also used to interpret the interaction between the drug and the silicas. The assymetry was assessed using the ratio of the leading slope (*a*) to that of the trailing slope (*b*). For drug mixes with silica-74 and silica-244 the *a*/*b* ratio was of the order 2.38–2.52 whereas for the drug itself and its mixture with silica-63 the ratio approached unity. In general where the drug was not bound the endotherm was symmetrical (*a*/*b*≈1) and where bonding occurred the peaks were assymetrical (Kim *et al.*, 1985). The fusion enthalpies were used as markers for dissolution for the silica-74 physical mixes and dispersions. The dissolution of the active drug from the mixes was inversely proportional to the transition energy of the sample. Thus the solid dispersion which displayed a zero heat of fusion displayed the most complete dissolution of the drug.

Waxy carriers may be used to provide a prolonged release of the drug because of their hydrophobic properties. Schroeder *et al.* (1978) examined the phase diagrams of tripelennamine hydrochloride and tolazoline hydrochloride with carnauba and castor waxes and stearyl alcohol. DSC of stearyl alcohol showed melting at 50–52°C and deviation from the baseline at above 100°C possibly due to decomposition or sublimation. Carnauba wax melted at 76–79°C and showed similar baseline changes at 160°C. These changes were taken as evidence that fusion at 85–90°C would not cause deleterious changes to their stability. Two-component systems showed that no significant interactions or eutectic interaction occurred between tolazoloine and the waxes. Three-component systems were prepared by fusion and solvent techniques. The DSC scans of the system (Fig. 7.15) containing carnauba wax, stearyl alcohol

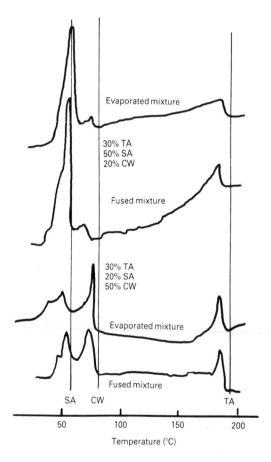

Fig. 7.15 — DSC scans for two compositions of tripelennamine–carnauba wax–stearyl alcohol. Samples of each composition were prepared by dispersing tripelennamine in a melt of carnauba wax and stearyl alcohol (fused mixture) and by dissolving all ingredients in chloroform and evaporating the solvent (evaporated mixture). Key: SA, stearyl alcohol; CW carnauba wax; and TA, tripelennamine hydrochloride (reproduced with permission from Schroeder *et al.*, 1978).

and tripelennamine showed that while minor differences occurred in the fused and evaporated samples the appearance of the endotherms corresponding to each individual component implied that no interaction occurred between the three components. This was confirmed by varying the wax concentrations at a constant 30% drug level. The melting points of each waxy component were again unchanged confirming that no interaction or eutectic formation took place in the ternary system. It was postulated therefore that other physical characteristics such as hardness, core composition and drug particle size would influence the release of drug from waxy matrices (Schroeder *et al.* 1978).

7.7 AGEING CHARACTERISTICS OF SOLID DISPERSIONS

Thermal analysis may be used in two manners to evaluate the likely ageing characteristics of solid dispersions. The obvious application is to study the artificially aged or devitrified dispersion and compare the scans before and after ageing. The other method utilizes the position of the glass transition temperature and compares it with theoretically derived values.

Ford & Rubinstein (1977) compared the DSC traces of one-hour-old and aged samples of chlorpropamide–urea solid dispersions. Ageing was achieved by accelerated storage at 60°C for 96 hours and by storage at ambient conditions for up to four weeks. Aged samples did not possess endotherms corresponding to the solidus line. Peaks corresponding to the liquidus line were generally 2–3°C higher than those obtained from the unaged samples. This was considered to be due to an increase in crystal size within the dispersion. Chiou & Niazi (1971) used the eutectic composition ($\equiv 52\%$ sulphathiazole) to predict the ageing of sulphathiazole–urea melt solid dispersions. Immediately after preparation the DTA scan showed an exotherm immediately prior to the endotherm corresponding to the fusion of the eutectic. This exotherm was thought to represent the conversion of amorphous sulphathiazole to a crystalline form. The exotherm was lost following storage under ambient conditions. X-ray diffraction studies confirmed this transformation. Thus thermal analysis may be used to predict age-induced changes because the loss of amorphousness or polymorphic modifications will be manifested in the thermal traces.

Summers (1978) examined the glass-forming properties of barbiturate–citric acid systems. Using equation (7.1) calculated T_g values were compared with experimentally determined values for citric acid dispersions with hexobarbitone, heptabarbitone and pentobarbitone. In each case the derived values were higher than theoretical values. It was supposed that this was due to a greater interaction between the drug and carrier molecules than between the drug molecules and the carrier molecules themsleves. Summers (1978) suggested that this resulted in 1:1 complexes between the barbiturates and citric acid in the glassy state but these were lost when the system devitrified to a simple eutectic. Timko & Lordi (1979) examined the ageing characteristics of bulk-prepared citric acid melts. Immediately after preparation the melt was a mixture of crystalline and amorphous states. On DSC analysis a broad recrystallization exotherm was apparent at $\approx 80°C$. Its area was examined during storage at 4, 23 and 37°C. Little change occurred at the lowest temperature and at 23°C a gradual decrease in area occurred. At 37°C an initial rapid decrease was followed by a slower change. The abrupt change corresponded to a loss of the T_g and it was thought that the initial rapid recrystallization was followed by a slower recrystallization, possibly resulting in a rearrangement of the primary crystallites (Timko & Lordi, 1979).

Timko & Lordi (1984) further considered that it would be possible to produce a glass of greater stability than either of its glassy components. This was demonstrated with a 1 : 1 citric acid : paracetamol glass whose DSC scans were unaltered after seven weeks of storage. Paracetamol itself formed a mechanical and thermally unstable glass which completely devitrified following 48-hours storage. Ford (1987) further developed the use of the T_g in predicting ageing problems of solid dispersions containing PEG 6000. The phase diagrams of seven drugs (chloramphenicol, glutethimide, griseofulvin, indomethacin, paracetamol, phenacetin and phenylbutazone) with PEG 6000 were examined following quench-cooling of the melt dispersions to either 4 or −120°C. With the exceptions of phenacetin and phenylbutazone the remaining five drugs produced at least partially stable glasses.

Typical diagrams include that for indomethacin (Fig. 7.13) and paracetamol (Fig. 7.16). These diagrams indicate the two extremes of paracetamol, where the

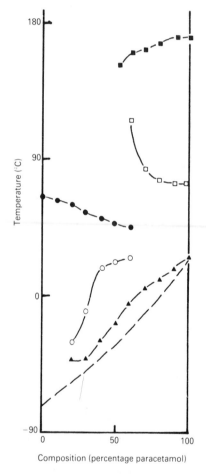

Composition (percentage paracetamol)

Fig. 7.16 — Phase diagram of paracetamol–PEG 6000, similarly constructed to Fig. 7.13. Key: ●, PEG fusion endotherm; ○, PEG recrystallization exotherm; ■, drug fusion endotherm; □, drug recrystallization exotherm; ▲, T_g values; broken line, predicted T_g values. (Reprinted from Ford (1987) p. 1762, by courtesy of Marcel Dekker, Inc.)

obtained glass transition temperatures were above those calculated, and indomethacin, where the transitions were below the calculated values. The severity of ageing on
the systems could be judged by such differences. PEG 6000 dispersions of chloramphenicol and paracetamol were the least effected by ageing. These two drugs
represented those diagrams where obtained T_g values were higher than the theoretical values. The strength of the interaction between the two components was
considered to be stronger than the mean strengths in PEG, paracetamol or chloramphenicol alone. Such interactions were deemed to protect the systems against ageing
by retarding drug association. Indomethacin, griseofulvin and glutethimide systems
were those most affected by ageing. The T_g values for the systems were lower than
the calculated values and indicated that a greater attraction for molecules of their
own species occurred than between the drug and PEG molecules. This resulted in a
greater association of the drugs or the PEG consequently making the systems more
sensitive to ageing. The rapid recrystallization of the phenacetin and phenylbutazone
systems rendered this method inappropriate.

7.6 CONCLUSIONS

Thermal analysis has played an important and valuable role in the analysis of solid
dispersions but requires the supportive evidence of other techniques such as IR
spectroscopy and X-ray diffraction studies to confirm the molecular state of the drug.
DSC is particularly useful in determining the solubility of a drug in a polymeric
carrier and both DSC and DTA are capable of detecting polymorphic modifications.
TG and EGA are important when decomposition may be a potential problem and
where the drug is volatile. Of particular interest is the role of both DSC and DTA in
predicting and examining age-induced changes. The usefulness of using glass
transition temperatures to predict ageing has still to be thoroughly evaluated but
phase transformations during storage are readily detected by either method.

REFERENCES

Andersen, F. M., Bundgaard, H. & Mengel, H. B. (1984) *Int. J. Pharm.*, **21**, 51–60.
Beaumont, R. H., Clegg, B., Gee, G., Herbert, J. B. M., Marks, D. J., Roberts, R.
    C. & Sims, D. (1966) *Polymer*, **7**, 401–417.
Beech, D. R., Booth, C., Dodgson, D. V., Sharpe, R. R. & Waring, J. R. S. (1972)
    Polymer, **13**, 73–77.
Bloch, D. W., El-Egakey, M. A. & Speiser, P. P. (1982) *Pharm. Acta Helv.*, **57**,
    231–235.
Chatham, S. M., Newton, J. M. & Walker, S. E. (1986) *Proceedings of 4th
    International Conference on Pharmaceutical Technology*, A.P.G.I., Paris, **2**,
    213–220.
Chiou, W. L. (1971) *J. Pharm. Sci.*, **60**, 1406–1408.
Chiou, W. L. & Niazi, S. (1971) *J. Pharm. Sci.*, **60**, 1333–1338.
Chiou, W. L. & Niazi, S. (1973) *J. Pharm. Sci.*, **62**, 498–501.
Chiou, W. L. & Riegelman, S. (1971) *J. Pharm. Sci.*, **60**, 1281–1302.
Chow, D. D, & Karara, A. H. (1986) *Int. J. Pharm.*, **28**, 95–101.
Collett, J. H., Flood, B. L. & Sale, F. R. (1976) *J. Pharm. Pharmacol.*, **28**, 305–308.

Corrigan, O. I. (1986) *Drug Dev. Ind. Pharm.*, **12**, 1777–1793.

Corrigan, O. I. & Stanley, C. T. (1980) *J. Pharm. Pharmacol.*, **34**, 621–626.

El-Gendy, G. A., Terada, K., Yamamoto, K. & Nakai, Y. (1986) *Int. J. Pharm.*, **31**, 25–31.

Ford, J. L. (1984) *Pharm. Acta Helv.*, **59**, 280–288.

Ford, J. L. (1986) *Pharm. Acta Helv.*, **61**, 69–88.

Ford, J. L. (1987) *Drug Dev. Ind. Pharm.*, **13**, 1741–1777.

Ford, J. L. & Rubinstein, M. H. (1977) *J. Pharm. Pharmacol.*, **29**, 209–211.

Ford, J. L. & Rubinstein, M. H. (1978) *Pharm. Acta Helv.*, **53**, 327–332.

Ford, J. L., Stewart, A. F. & Dubois, J.-L. (1986) *Int. J. Pharm.*, **28**, 11–22.

Gordon, M. & Taylor, J. S. (1952) *J. Appl. Chem.*, **2**, 493–500.

Hoelgaard, A. & Møller, N. (1975) *Arch. Pharm. Chemi. Sci. Ed.*, **3**, 34–47.

Jones, S. P., Grant, D. J. W., Hadcraft, J. & Parr, G. D. (1984) *Acta Pharm. Technicol.*, **30**, 213–223.

Kambe, Y. (1980) *Polymer*, **21**, 352–355.

Kaur, R., Grant, D. J. W. & Eaves, T. (1980a) *J. Pharm. Sci.*, **69**, 1317–1321.

Kaur, R., Grant, D. J. W. & Eaves, T. (1980b) *J. Pharm. Sci.*, **69**, 1321–1326.

Kim, K. H., Frank, M. J. & Henderson, N. L. (1985) *J. Pharm. Sci.*, **74**, 283–289.

Lang, M. C., Noel, C. & Legrand, A. P. (1977) *J. Polym. Sci.*, **15**, 1319–1327.

Nakai, Y., Yamamoto, K., Terada, K. & Ozawa, K. (1981) *Chem. Pharm. Bull.*, **101**, 1016–1022.

Nielson, L. E. (1974) *Mechanical properties of polymers and composites*, vol. 1, M. Dekker, New York, 25.

Otagiri, M., Imai, T., Hirayama, F., Uekama, K. & Yamasaki, M. (1983) *Acta Pharm. Suec.*, **20**, 11–20.

Schroeder, H. G., Dakkuri, A. & DeLuca, P. P. (1978) *J. Pharm. Sci.*, **67**, 350–353.

Sekiguchi, K. & Obi, N. (1961) *Chem. Pharm. Bull.*, **9**, 866–872.

Shefter, E. & Cheng, K. C. (1980) *Int. J. Pharm.*, **6**, 179–182.

Summers, M. P. (1978) *J. Pharm. Sci.*, **67**, 1606–1610.

Summers, M. P. & Enever, R. P. (1976) *J. Pharm. Sci.*, **65**, 1613–1617.

Summers, M. P. & Enever, R. P. (1977) *J. Pharm. Sci.*, **66**, 825–828.

Summers, M. P. & Enever, R. P. (1980) *J. Pharm. Sci.*, **69**, 612–613.

Szejtli, J., Bolla-Pusztai, E., Szabo, P. & Ferenczy, T. (1980) *Pharmazie*, **35**, 779–787.

Szeman, J., Stadler-Szoke, A. & Szejtli, J. (1987) *Acta Pharm. Technol.*, **33**, 27–30.

Sztatisz, J., Gal, S., Komives, L., Stadler-Szoke, A. & Szejtli, J. (1980) Thermoanalytical investigations on cyclodextrin inclusion compounds, in *Proceedings of an International Conference on Thermal Analysis*, Hemminger, E. (ed.), Birkhaeuser, Basel, 487–493.

Takayama, K., Nambu, N. & Nagai, T. (1977) *Chem Pharm. Bull.*, **25**, 2608–2612.

Tan, Y. Y. & Challa, G. (1976) *Polymer*, **17**, 739–741.

Theeuwes, F., Hussain, A. & Higuchi, T. (1974) *J. Pharm. Sci.*, **63**, 427–429.

Timko, R. J. & Lordi, N. G. (1979) *J. Pharm. Sci.*, **68**, 601–605.

Timko, R. J. & Lordi, N. G. (1982) *J. Pharm. Sci.*, **71**, 1185–1186.

Timko, R. J. & Lordi, N. G. (1984) *Drug Dev. Ind. Pharm.*, **10**, 425–451.

Turner, D. T. & Schwartz, A. (1985) *Polymer*, **26**, 757–762.

Uekama, K., Hirayama, F., Esaki, K. & Inque, M. (1979) *Chem. Pharm. Bull.*, **27**,

76–79.

Uekama, K., Narisawa, S., Hirayama, F. & Otagiri, M. (1983a) *Int. J. Pharm.*, **16**, 327–338.

Uekama, K., Oh, K., Otagiri, M., Seo, H. & Tsuruoka, M. (1983b) *Pharm. Acta Helv.*, **58**, 338–342.

Uekama, K., Narisawa, S., Hirayama, F., Otagiri, M., Kawano, K., Ohtani, T. & Ogino, H. (1983c) *Int. J. Pharm.*, **13**, 253–261.

Uekama, K., Hirayama, F., Fujise, A., Otagiri, M., Inaba, K. & Saito, H. (1984) *J. Pharm. Sci.*, **73**, 382–384.

Uekama, K., Sakai, A., Arimori, K., Otagiri, M. & Saito, H. (1985) *Pharm. Acta Helv.*, **60**, 117–121.

Winfield, A. J. & Al-Saidan, S. M. H. (1981) *Int. J. Pharm.*, **8**, 211–216.

Yakou, S., Umehara, K., Sonobe, T., Nagai, T., Sugihara, M. & Fukuyama, Y. (1984) *Chem. Pharm. Bull.*, **32**, 4130–4136.

8

The use of thermal analysis in polymeric drug delivery systems

The use of polymers in the controlled release of drugs is a very large and diverse area of pharmaceutical research and development. Water-soluble carriers may be used to increase the dissolution rate and extent of absorption of poorly water-soluble drugs. Thermal analytical treatment of these rapidly dissolving systems can be found in Chapter 7. Polymers whose solubility is pH-sensitive, e.g cellulose acetate phthalate, are used as coatings for tablets and are discussed in Chapter 9. This chapter concentrates on those polymers which are used to provide a controlled release. Only examples of representative polymers will be illustrated. Consideration is given to polymer characterization prior to formulation and thermal analysis of the formulated products.

8.1 INTRODUCTION

Barton *et al.* (1978) reviewed the use of thermal analytical techniques in polymer analysis. TG is useful in the measurement of thermal stability, the efficiency of antioxidants and the curing properties of some resins. DSC and DTA measure both glass and first-order transitions, the efficiency of antioxidants and purity (DSC). TMA provides information on transitions, cure, modulus, creep, thermal expansion and swelling behaviour, although care is required in sample preparation because the transition measured vary according to the stress applied. TBA provides measurements of transitions, cure, modulus and the effect of environment on physical properties. Barton *et al.* (1978) considered that TBA could not distinguish between melting and glass transitions.

The important transitions for the formulation scientist to measure are glass transition and melting, crystallization and decomposition transitions. The glass transition occurs at a specific temperature, the glass transition temperature (T_g). Its value can be affected by both instrumental variations and polymer history. Below the T_g the polymer atoms and chains can only undergo low amplitude vibratory motion and the polymer is usually in an amorphous and brittle inelastic state. Above the T_g

the polymer chains undergo rotational and diffusional movements and the polymer is in the rubbery state. The transition can be effected by molecular weight, position of pendant side-chains, chain stiffness and the polarity of the chain. Additionally branching, cross-linking, copolymerization and plasticizers will alter its value. Above the T_g, and depending on its value and polymer properties, crystallization and subsequent melting can occur, with or without degradation. It should be remembered that even crystalline polymers are never 100% crystalline due to steric hindrance provided by their long polymer chains. Instrumental factors, including heating and cooling rates and sample size, also influence transition temperatures.

Polymers are synthesized by either addition reactions or condensation reactions (Wood, 1980). Addition produces backbones consisting of $C-C$ bonds which are normally resistant to enzymic and hydrolytic attack, e.g. hydrogels. Condensation reactions produce byproducts, e.g. water molecules or hydrogen chloride, and examples include polyamides, polyesters and polyurethanes. Generally these polymers will be biodegradable and examples include polylactic acids, polyglycolic acid, poly-ε-caprolactone, polyglutamic acid and poly(alkyl-2-cyanoacrylates). Biodegradation occurs by four mechanisms (Gilding, 1981). These are solubilization or dissolution (e.g. polyvinyl alcohol, polyethylene oxide, pluronics and polyvinylpyrrolidone), ionization degradation (e.g. alkylvinylether–maleic anhydride copolymer and acrylic copolymers), enzyme-catalysed biodegradation (e.g. gelatine) and simple hydrolysis (e.g. polyglycolic acid, polylactic acids and polyethylene oxide–polyethylene terephthalate copolymers). An ideal biodegradable polymer (Pitt *et al.* 1976) should have sufficient strength for handling, be soft and pliable to minimize tissue irritation, possess little crystallinity and a T_g no greater than body temperature and be compatible with the drug. Homopolymers are unlikely to satisfy all these requirements and therefore copolymers provide more suitable bases for polymeric, biodegradable drug delivery systems.

There is a large pool of polymers which may be considered as suitable for use in prolonged release systems. These polymers may be classified by several methods including on the basis of their biodegradability. Those which are not biodegradable include silastic (polydimethylsiloxane), a synthetic rubber. Their major limitation is that following drug depletion the delivery device must be surgically removed if designed as an implant. Obviously this is not a problem for non-parenteral routes.

8.2 POLYMER CHARACTERIZATION

8.2.1 Polyacids (polyesters)

Although it is possible to prepare polyesters by condensation of the monomeric acids, the yield is invariably a low molecular weight product. The molecular weight of the product may be increased by polymerization via the cyclic diesters. The precursors of polyglycolic acid, glycolide and lactide, respectively possessed melting points of 80 and 98°C when examined by DSC (Gilding & Reed 1979). It is necessary to provide a constant thermal history to the polymer before examination to standardize previous thermal history. For instance copolymers of lactic and glycolic acids were heated and quenched on cold aluminium blocks prior to analysis to provide such a standard thermal history (Gilding & Reed, 1979) and remove the effects of previous treatments from the polymer structure.

TG may be used to examine the conversion of monomers to the polymeric chain. The copolymers of lactic and glycolic acids were prepared from their respective dimers, glycolide and lactide (Gilding & Reed, 1979). Samples from the reaction mixture at 200°C were quenched and extracted with solvent and the copolymer composition and percentage conversion estimated from TG. Fig. 8.1 illustrates

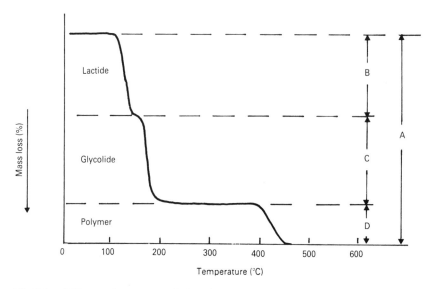

Fig. 8.1 — TG scan of pre-extracted glycolic acid/lactic acid polymerization mixture (reproduced from Gilding, D. K. & Reed, A. M., Fig. 2b p. 1461, *Polymer* 1979, by permission of the publishers, Butterworth & Co. (Publishers) Ltd ©).

representative results and from a knowledge of the volatility of the ingredients each segment of the weight loss curves was used to determine the composition of the reaction blend. The values of A, B, C and D in Fig. 8.1 allowed this calculation. The composition was confirmed by NMR. The data were also used to show that at the start of the polymerization process the glycolide dimers were preferentially incorporated at the expense of the lactide dimers.

8.2.1.1   Polylactic acids

The glass transition represents a measure of the flexibility of the polymer chain and may be an indication of the ease of hydrolysis of the ester bonds. The properties of the polymers vary with the form of lactic acid used in the manufacture and previous sample history. Poly-*d*-lactic acid (P-dLA) has T_g and T_m values of 67 and 180°C respectively (Pitt *et al.*, 1980) and is a crystalline, tough, inelastic polymer. Alternatively poly-*l*-lactic acid (P-lLA) showed endotherms at 66 and 165°C and an exotherm at 111°C with concomitant heats of transition of 0.7, 10.3 and 8.1 cal/g but on annealing only the higher endotherm was apparent (Pitt & Gu, 1987). The exotherm was due to recrystallization of the quenched film. The difference in enthalpy between this exotherm and the endotherm corresponding to melting was therefore equivalent to the crystallinity of the film. Based on an assumed heat of fusion of 22 cal/g, the crystallinity in the P-lLA of Pitt & Gu (1987) was ≈ 10%. Exposure of the film to water for 2000 hours did not result in any change in crystallinity but exposure to ethanol resulted in a single endotherm at 166°C of 10.1 cal/g and an apparent 50% crystallinity.

Other physico-chemical values for P-lLA obtained by thermal analysis were a crystallinity of 37% and a T_g at 57°C (Gilding & Reed, 1979) and a melting point at 174°C for the crystalline material (Christel *et al.*, 1980). Kulkarni *et al.* (1971) examined P-lLA of unknown molecular weight and found a melting point of 170°C and second-order transitions at 110–115°C. Decomposition occurred at above 300°C and a high crystallinity was claimed for drawn fibres.

Kalb & Pennings (1980) examined the crystallization kinetics of P-lLA. The melting temperatures of isothermally crystallized polymer were determined by DSC as the peak temperature. Fig. 8.2 shows that the intercept of the experimental data

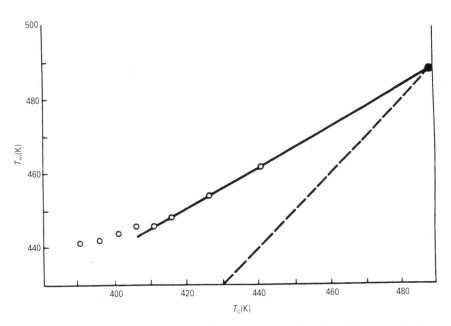

Fig. 8.2 — The crystalline melting point (T_m) of P-lLA as a function of the crystallization temperature, obtained from DSC measurements. The extrapolation of the data to $T_m = T_c$ – – –, yields an equilibrium melting point of about 488 K, shown as ● (reproduced from Kalb, B. & Pennings, A. J., Fig. 2 p. 608, *Polymer* 1980, by permission of the publishers, Butterworth & Co. (Publishers) Ltd ©).

with the $T_m = T_c$ line predicted an equilibrium melting temperature of 215°C. The estimated error was ± 10% due to lammellar thickening of the polymer crystals. The heat of fusion of a sample isothermally crystallized at 149°C was 50.7 J/g. The T_g was 55°C. Samples annealed at 45°C showed an endotherm at their T_g but samples which had been rapidly quenched showed only an endothermic inflection. The endotherm was attributed to a stress relaxation phenomenon that occurred within the polymer when allowed to stabilize. Such occurrences highlight the importance of removing the previous history of polymers.

Poly-dl-lactic acid (P-dlLA) was thought by Kulkarni $et\ al.$ (1971) to have a melting point of 60°C and no detectable glass transition when examined by DTA at 20°C/min. Decomposition occured at above 300°C. The polymer was claimed to be amorphous (Kulkarni $et\ al.$, 1971; Gilding & Reed, 1979). The transition of Kulkarni $et\ al.$ (1971) is probably a T_g. Christel $et\ al.$ (1980) gave it a value of 59.5°C and Pitt $et\ al.$ (1980) confirmed the T_g as 57°C, describing the polymer as amorphous, tough and inelastic. Rak $et\ al.$ (1985) used DSC to characterize the d,l-lactide and subsequently formed polymers. The lactide displayed only one endotherm at 126–126.4°C with a heat of fusion of 33.7 cal/g. The influences of molecular weight were examined on the T_g of P-dlLA. Values of T_g were quoted as 49.3, 41.7 and 51.0°C for samples of average molecular weights 4810, 8740 and 36 970 respectively. Samples with higher molecular weights displayed T_gs in the range 51.0–54.6°C with no apparent molecular weight dependence. The associated heats of transition displayed no molecular weight dependence and were similar with a mean value of 1.43 cal/g. There were no further transitions in the DSC traces except for a gradual endothermic drift commencing at 270–295°C which was due to carbonization of the samples.

Siemann (1985) reported the influence of moisture content on the T_g values of P-dlLA using DSC. Samples were dried for 24 hours at 40°C before use and before analysis were heated to 20°C above T_g then quenched to 0°C. The T_g was reduced by 12 K following exposure to moisture for six hours and remained unchanged even after 144 hours' exposure. Incorporation of salicylic acid into the P-dlLA gradually reduced the T_g to a minimum value at 6% salicylic acid which was 10 K lower than the untreated material. Exposure of this dispersion to moisture resulted in a total depression of the T_g of 28 K.

Wakiyama $et\ al.$ (1982) reported the thermal behaviour of these batches of P-dlLA. The polymer with nominal molecular weight of 9100 gave values for T_g, melting point and decomposition of 55, 270–280 and 280°C respectively. The values for similar transitions were 58, 266 and 274°C for a molcular weight of 17 000 and 58, 294 and 324°C for a molecular weight of 25 000.

8.2.1.2   Polyglycolic acids

Pitt $et\ al.$ (1980) gave the T_g of polyglycolic acid (PGA) as 36°C with a melting point at 230°C describing the polymer as crystalline, tough and inelastic. Gilding & Reed (1979) confirmed the T_g value. The degree of crystallinity is important because hydrolysis is generally recognized as proceeding through the amorphous phase only. Its crystallinity, in the form of the suture material 'Dexon', was reported by Gilding & Reed (1979) to be 46–52%. Chistel $et\ al.$ (1980) recognized the melting point for PGA as 218°C. Poly-d,l-ethyl glycolic acid is amorphous and rubbery and has a T_g of 16°C and polydimethyl glycolic acid is crystalline, tough and inelastic with a T_m at 240°C (Pitt $et\ al.$, 1980).

8.2.1.3   Copolymers containing lactic acid

Copolymers intended for use as sutures require an element of crystallinity to maintain their physical integrity and strength but when used as polymeric release systems the matrix should be monophasic to ensure heterogeneous drug dispersion and a regular controlled release. The thermal properties of copolymers of glycolic acid with various lactic acids have been studied by many workers.

Weight fractions of polyglycolic acid : Poly-*l*-lactic acid at 10:0, 9:1, 1:9 and 0:10 each possessed a T_g followed by a crystallization exotherm and a subsequent melting endotherm (Gilding & Reed, 1979). Typical DSC scans, obtained at 10°C/min, are shown in Fig. 8.3. Ratios of 3:7, 5:5 and 7:3 displayed a glass transition only and

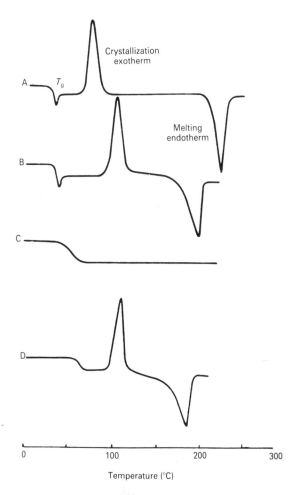

Fig. 8.3 — DSC scans of glycolic acid/lactic acid polymers. A, PGA; B, 90:10 GA:LA; C, 50:50 GA:LA, D, PlLA (reproduced from Gilding, D. K. & Reed, A. M., Fig. 5 p. 1462, *Polymer* 1979, by permission of the publishers, Butterworth & Co. (Publishers) Ltd ©).

consequently compositions in the range 25–70 mol% glycolic acid were amorphous. The 3:7 copolymer (Pitt & Gu, 1987) showed a single endotherm at 56°C with a heat of fusion of 1.5 cal/g. This was calculated to be equivalent to a degree of crystallinity of only 7% and confirms the amorphous nature of this particular composition. On exposure of a film of this composition to water (Pitt & Gu, 1987) the endotherm was replaced by a new endotherm at 73°C and multiple endotherms in the range 125–135°C. The heats of transition of these endotherms increased on storage and were consistent with hydrolysis of the polymers to smaller chains of glycolic or lactic acids which were capable of forming crystallites. Christel *et al.* (1980) reported the T_g values of the copolymers to be 52, 44 and 44°C at P-*l*-LA weight fractions of 75, 50 and 25% respectively. The thermal stability of PGA–P–*l*LA copolymers was assessed by TG (Gilding & Reed, 1979). Weight loss occurred only at above 200°C and indicated that the copolymers are stable to melt processing without degradation provided moisture is absent. The degradation products were the glycolide and lactide subunits.

Examination of the copolymer of PGA and P-dlLA revealed that compositions containing 0–70% glycolic acid were amorphous (Gilding & Reed, 1979). Christel *et al.* (1980) reported the T_g values of the copolymers to be 43 and 54°C at PGA weight fractions of 75 and 25% respectively.

Christel *et al.* (1980) used DTA to also determine the T_gs of copolymers of P-*l*LA and P-dlLA. The copolymers containing 25, 50, 75 and 100% P-dlLA gave relatively invarient T_gs at 60, 60, 59 and 59.5°C respectively.

The properties of copolymers of P-dlLA and poly-ε-caprolactone were examined by Pitt *et al.* (1976) to evaluate their potential as implant material. The T_gs of the various monomer ratios are determined as a function of P-dlLA content. All copolymers containing less than 80% *dl*-lactide were suitable for implementation because the T_gs were below 37°C (Fig. 8.4) and the copolymers would be flexible enough for use. The solid line on Fig. 8.4 is the data calculated from equation (8.1) (see also equation (7.2)).

$$\frac{1}{T_g} = \frac{W_A}{T_{gA}} + \frac{W_B}{T_{gB}} \qquad (8.1)$$

where T_g is the glass transition of the copolymer, W is the weight fraction of each of the monomers A and B and T_{gA} and T_{gB} are the glass transition temperatures of each homopolymer. Fig. 8.4 indicates the closeness of fit of the experimentally derived points to the calculated values. Schindler *et al.* (1977) further commented about the copolymer that those ratios with a T_g below room temperature were very tough but of good elasticity. The cross-links required for toughness were supplied by the crystalline phase formed by the longer caprolactone units.

8.2.2 Poly-ε-caprolactone
This polymer was described (Pitt *et al.*, 1980) as crystalline, tough but flexible. The values for the T_m and T_g were 63 and −65°C respectively. Dubernet *et al.* (1987) confirmed the values of the T_g but gave a melting point 3°C lower and a decomposition exotherm in the range 200–210°C.

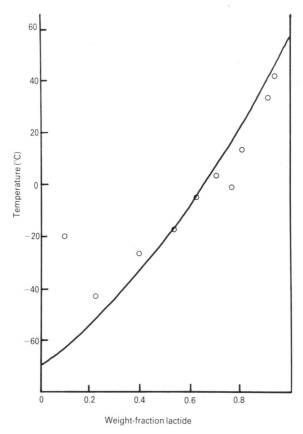

Fig. 8.4 — Variation of the glass transition temperature (T_g) of copolymers of caprolactone and *dl*-lactide as a function of *dl*-lactide content. The solid line is the calculated relationship, based on equation (8.1) (from Pitt *et al.*, 1976).

8.2.3   Polyethylene terephthalate (PET)

This polymer is crystalline, tough and inelastic with a T_g and T_m at 67 and 256°C respectively (Pitt *et al.*, 1980). Gillham & Schwenker (1966) performed a full thermal analysis on this polymer using DTA, TG and TBA (Fig. 8.5). DTA indicated a T_g in the range 65–88°C and the TMA spectrum indicated a decrease in modulus at above 65°C and the mechanical dampening showed a broad peak centred at 100°C which indicated appreciable crystallinity ($\approx 33\%$). A secondary recrystallization process was manifested by the appearance of a decrease in slope of the modulus curve above 160°C, a shoulder in the dampening curve at 175°C and a DTA exotherm commencing at 170°C. The melting transition was noted by a modulus drop to a minimum at 256°C, a sharp peak in the dampening curve at 252°C and a sharp DTA endotherm commencing at 228°C and peaking at 249°C. The changes in the curves above 350°C were indicative of chemical change and decomposition. At about 400°C the modulus curve rose steeply before levelling off at 430°C, the dampening curve displayed a

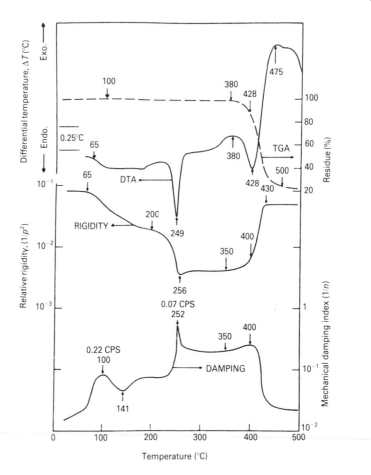

Fig. 8.5 — Thermoanalytical traces of poly(ethylene terephthalate) in nitrogen (reproduced with permission from Gillham, J. K. & Schwenker, R. F., *Applied Polymer Symposia*, **2**, 59–75) Copyright © 1966, John Wiley & Sons, Inc.).

broad peak at 400°C followed by a steep drop and the DTA curve showed an endotherm around 400°C peaking at 428°C and a subsequent large exotherm peaking at 475°C. Substantial weight loss in the TG curve occurred in the range 390–500°C with major weight loss around 400–475°C. Techniques such as gas chromatography and IR spectroscopy would be required to identify the decomposition products but Gillham & Schwenker (1966) suggested that the mechanical data indicated that the residue was highly cross-linked.

Miller (1968) used this polymer to illustrate the differences in transition temperatures that could be derived by different methods of thermal analysis. Thus DTA located transitions at − 130, − 62, − 5, 77, 207 and 225°C. TMA was completed using three separate modes. Tension data gave transitions at − 126, − 7, 33, 76, 155 and 242°C, linear expansion located transitions at − 125, 86, 130, 200 and 252°C whilst penetration studies indicated changes at only − 130, 83 and 243°C. Many of these transitions are not coincident and not detected by all techniques. Some events were unassigned, a T_g explained the trends at $\simeq 70°C$, and melting occurred at 226–252°C.

8.2.4 Polyvinyl alcohol (PVA)

PVA is often used to plasticize other more rigid polymers. Gillham & Schwenker (1966) identified its thermal properties (Fig. 8.6). The T_g at 77°C was identified by a

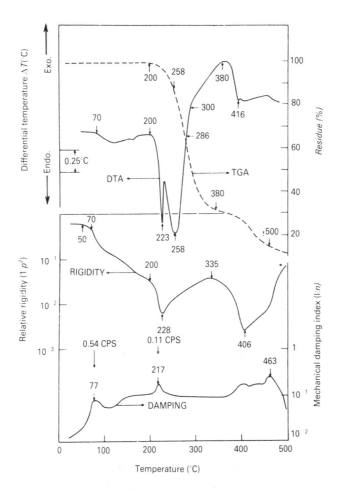

Fig. 8.6 — Thermoanalytical traces of poly(vinyl alcohol) in nitrogen (reproduced with permission from Gillham, J. K. & Schwenker, R. F., *Applied Polymer Symposia*, **2**, 59–75, Copyright © 1966, John Wiley & Sons, Inc.).

dampening maximum and a DTA endothermic trend. Subtle changes in the dampening and mechanical curves at 100–160°C were attributed to molecular movements in the crystalline portions of the polymer. The melting transitions of the crystalline regions were represented by a minimum in the modulus at 228°C, a dampening maximum at 217°C and a DTA endotherm at 223°C. The subsequent broad large endotherm at 258°C and changes in the TG curve indicated that the melting process was accompanied by decomposition and weight loss.

8.2.5 Poly(amino acids)

Crighton & Findon (1981) examined the DTA transitions of some simple poly(amino acids). Poly(glutamic acid) showed a melting/decomposition endotherm at 250°C. Poly(asparagine) showed endotherms at 200 and 240°C and poly(aspartic acid) at 200 and 253°C. Both of these higher temperature endotherms reflected some characteristics of melting. Endotherms potentially indicative of decomposition were noted at higher temperatures. Crighton & Findon (1981) attributed these to overlapping chemical degradations but suggested that other techniques such as TG and EGA were required to identify their characteristics.

8.3   DRUG DELVIERY SYSTEMS

8.3.1   Terminology

Implants are polymer drug preparations usually placed subcutaneously. The polymers chosen are either non-biodegradable, e.g. polydimethylsiloxane, or biodegradable, e.g. poly(caprolactone), poly(lactic acid) or poly(glycolic acid).

Two major mechanisms exist for implantation. The polymer may coat the core containing the drug as a membrane in which case drug release is controlled by diffusion through the membrane. The alternative system involves a dispersion of the drug, either in the molecular or fine crystalline state throughout the polymeric matrix. This may be formulated as either a single slab or matrix system (termed a monolith) or as a multiparticulate devices whose components may be known as microspheres, beads, microscapsules or nanoparticles. General preparative techniques include filmcasting, microencapsulation, moulding, compression, spray-drying, melt extrusion, micelle polymerization of the monomer units in the presence of the drug substances by irradiation. In the majority of these systems the drug can be released by diffusion through the polymer or erosion of the polymer by enzymes or hydrolysis. This depends on the physico-chemical interaction between the drug and the carrier.

Microencapsulation is a process by which the drug is coated by a polymeric coating in which the drug may show a partial solubility. The methods of preparation include coacervation, interfacial polymerization, precipitation, hot melt and solvent evaporation techniques. Microencapsulation is generally used for oral presentation, for instance to mask the taste of bitter drugs or to protect the gastro-intestinal tract from irritation. *Microcapsules* for parenteral use are spherical particles containing drug concentrated in a centre core (Leung *et al.*, 1987). Although many workers use different terminologies to describe their dispersed polymeric drug delivery systems and they have been variously described as microcapsules, microspheres, pseudolattices and nanoparticles, for the purpose of this chapter they are grouped together as microspheres bacause of similarities in their method of preparation. Nanoparticles are generally considered to be smaller than the other systems. *Microspheres* are spherical particles containing dispersed drug molecules in either in solution or crystal form (Leung *et al.*, 1987). The drug is either suspended or dissolved in a biodegradable polymer, e.g. poly(isobutyl cyanoacrylate) or poly(lactic acid). Their size is in the range 100 nm–1 μm. *Nanoparticles* were defined (Leung *et al.*, 1987) as transport carrier compartments for drugs of active molecules of non-liposomal character in the nanometer size range (10–1000 nm). Methods of preparation include colloidal coacervation, coating by polymerization of interfacial polymerization. *Films* are often preclinically prepared dispersions of the drug in the carrier prepared by solvent evaporation techniques.

Commerical polymeric devices include 'Ocusert' and 'Progestasert' which are insert reservoir devices which use ethylene–vinyl acetate copolymer membranes, and 'Septopal' and 'Garamycin' chains which consist of implantable beads composed of methyl methacrylate–methyl acrylate copolymers containing gentamycin.

Gilding (1981) considered that 25% w/w was probably the acceptable upper limit of drug loading and that the T_g should be outside the normal operating temperature range, i.e. 37°C, because otherwise discontinuity of drug release may occur. It is possible that a drug may plasticize a polymer and that as drug depletion or moisture absorption occurs the glass transition temperature may pass through body temperature and thereby seriously alter the drug release profile.

8.3.2 Applications of thermal analysis to drug–polymer interactions

Apart from determining transition temperature DTA and DSC are particularly useful in evaluating the form of the drug in the dispersion, The absence of an endotherm corresponding to drug fusion is indicative of the drug being dispersed as either a molecular dispersion or a solid solution. Bissery *et al.* (1983) considered that a lowering of the polymer T_g was indicative of solid solution formation. Various forms of drug dispersion may occur. The drug may be totally soluble in the organic solvent and therefore, when the microspheres are deposited, may recrystallize within the carrier in the same or a different polymorphic form to the starting material. There is the added possibility that the drug will be left in the microspheres in the molecular or amorphous state. If the drug is insoluble in the solvent than it is likely that the dispersed drug will be in the same polymorphic form and crystal size as the starting material.

Interactions between drug and carrier may also occur. Benoit *et al.* (1986a) used a P-dlLA around molecular weight of 60 000 and DSC at 20°C/min and DTA and TG at 10°C/min to examine its microspheres with 23% progesterone. Although the T_g was generally evident, depending on the dispersion it was classified as (a) too diffuse to measure, (b) diffuse or (c) apparent with an endotherm. On storage at 22°C the endotherm became more defined. Although the microspheres were stored under vacuum, weight loss, determined by TG as 1.8–4.7%, was attributed to methylene chloride i.e. residual solvent. The lack of endotherms to melting of cyrstalline drug suggested that drug was molecularly dispersed throughout the polymer. Polymer––drug interactions may be estimated following accelerated storage. By storage at 110°C, a temperature above the T_g but below the melting point of the progesterone, Benoit *et al.* (1986a) showed that the drug and carrier were immiscible. Their argument was that the treatment would not cause a precipitation of a miscible drug because the drug would remain molecularly dispersed whereas molecular movement of progesterone would be facilitated by holding the carrier at a temperature in excess of its T_g and allowing free movement of the drug molecules to effect crystallization.

Microspheres of.P-dlLA containing (chloro-2-ethyl)-1-cyclohexyl-3-nitroso-1-urea (CCNU) were examined by Benoit *et al.* (1986b). DTA scans from 30°C did not display the characteristic T_g at 57°C for the polymer or an endotherm at 93°C equivalent to the drug melting. A broad endotherm at 115–140°C was confirmed by TG to be the loss of volatile ingredients, attributable to the drug. Weight loss in the temperature range 20–180°C was taken to be equivalent to the total amount of drug loading. DSC from 0°C gave a broad endothermic change at 25°C equivalent to a T_g. The type of distribution of CCNU was discussed by Benoit *et al.* (1986b). The T_g of the polymer was 49 ± 5°C when the drug was dispersed in the crystalline state but was reduced to 21 ± 4°C when the drug was in the amorphous state, indicating that the drug plasticized the polymer. No changes in the curves were apparent following storage for 35 months, indicating physico-chemical stability.

Poly-ε-caprolactone microspheres containing 10% nitrofurantoin were examined by DTA analysis from well above the T_g of the polymer (Dubernet *et al.*, 1987). Two thermal events were apparent: an endotherm at 60°C corresponding to the melting point of the crystalline portions of the polymer and an exotherm at 270°C associated with nitrofurantoin degradation. Because the drug degraded before it could show its characteristic melting endotherm and associated heat of fusion it was not possible to calculate its precise degree of crystallinity. X-ray diffraction was required to confirm the crystalline nature of the drug which was assumed due to its insolubility in the casting fluid. Poly-ε-caprolactone did not protect the drug against thermal decomposition but the possibility was mooted that nitrofurantoin protected the polymer since its decomposition exotherm at above 200°C was inapparent in analysis of the microspheres.

Similar microspheres based on polycaprolactone and containing 15.6% progesterone were scanned by DSC and displayed the typical melting point of the polymer but not an endotherm corresponding to progesterone fusion (Chang *et al.*, 1987). This again indicated that the drug was molecularly dispersed or present in the amorphous state.

Knowledge of the T_g is important in assessing microsphere properties. Bissery *et al.* (1983) prepared microspheres containing various drugs using polymers such as poly(ε-caprolactone), P-dlLA and poly(β-hydroxybutyrate) (PHB). The latter had a T_g at 15°C and a melting point of 170–185°C. Microspheres were cast from methylene chloride or from chloroform using solvent evaporation. The casting temperature was 22°C and thus P-dlLA had a T_g above the casting temperature. The P-dlLA microspheres were uniform, smooth, glassy and defect-free because the T_g was below the casting temperature. PHB and poly(caprolactone) were relatively crystalline polymers with melting temperatures above the casting temperature. Because the PHB crystallized slowly the microspheres were relatively smooth but very porous. However, the poly(caprolactone) crystallized rapidly giving crumpled irregular microspheres.

8.3.2   Internal structure of microspheres

Since many microspheres are very porous thermal analysis has proved useful in characterizing their internal pore structure. Ishizaka *et al.* (1979) examined the permeability of polyamide microspheres using DSC from -30°C. Pure water displayed a single endotherm at 0°C. A microcapsule dispersion in water displayed three endotherms. The first of these, at -15°C, was lost in the presence of 1 M urea and was attributed to the melting of highly structured water on and around the hydrophilic polymers making up the microcapsule membrane. The second, somewhat below the melting point of ice, was attributed to the melting of free water and lowered by the possible presence of impurities. A third endotherm at about 7°C was thought to be due to melting of organic solvent remaining in the dispersion medium which was not removed on dialysis after preparation.

Vidmar *et al.* (1982) used DSC to study water structure in pores in ethylcellulose microcapsules. Only one endotherm was apparent between 0 and 5°C and was attributed to melting of the dispersant. No other endotherm coresponding to melting of water associated with the microsphere structure was apparent and it was concluded that no water existed in the members of the ethylcellulose microspheres examined.

8.3.3 Drug solubility in polymers

Irrespective of the method of preparation, any drug will be soluble in a polymer even to a small and limited extent. DSC has proved a useful tool in determining the degree of solubility. Theeuwes *et al.* (1974) studied the solubility and crystal characteristics of cholesterol and progesterone dispersed in polydimethylsiloxane by film casting. The method detected the heat of mixing of the drugs in the polymer, polymorphic modifications, the drug melting point and, most importantly, the fraction of the drug present in the crystalline form. Prior to solubility determination the film samples were held in the thermal analyser a few degrees below the melting point of the drugs. The areas of the melting endotherms on subsequent analysis were compared with those of samples scanned from room temperature without prior temperature holding. Because the areas were identical Theeuwes *et al.* (1974) assumed that the polymer and the drugs were in equilibrium with each other at the scanning speed employed and the area corresponding to the amount of crystalline drug. Recycling the samples was considered a requirement to check if polymorphic modifications occurred. The data in (Table 8.1) gives the observed endothermic heats of transition at various loadings. Since the heats were similar, the same crystal form of cholesterol was present at each treatment. The plot of the data of film drug loading and heats of transition (Fig. 8.7) gave a straight-line intercept at 35 mg/g polymer film and this was equivalent to cholesterol solubility. The endothermic area corresponded not only to a heat of melting but also has a contribution from the heat of mixing of the drug within the polymer. Theeuwes *et al.* (1974) calculated this heat of mixing from equation (8.2)

$$q_o = (m_t - m_s)q_m - (m_t - m_s)q_d \qquad (3.2)$$

where q_o is the observed heat per unit film weight, q_m is the heat of fusion of the drug, q_d is the heat of mixing, m_t is the total mass and m_s is the dissolved mass of the drug, each per unit weight of film. The heat of mixing was 0.6 cal/g. This slight difference is depicted in Fig. 8.7 as the difference between the derived and obtained heats of fusion (Theeuwes *et al.*, 1974). Progesterone melted at 402 K with a heat of fusion of

Table 8.1 — Observed endothermic heats of melting of silicone rubber samples containing dispersed cholesterol (reproduced with permission from Theeuwes *et al.*, 1974)

Concentration of cholesterol in the film (mg/g of film)	Observed endotherm (first heating) (cal/g)	Observed endotherm (second heating) (cal/g)
7.96	not observed	—
43.7	0.088	—
61.2	0.328	0.287
83.6	0.62	0.66
160.0	1.54	1.56
213.8	2.21	2.34
222.5	2.47	—
322.9	3.65	—

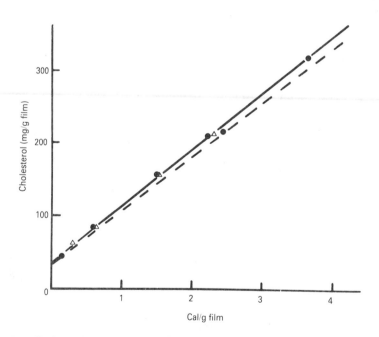

Fig. 8.7 — Cholesterol concentration of film samples as a function of the observed heat of fusion. Key: ●, observed heats at first melting; △, observed heats at second melting, ——, average line through experimental data; and –––, calculated line from drug solubility intercept and heat of melting of the pure drug (reproduced with permission from Theeuwes *et al.*, 1974).

19.1 cal/g. This was equivalent to the α-form. Recycling produced a scan with a peak at 395 K which was the β-form. The heat of fusion versus drug concentration diagram yielded an intercept, and hence solubility, of 6 mg/g film and a heat of mixing of 1.3 cal/g. Unlike the cholesterol films, reheating produced different scans with peaks at 378, 381 and 396 K. These were assumed to be progesterone polymorphs because altering the cooling conditions altered their intensity but not their location. It must be emphasized that the solubilities correspond to the solubility at the melting point and may not reflect the true solubility of the drug in the polymer at ambient or body temperatures. Rapid heating rates, beacuse they result in non-equilibrium conditions, may produce a better approximation to these lower temperature solubilities.

8.3.4 Microsphere degradability

Drug release from microspheres is often a function of their resistance to erosion and related to the amount of cross-linkage between the polymer chains, which controls its susceptibility to enzymatic attack. This is particularly the case with albumin-based microbeads cross-linked with glutaraldehyde. The greater the content of glutaldehyde the greater the resistance to erosion (Sheu *et al.*, 1986). The untreated product displayed a decomposition endotherm at 214°C. Use of 0.5, 1 and 1.5% glutaraldehyde increased the endotherm temperature to 219, 220.7 and 223°C respectively. This represented an increased modification of the lysine residues and produced a product that was less susceptible to digestion by α-chymotrypsin and possessed increased physical strength and durability. Taken to its extreme it was postulated that large levels of cross-linker would produce a non-erodable matrix. DSC was therefore used as a predictor of microsphere degradability.

8.3.5 Drug release mechanisms

Thermal analysis may be used to examine release mechanisms from polymer matrices. Carelli *et al.* (1986) described the feasibility of the techniques using the release mechanisms of progesterone from polydimethylsiloxane matrices as an example. The matrices contained generally 20% progesterone and 14% water-soluble carrier chosen from glycerol, ethylene glycol or polyethylene glycol 200. Polymer disks, following elution with 0.13 N phosphate buffer were examined by DSC either at 2°C/min from -20°C to evaluate the quality of fluid in the discs or at 8°C/min from 50°C to determine the polymorphic form of progesterone present at various distances from the disc surface. The study consisted of several experiments.

(a) The eluting buffer, or solutions containing 2.6% (iso-osmotic) or 10% glycerol, 1.7% (iso-osmotic) or 10% ethylene glycol, and 4% (iso-osmotic) or 10% PEG 200 were characterized to produce a melting range (ΔT_m) based on the purity correction method of Sondack (1972) (see Chapter 5) and equivalent to the difference in temperature between the final melting temperature and that temperature when 10% of the solid ice solution had melted. This melting range increased with increased solute levels and was used qualitatively to estimate the fluid that had replaced the soluble entities within the eluted matrix. In matrices containing 20% drug and no carrier no endotherm corresponding to moisture was detected. When progesterone was released from the matrix it was not replaced by eluting fluid. Matrices containing glycerol and ethylene glycol displayed ΔT_m values which indicated that the soluble content of the matrices had been replaced to yield fluid of near iso-osmolarity after one day of elution and below iso-osmolarity after 60 days. In matrices containing PEG 200 the value exceeded the iso-osmotic value after one day and indicated a delayed clearance of PEG.

(b) The water content of the matrices was determined on the basis of the observed heats of fusion. By determining the water content (volume/ unit polymer weight) (V_w) expressed as a ratio to V_r (the total volume of drug and water carrier released per unit polymer weight) it was possible to derive Figs 8.8 and 8.9. V_r is

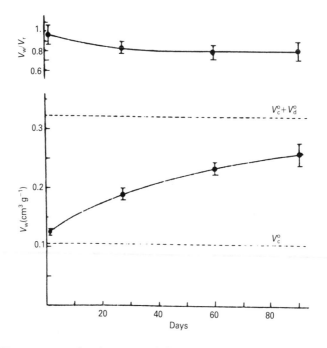

Fig. 8.8 — Water content and replacement ratio for matrix containing 20% progesterone and 14% glycerol as functions of time. V_w, volume of absorbed water per unit polymer weight; V_r, total volume of drug and water carrier released per unit polymer weight; V_c^o, volume of water carrier formulated within matrix per unit polymer weight; V_d^o, volume of drug formulated within matrix per unit polymer weight. Each point represents the mean of three samples from distinct batches. Vertical bars represent the standard deviation (reproduced with permission from Carelli *et al.*, 1986).

a summation of the volumes of carrier and drug in the matrix. For the matrix containing glycerol at day 1 the ratio was near unity indicating complete replacement of the carrier by the eluting solvent (Carelli *et al.*, 1986). The solvent absorption thereafter continued as evidenced by the decrease of the ratio from unity and the gradual increase in V_r. This was equivalent to direct replacement of the drug by eluant. This value did not reach the summated volume and indicated that all the prednisolone could not be replaced within the

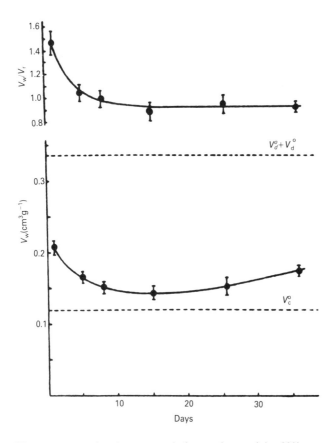

Fig. 8.9 — Water content and replacement ratio for matrix containing 20% progesterone and 14% PEG 200 as functions of time. V_w, volume of absorbed water per unit polymer weight, V_r, total volume of drug and water carrier released per unit polymer weight; V_c^0, volume of water carrier formulated within matrix per unit polymer weight; V_d^0, volume of drug formulated within matrix per unit polymer weight. Each point represents the mean of three samples from distinct batches. Vertical bars represent the standard deviation (reproduced with permission from Carelli *et al.* 1986).

matrix. This represented the difference between prednisolone particles which were wettable by pore fluids and particles that were incapable of being wetted and were released by diffusion through the polymer. Wettable particles should contribute mostly to initial stages of rapid drug release. The differences in Fig. 8.9 (the ratio exceeded unity and there was a high initial value of V_r) confirmed the delayed removal of PEG.

(c) Samples taken from different levels of the matrix containing 14% glycerol were examined following elution (Fig. 8.10). The approximately 0.05-cm thick inner section showed conversion of progesterone to its hydrate (curve C). A shift in the onset of the water-release peak occurred and the decomposition/melting endotherm moved to a lower temperature; the latter occurred in the uneluted control matrix (curve G). The hydrated form was absent in DSC scans of the

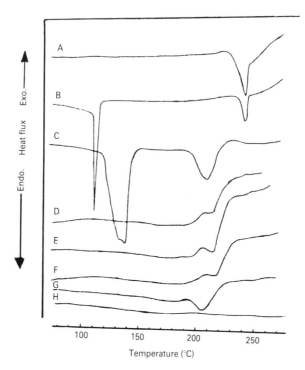

Fig. 8.10 — DSC scans of (A) commerical prednisolone, (B) prednisolone hydrous crystal form, (C) inner section of matrix containing 20% progesterone and 14% glycerol and eluted 4 days, (D) outer section of matrix containing 20% progesterone and 14% glycerol and eluted 4 days, (E) inner section of matrix containing 20% progesterone and 14% glycerol and eluted 7 days, (F) outer section of matrix containing 20% progesterone and 14% glycerol and eluted 7 days (G) control matrix and (H) vulcanized poly(dimethylsiloxane) (reproduced with permission from Carelli *et al.*, 1986).

0.02-cm thick outer layer of the same eluted matrix and both inner and outer segments (curves E and F) of samples eluted for seven days.

Prednisolone hydrate crystals are known to precipitate from saturated solutions of the anhydrous material. Consequently Carelli *et al.* (1986) confirmed the complex nature of the release of prednisolone from polydimethylsiloxane matrices. No account was given to the influence of the T_g on drug mobility through the rubber matrix; it appeared that prednisolone release was complicated by factors involving drug dissolution, precipitation of the hydrate and the subsequent redissolution of these crystals. Additionally the slower movement by diffusion of prednisolone through the matrix contributed to the release rates.

The treatment of Carelli *et al.* (1986) is very useful in assigning release mechanisms in matrices from which the drug can dissolve following pore formation. It is unlikely that similar release mechanisms can be validated by thermal analysis when biodegradation of the polymer occurs.

8.3.6 Ophthalmic delivery systems

DSC has been used to assess the suitability as carriers of pilocarpine and its nitrate salt (Saettone *et al.*, 1984). Grades of PVA and hydroxypropylcellulose (HPC) were examined as vehicles and polyacrylic acid as a potential cation for occular delivery systems containing pilocarpine. Pilocarpine nitrate displayed a melting point at 176°C and decomposition exotherms at 190 and 220°C. The position of the nitrate decomposition exotherm varied with the grade of polymer used. For PVP 90 000 the transition was centred at 230°C whereas for PVP 25 000 the transition was displaced to 190°C. These may be degradation products interacting with the polymer or nitrate decomposition. Examination of the drug–polymer systems on cooling produced two exotherms at 160-110°C which were not considered significant. PVA inserts containing the polyacrylic acid salt showed different behaviour. The salt alone displayed a single melting endotherm at 260°C. On examination of the inserts containing the polyacrylate salt the transitions due to polyvinyl alcohol were lowered in temperature and no transitions corresponding to the melting of the salt were apparent. On cooling the carrier crystallized over a narrow temperature range with only a low level of supercooling. These features were indicative of a tight binding of the components within the matrix and Saettone *et al.* (1984) claimed that this could produce a restrained mobility and release of the pilocarpine. This was confirmed in *in vivo* experiments.

REFERENCES

Barton, J. M., Lee, W. A. & Wright, W. W. (1978) *J. Therm. Anal.*, **13**, 85–89.

Benoit, J. P., Courteille, F. & Thies, C. (1986a) *Int. J. Pharm.*, **29**, 95–102.

Benoit, J. P., Madelmont, G., Puisieux, F. & Thies, C. (1986b) Polymer Prep., **27**, 27–28.

Bissery, M. C., Puisieux, F. & Thies, C. (1983) *Proceedings of 3rd International Conference on Pharmaceutical Technology*, A.P.G.I., Paris, **1**, 233–239.

Carelli, V., Di Colo, G. & Nannipieri, E. (1986) *Int. J. Pharm.*, **30**, 9–16.

Chang, R-K., Price, J. C. & Whitworth, C. W. (1987) *Drug. Dev. Ind. Pharm.*, **13**, 249–256.

Christel, P., Chabot, F., Leray, J. L., Morin, C. & Vert, M. (1980) *Biomaterials*, **3**, 271–280.

Crighton, J. S. & Findon, W. M. (1981) In '*Proceedings of the 2nd European Symposium on Thermal Analysis*, Dollimore, D. (ed.), Heyden, London, 247–250.

Dubernet, C., Benoit, J. P., Gouarraze, G. & Duchene, D. (1987) *Int. J. Pharm.*, **35**, 145–156.

Gilding, D. K. (1981) Biodegradable polymers. In. *Biocompatibility of clinical implant materials*, Vol. II, Williams, D. F. (ed.), C.R.C. Press, Florida, U.S.A., 209–232.

Gilding, D. K. & Reed, A. M. (1979) *Polymer*, **20**, 1459–1464.

Gillham, J. K. & Schwenker, R. F. (1966) *Applied polymer symposia*, **2**, 59–75.

Ishizaka, T., Koishi, M. & Kondo, T. (1979) *J. Membrane Sci.*, **5**, 283–294.

Kalb, B. & Pennings, A. J. (1980) *Polymer*, 607–612.

Kulkarni, R. K., Moore, E. G., Hegyeli, A. F. & Leonard, F. (1971) *J. Biomed. Mater. Res.*, **5**, 169–181.

Leung, S-H., S., Hui, H-W., Robinson, J. R. & Lee, V. H. L. (1987) In *Controlled drug delivery*, Robinson, J. R. & Lee, V. H. L. (eds.), Marcel Dekker, N.Y., Chapter 10, 433–480.

Miller, G. W. (1968) In *Analytial calorimetery*, Porter, R. S. & Johnson, J. E. (eds.), Plenum Press, New York, 71–82.

Pitt, C. G. & Gu, Z-W. (1987) *J. Controlled Release*, **4**, 283–292.

Pitt, C., Christensen, D., Jeffcoat, R., Kimmel, G. L., Schindler, A., Wall, M. E. & Zweidinger, R. A. (1976). In: *Proceedings drug delivery systems*, Ed: Gabelnick, U.S. Department of Health & Welfare, 141–192.

Pitt, C. G., Marks, T. A. & Schindler, A. (1980) In *Biodegradable drug delivery systems*, Baker, R. (ed.), Academic Press, London, 19–43.

Rak, J., Ford, J. L., Rostron, C. & Walters, V. (1985) *Pharm. Acta Helv.*, **60**, 162–169.

Saettone, M. F., Giannaccini, B., Chetoni, P., Galli, G. & Chiellini, E. (1984) *J. Pharm. Pharmacol.*, **36**, 229–234.

Schindler, A., Jeffcoat, R., Kimmel, G. L., Pitt, C. G., Wall, M. E. & Zweidinger, G. L. (1977) In *Contemporary topics in polymer science*, Pearce, E. M. & Schaefgen, J. R. (eds), Plenum Press, New York, 251–286.

Sheu, M-T., Moustafa, M. A. & Sokoloski, T. D. (1986) *J. Parent. Sci. Technol.*, **40**, 253–258.

Siemann, U. (1985) *Thermochim. Acta*, 513–516.

Sondack, D. L. (1972) *Anal. Chem.*, **44**, 888, 2089.

Theeuwes, F., Hussain, A. & Higuchi, T. (1974) *J. Pharm. Sci.*, **63**, 427–429.

Vidmar, V., Jalsenjak, I. & Kondo, T. (1982) *Int. J. Pharm.*, **34**, 411–414.

Wakiyama, N., Juni, K. & Nakano, M. (1982) *Chem. Pharm. Bull.*, **30**, 2621–2628.

Wood, D. A. (1980) *Int. J. Pharm.*, **7**, 1–18.

9

The use of thermal analysis in the development of solid dosage forms

The uses of thermal analytical techniques have increased immensely in the development of solid dosage forms. The uses in the analysis and characterization of drugs are covered in Chapter 6. This chapter develops these uses to include the characterization of excipients and how formulation may be developed from a knowledge of thermal analysis. Additional specialized operations, e.g. film coating, can be evaluated by thermal analysis.

9.1 EXCIPIENT CHARACTERIZATION

Excipients may be characterized in a manner similar to drugs. It is not the intention of this section to review fully the literature dealing with excipient characterization but to indicate the potential of thermal analysis in this field. Reference to other excipients is made in Chapters 7, 8 and 10.

9.1.1 Starch and gelatinization

Starch products are widely used in tabletting for their disintegration and binding properties. Starches are used both extra- and intra-granularly to effect disintegration and their aqueous gels are used as wet binders. Information on the thermal properties of starches and their gels is found in food technology literature. Care should be exercised in using thermal analysis in the determination of the gelatinization process of starches (Lund, 1984), not because of a limitation of the technique but because of the inhomogeneity of starches due to their natural origin.

    DTA scans of potato and wheat starches (Morita, 1956) were characterized by a number of endotherms in the range 135–310°C and two endotherms in the range 375–520°C. Pretreatment by solvent extraction affected the scans. Endotherms at 145 and 275°C in wheat starch were due to selective dehydration and dextrinization and one at 310°C corresponded to the elimination of polyhydroxyl groups accompanied by depolymerization and decomposition which accounted for later decomposition exotherms. The trapped effluent was analysed by ultra-violet and IR spectroscopy. Many starches display a broad endotherm at 50–120°C corresponding to the volatilization of adsorbed water.

The process of gelatinization (heating starch with water) includes the following stages (Olkku & Rha, 1978): (a) the granules hydrate and swell, (b) loss of birefrigence, (c) increase in clarity of the mixture, (d) rapid increase in consistency, (e) dissolution and diffusion of linear molecules out of the ruptured starch granules and (f) retrogression of the mixture to a paste-like mass of gel. The phenomenon is linked to amylose concentration, the length of its molecular chains and their state of dispersion. In wheat starch the behaviour can be represented by a small amount of swelling at 60–70°C involving disruption of weakly bound or readily accessible amorphous sites and followed at higher temperatures (80–90°C) by the disruption of more highly bound or less accessible sites and eventual fragmentation of the starch granules (Bechtel et al., 1964). Methods of analysis of starch gelatinization include studying consistency changes during gelatinization (viscoamylography, plastography) and proton magnetic resonance.

DSC is a useful tool in the investigation of phase transitions of starch–water systems (Biliaderis et al., 1980) allowing investigation over a wide starch:water ratio range and at gelatinization temperatures above 100°C. Samples in the range 10 ± 2 mg may be analysed in hermetically sealed pans which are essential in preventing moisture loss during analysis. Corn and potato starches were studied using a water content of about 47% (Biliaderis et al., 1980) and two endothermic transitions were apparent below 100°C. DSC was used to describe temperatures T_{p1} and T_{p2} corresponding to the peaks or shoulders of the endotherms in addition to T_0 and T_m, the temperatures corresponding to the onset temperature and the melting temperature, i.e. the intercept of the tailing edge of the endotherm with the baseline. The values of T_0, T_{p1}, T_{p2} and T_m were 55, 60, 68, and 85°C for potato starch and 60, 67, 78 and 89°C for corn starch. The respective heats of transition for these starches were 4.4 and 3.3 cal/g (Biliaderis et al., 1980).

Variation in the water content will produce differences in the shape of the DSC scans. Fig. 9.1 demonstrates the effect of increasing moisture content of the starch. At high water levels (Biliaderis et al., 1980) only one endotherm (T_{p1}) was apparent but as the moisture content was reduced, although this endotherm remained, a second endotherm (T_{p2}), whose peak temperature increased as the moisture level decreased, became apparent. The moisture content clearly effected the gelation processes as follows (Biliaderis et al., 1980). At high levels extensive hydration and swelling of the amorphous portions of the grains encouraged melting of the starch crystallites, the whole process taking place over a narrow temperature range giving a narrow endotherm. As the starch content increases, the destabling effect of the amorphous regions decreases and only partial melting (dissolution) of the crystallites occurred (Biliaderis et al., 1980). The water is however distributed throughout the partially formed gel with the net result that melting of the unmelted crystallites would occur at higher temperatures giving the biphasic scans noted in Fig. 9.1. Without water only heat is available as a grain denaturant and consequently melting occurred at higher temperatures than in the presence of water. At low water values the heat of transition represented the enthalpy of fusion of the starch crystallites but at high moisture levels the heat of transition corresponded to a variety of reactions such as granule swelling, crystallite melting and hydration phenomena.

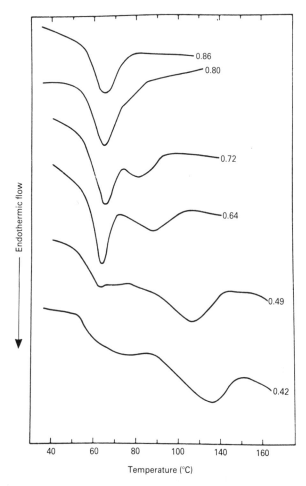

Fig. 9.1 — DSC scans of smooth pea starch heated at different water concentrations; numbers represent volume fraction of water. Percentage concentrations of starch (w/w) for these experiments from top to bottom were 19.0, 26.6, 37.0, 45.6, 60.8 and 67.5. Reprinted from Biliaderis *et al.*, *Journal of Food Science*, 1980 45 (6): 1671. Copyright © Institute of Food Technologists.

Pretreatment will markedly alter the properties of starches. Commonly water-binding capacities and enzyme susceptibilities increase but swelling powers decrease. Donovan *et al.* (1983) studied the effects of treatment methods, including defatting under methanolic reflux, heat–moisture treatment and their combination, on gelatinization. Fig. 9.2 indicates that the gelatinization temperature range broadened and shifted to higher temperatures with heat–moisture treatment. The enthalpy change for the transition decreased with heat–moisture treatment and the temperature of gelatinization increased with an increase in the moisture content during heat treatment, the transition becoming biphasic. These DSC scans were obtained at heating rates of 5°C/min but Donovan *et al.* (1983) also used a rate of 1°C/min to

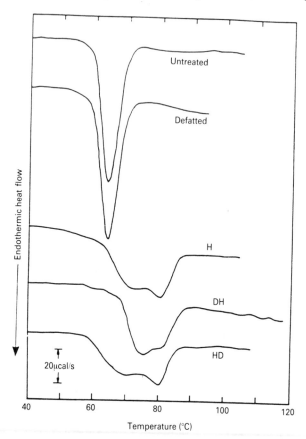

Fig. 9.2 — Differential scanning calorimeter scans for untreated, defatted and heat-moisture (27%) treated potato starch at 5°C/min. Sample sizes (d.b., top to bottom): 1.92 mg, 1.77 mg, 2.4 mg, 2.22 mg, 1.99 mg, with 12–13 mg of water. H, heat-moisture treated; DH, defatted, then heat-moisture treated; HD, heat-moisture treated, then defatted. (Reproduced with permission from Donovan *et al.*, 1983).

allow comparison with results of birefringence studies. A comparison of the two methods (Fig. 9.3) showed a greater drift in DSC baseline due to the increase in sensitivity required at the lower heating rate. Birefringence loss corresponded to the second endotherm peak for the heat–moisture-treated potato starch. For untreated potato starches the temperature of birefringence was lower and corresponded to the endotherm peak. Some structure was retained following treatment but was detectable only by DSC and not optical studies.

The birefringence temperatures of wheat starch (Donovan *et al.*, 1983) matched those of a lower melting endotherm for treated starches and were similar to the temperatures of the untreated starch. This indicated that some structure was retained following heat–moisture treatment which may have produced types of crystallites different from those in the untreated product in having a higher gelatinization temperature because of a greater stability. The first transition corresponded to

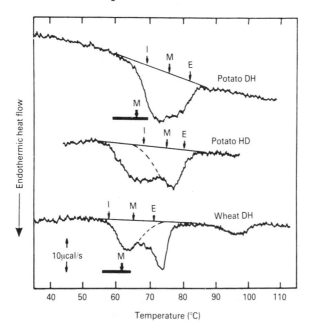

Fig. 9.3 — Comparison of scanning calorimeter measurements with loss of birefringence for
samples heat-treated at 27% moisture. Heating rate 1°C/min for both methods. I, M, and E
indicate initiation, midpoint and end of the birefringence change observed visually. The thick
bars show the gelatinization temperature ranges of the untreated starches. Dashed curves are
drawn to represent the probable course of birefringence loss (arbitrary vertical scale). Dry
weight of DSC samples, from top to bottom; 4.29 mg, 4.33 mg, 4.5 mg with 15 mg of water.
Reference 151 mg of aluminium. H, heat-moisture treated; DH, defatted, then heat-moisture
treated; HD, heat moisture treated then defatted. (The marked slope of the upper DSC trace is
due to error in setting the instrument baseline slope adjustment.) (Reproduced with permission
from Donovan *et al.*, 1983.)

the transition of normal crystallites in the granule. The second endotherm corresponded to transitions of crystalline regions perfected by the heat treatment. This
allowed polymer chain motion and crystal growth of crystalline regions with perhaps
intermolecular hydrogen bonds being replaced by hydrogen bonds with water
molecules (Donovan *et al.*, 1983).

DSC may be used to determine the amount of gelatinization in starch suspensions
(Lund, 1984). Since the endotherm area corresponds to the amount of gelatinization,
it follows that this will decrease if gelatinization has previously occurred. Thus by
exposing a suspension to heat and analysing at different time intervals it is possible to
determine the ungelatinized content of the starch. Although first-order kinetics may
describe gelatinization, Lund (1984) thought it highly questionable to develop a
kinetic model for gelatinization but cited activation energies for gelatinization as
between 14 and 25 Kcal/mol. The large variation was caused by differences in nature
of the raw material. Muhr *et al.* (1982) used high-pressure DTA to examine the
gelatinization of starches prepared as 33% starch slurries. The heats of fusion
decreased with increasing pressures. Thus values were obtained of 3.1, 2.0, 1.8, 1.7,

1.2 and 1.1 cal/g at pressures of 0.1, 88, 143, 156, 272 and 329 MPa for wheat starch. The peak decreased at higher pressures. The decrease in area was not as apparent for potato starch (Muhr *et al.*, 1982) but the temperature corresponding to the endotherm peak increased from 62.4 to 66.8°C as the pressure was increased from 0.1 to 285 MPa. Considerably greater pressures are therefore required to gelatinize potato starch at room temperature than wheat starch.

9.1.2 Gelatine and gelatization

A solution of gelatine is used as a wet granulating agent. Because gelatine is naturally based it suffers from batch to batch variation. Literature values of its glass transition temperatures vary by as much as 100°C and possibly due to differences in the residual water levels of samples. Marshall & Petrie (1980) examined the DSC scans of gelatine films prepared by hot casting at 49°C or cold chilling at 5°C for 3 min prior to drying at about 20°C. The amount of gelatine structure was related to an endotherm associated with structural loss on heating. Cold-dried films had a structure equivalent to 5 cal/g gelatine (Marshall & Petrie, 1980). DSC indicated moisture loss up to about 200°C and a small endotherm at around 230°C corresponded to loss of structure. Immediate reheating produced an endothermic shift corresponding to the T_g at 220°C. Water plasticizes gelatine films and Fig. 9.4 indicates the changes induced by

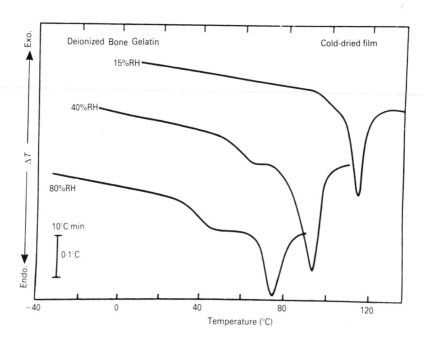

Fig. 9.4 — DSC scan for a cold-dried gelatine film equilibriated at 15, 40 and 80% relative humidity (RH) respectively (reproduced with permission from Marshall & Petrie, 1980).

storage. At higher humidity levels the T_gs were well separated from the melting endotherms but, following storage at 15% relative humidity, the two events became superimposed. Immediate recycling obliterated the structural loss endotherm and enabled T_g determination, confirming that it was reduced increasingly at higher humidities through the plasticization of gelatine by water. The endotherm represented a helix-to-coil transition (Marshall & Petrie, 1980). The DSC of a solution containing 10% gelatine chill set at 4°C revealed the helix-to-coil structural loss at 32°C. Annealing the gels produced greater stability in the gel structure.

9.1.3 Celluloses
Celluloses are used for a variety of reasons, e.g. microfine or microcrystalline cellulose is a direct compression adjunct and hydroxypropylmethylcellulose is used, because of its readiness to form water-soluble gels, in sustained-release matrices. Most grades of cellulose without prior treatment display a broad endotherm in the range 75–120°C which corresponds to loss of unbound water. Heating celluloses to high temperatures produces pyrolysis (Kilzer & Broido, 1965) which was represented by (a) a small endotherm at around 220°C corresponding to loss of water to yield a dehydrocellulose, (b) an endothermic depolymerization at about 280°C often running in competition with dehydration producing a tar probably involving laevoglucasan and (c) an exothermic degradation producing gaseous products (water, carbon dioxide and carbon monoxide) and tarry byproducts. The degradation routes were confirmed (Kilzer, 1971) using a combination of TGA with effluent gas analysis and DSC. Dollimore & Hoath (1981) used IR spectroscopy in conjunction with TMA to confirm the degradation of cellulose. Gas evolution commenced at 250°C but the temperatures corresponding to maximum evolution of carbon dioxide, carbon monoxide and water differed.

DTA was used (Nyqvist *et al.*, 1978) to characterize Avicel pH101® which displayed three exotherms at −30, −9 and 0 to 5°C when examined at a fast (undisclosed) heating rate and under vacuum. These transitions were attributed to some form of adsorption phenomenon.

9.1.4 Lactose
Lactose is widely used as an excipient in tablets and capsules and exists in several forms which vary in their response to water. α-Lactose is an unstable anhydrate which absorbs one water molecule per lactose molecule to form α-lactose monohydrate. β-Lactose exists as a stable anhydride and additionally as a glassy form which is stable at low relative humidities but is hygroscopic at higher humidities.

α-Lactose monohydrate is particularly sensitive to sample encapsulation (Berlin *et al.*, 1971). In crimped containers a dehydration endotherm commenced at 97°C, peaked at 144°C and returned to baseline at 167°C. A melting endotherm commenced at 202°C, peaked at 223°C and returned to baseline at 227°C. Heats of dehydration and fusion were 12.3±0.7 kcal/mol and 35.9±1.7 cal/g. Charring and decomposition occurred at higher temperatures. Heating in sealed containers split the dehydration endotherm and an additional exotherm and endotherm developed. Berlin *et al.* (1971) considered that the exotherm in the sealed capsule was due to the solution of lactose in the released water and subsequent conversion of α-lactose to the β-form. A concentrated solution of lactose in the water would form and be unable to escape from the sealed container, resulting in mutarotation. The resultant endotherms are due to melting of the α- and β-forms.

Itoh *et al.* (1977), examining α-lactose hydrate, found DTA endotherm peaks at 154 and 221°C and a small, unassigned exotherm at 181°C. Treating with methanol or reflux resulted in a loss of this exotherm at 181°C and the dehydration endotherm, indicating that treatment had yielded and anhydrous stable α-form. Itoh *et al.* (1977) also attempted to dry α-lactose monohydrate by heat (Fig. 9.5). The DTA changes

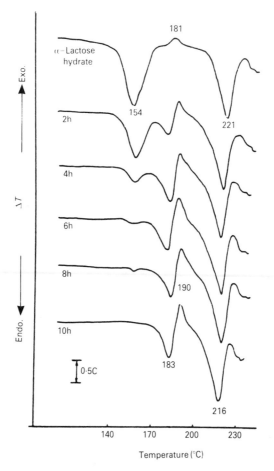

Fig. 9.5 — DTA scans of α-lactose hydrate dried by heating for various times at 110.6°C (reproduced with permission from Itoh *et al.*, 1977).

were an increase in an exotherm at 181–190°C and the development of a new endotherm at 183°C. Thus the heat-prepared anhydrate was unstable and different to the methanol-prepared anhydrate. The unstable anhydrate rehydrated to give a DTA scan that showed the restoration of a dehydration endotherm but endotherm and exotherm peaks were still apparent although their peaks transferred from 183 and 190°C to 175 and 181°C respectively, indicating that the physico-chemical properties were not fully restored. Lerk *et al.* (1984a), using X-ray diffraction to quantify the scans of α-dextose monohydrate, found that dehydration produced both the stable and unstable anhydrous lactose, the latter converting to a β/α-compound at temperatures around 180°C. The compound formed at about 150°C was confirmed by X-ray diffraction as a 1:1 of α- and β-lactose (Lerk *et al.*, 1984a). The initial endotherm of the unstable anhydrous α-lactose was due to its melting and the subsequent exotherm and endotherm were due to crystallization and fusion of the α/β-compound.

Lerk *et al.* (1980) examined the DSC of α-lactose monohydrate following compression or grinding using open pans. Untreated α-lactose monhydrate dis-played an additional endotherm at 164°C which was followed by an endotherm at 175°C. Since drying the lactose at 100°C produced an endotherm at 173°C it was anticipated that the extra peak in the α-lactose monohydrate was due to unstable anhydrous α-lactose or β-lactose. Stable anhydrous α-lactose displayed endotherms at 211 and 224°C. Grinding and compression of stable anhydrous α-lactose decreased the peak melting temperatures (Lerk *et al.*, 1980). The α-lactose monohydrate dehydration peak broadened after 5 minutes' grinding or 20 minutes' compaction and decreased after 20 minutes grinding to 80°C. TG and DTA of α-lactose monhydrate following 20 hours' milling confirmed the broadening of the dehydration endotherm (Krycer & Hersey, 1981). No water was released at about 100°C due to a build up of bound water. Surface energy increases induced by particle size reduction were dissipated by the moisture adsorbed.

Intensely ground α-lactose monohydrate showed a similar scan to amorphous lactose (Fig. 9.6) with an exotherm at ≃160°C and an endotherm at 210°C (Lerk *et al.*, 1984b) and only a slight dehydration endotherm compared with the untreated material. Mechanical treatment had reduced the crystallinity and crystallization, marked by the exotherms, occurred. α-Lactose monohydrate formed the β/α-compound. X-ray diffraction showed that the β-lactose content of the α-lactose sample rose from 2% before treatment to 42% on grinding and thermal treatment.

β-Lactose displayed a single endotherm beginning at 220°C, peaking at 232°C and returning to the baseline at 241°C with a heat of fusion of 48.7±1.3 cal/g (Berlin *et al.*, 1971). β-Lactose is usually prepared by crystallization from an aqueous lactose solution above 93.5°C, however the product is often contaminated with α-lactose. The DTA scans of β-lactose at different heating rates (Itoh *et al.*, 1978) are given in Fig. 9.7. The melting point increased with an increase in heating rate. Since no prior transitions were apparent, the anomaly was attributed to decomposition on fusion. Prior treatment raising the temperature to near the melting point before scanning reduced the depression of melting point. β-Lactose, following preheating at 40°C/min to 224°C to avoid decomposition, possessed a melting point of 229.5°C when subsequently scanned at 0.5°C/min. Grinding and compression of β-lactose resulted in a decrease in the peak melting temperatures (Lerk *et al.*, 1980) and intensively ground β-lactose developed an exotherm at 140°C (Lerk *et al.*, 1984b) (Fig. 9.6).

Berlin *et al.* (1971) freeze-dried an aqueous solution of lactose to prepare amorphous lactose which displayed a broad endotherm between 10 and 120°C equivalent to an enthalpy change of 10.8±0.5 kcal/mol. Reheating the lactose glass

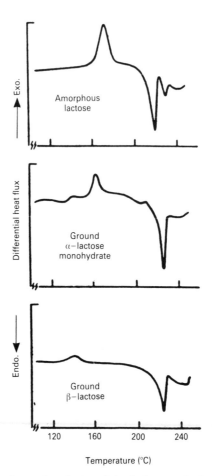

Fig. 9.6 — DSC curves of amorphous lactose, intensively ground α-lactose monohydrate and intensively ground β-lactose, respectively, recorded at a heating rate of 10°C/min (reproduced with permission from Lerk *et al.*, 1984b).

in a sealed container gave a small exotherm peaking at 110°C with an energy of −4.7 kcal/mol anhydrous lactose (Berlin *et al.*, 1971). Storage of the amorphous solid at 70% relative humidity resulted in water sorption and a gradual conversion, DSC showing that α-lactose monhydrate was formed. TG of the glass and α-lactose monohydrate indicated that water loss for the hydrate occurred over a discrete temperature interval between 97 and 165°C whilst the water in the glass was lost over a wide range indicating that the water was not bound in the same manner as in the crystal (Berlin *et al.*, 1971). The similarity in the heats of dehydration between the two forms indicated that the manner of interaction between water and the two forms was similar (probably through hydrogen bonding). Lerk *et al.* (1984b) showed that amorphous lactose displayed an exotherm at about 160°C and an endotherm at 210°C as indicated in Fig. 9.6.

Amorphous lactose, prepared by grinding or freeze-drying, prior to storage possessed a DSC exotherm which did not correspond to a TG weight loss (Morita *et*

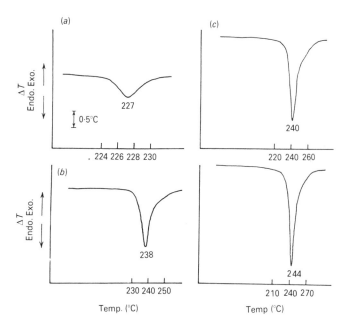

Fig. 9.7 — DTA scans of β-lactose recorded at various heating rates. Rates of heating remained unchanged throughout each measurement and were (a) 1°C/min (b) 5°C/min (c) 10°C/min (d) 15°C/min. (Reproduced from Itoh, T. *et al.* An improved method for the preparation of crystalline β-lactose and observations on the melting point, *Journal of Dairy Research* (1978), with permission of Cambridge University Press.)

al., 1984). Amorphous lactose was stable for 30 days when stored over desiccant at 30°C, but at 60% relative humidity the exotherm of the freeze-dried sample almost disappeared whereas that of the ground sample was still apparent. Dehydration endotherms became clear for the stored samples; the peak for the freeze-dried samples was larger and occurred at a higher temperature than that of the ground sample. This provided evidence that the freeze-dried sample was more completely amorphous than the ground sample (Morita *et al.*, 1984) and possessed a larger surface area, indicating that water was more favourably bound in the former sample facilitating conversion back to the stable monohydrate.

These different lactose forms possess varying tabletting properties. The α-lactose anhydrate has superior binding properties to the monohydrate (Vromans *et al.*, 1985). Moisture release from α-lactose monohydrate was determined by particle size since the dehydration endotherm became broader at the smaller particle sizes. This broadening was also apparent in milled samples and those subjected to compression (Figs 9.8, 9.9). Isothermal TG confirmed that water release was constant during the first stages of release. The smaller particles which have a larger surface area released their moisture at a faster rate than larger particles (Vromans *et al.*, 1985). The shift to lower temperatures of the endotherms of the smaller particles is explained by their surface area and by comminution causing a release of the water of crystallization. The heats of dehydration (about 144 J/g) were independent of particle size, indicating that the dehydration processes were similar.

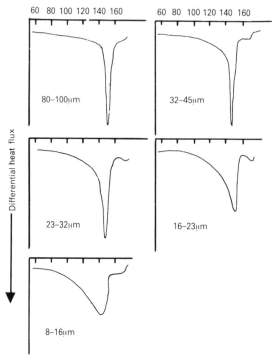

Fig. 9.8 — Dehydration peaks of particle size fractions of α-lactose monohydrate, recorded by DSC at a heating rate of 10°C/min (reproduced with permission from Vromans *et al.*, 1985).

Knowledge of the thermal properties of lactose has led to production of direct compression lactose products (Anon, 1984). Spray-drying a dispersion containing α-lactose monohydrate crystals in a lactose solution resulted in the formation of aggregates held together by amorphous lactose which is responsible for its good bonding characteristics. The product is marketed as DCLactose®11 and DSC is used to assay the amorphous lactose content based on the ability of the amorphous moiety to crystallize under certain humid storage conditions (Anon, 1984). The heat of crystallization depended on the amount of α- and β-lactose present and the optimum amount of amorphous lactose with respect to formulation properties was 15–20%. DCLactose®21 is a commercial β-lactose product prepared by roller-drying pure lactose solution. The product contains 80% β-lactose and 20% α-lactose and no amorphous lactose. DSC may be used to determine its α-lactose monhydrate content from the size of the dehydration endotherm at 150°C, amorphous lactose by the method discussed above and the stability of the anhydrous material by examining the recrystallization endotherms following storage under humid conditions (Anon, 1984). A third product (DCLactose®30), based on anhydrous α-lactose, may also be characterized by DSC to determine the presence of any non-amorphous forms: α-lactose monohydrate by the dehydration endotherm and unstable anhydrous α-lactose by melting/recrystallization in the range 160–180°C giving the β/α-compound or by DSC following storage under humid conditions (Anon, 1984).

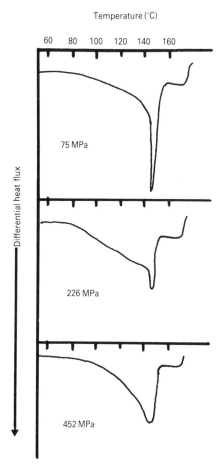

Fig. 9.9 — Dehydration peaks of α-lactose monohydrate, 100 mesh compacted at different compaction loads, recorded by DSC at a heating rate of 10°C/min (reproduced with permission from Vromans *et al.*, 1985).

9.1.5 Magnesium stearate

Magnesium stearate is used as a lubricant and glidant. It is not a single chemical entity but often consists of different proportions of several different chain-length fatty acids. Commercial samples may contain up to 50% impurities such as free fatty acids and fatty acid salts. The United States Pharmacopoeia and the British Pharmacopoeia allow a mixture of magnesium stearates and palmitates which may give a product of variable bulk density. The inhomogeneity, coupled with variation in moisture content, results in a problem in standardizing a specification for it. Any variation may result in vast changes in its lubricating properties.

Several pseudopolymorphs and hydrates of magnesium stearate were characterized by DTA and TG in combination with X-ray and IR techniques (Muller, 1977). The presence of different crystal habits in the sample indicates that different drying conditions were used during manufacture and may lead to forms with poorer lubrication properties.

Miller & York (1985) emphasized the importance of thermal analysis as a technique in the characterization of magnesium stearate and palmitate. Chemical assays, moisture content, surface area measurements, IR spectroscopy and X-ray diffraction were used to characterize the powders in conjunction with HSM, TG and DSC. DSC and TG were completed on samples stored at 20°C (Fig. 9.10) and at less

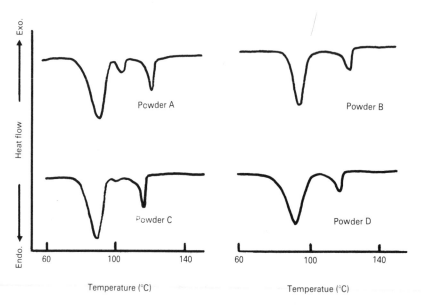

Fig. 9.10 — DSC scans at 2°C/min of undried magnesium stearate (A and B) and magnesium palmitate (C and D) (reproduced with permission from Miller & York, 1985).

than 45% relative humidity, DSC being also used on samples dried at 90°C at less than 0.1 torr (Fig. 9.11). Two batches of magnesium stearate (A and B) and two of magnesium palmitate (C and D) were produced (Miller & York, 1985). Batches A and C were prepared after forming the sodium salt as an intermediate by the addition of sodium hydroxide and produced irregular crystals. These had a stoichiometric ratio of 1:<2 fatty acid salt:water. Four endotherms were apparent (Fig. 9.10 and 9.11) of which three were immediately apparent without recourse to drying. The fourth was obscured by the endotherm due to moisture loss and was only apparent in the pre-dried sample. Only the latter two endotherms were apparent on drying. The peak temperatures corresponded to 90, 103, 121 and 132°C for the magnesium stearate sample and to 89, 101, 116 and 126°C for the palmitate sample. The presence of two endotherms equivalent to water loss was indicative that the moisture was bound in two different manners in the samples A and C (Miller & York, 1985). Batches B and D were prepared via the ammonium salt and gave thin, plate-like particles and had a stoichiometric ratio of 1:2 fatty acid salt:water. Only two

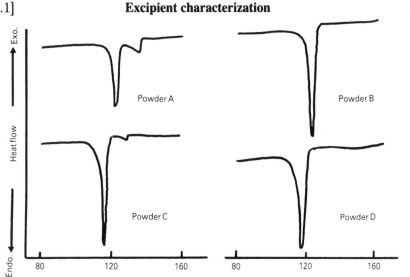

Fig. 9.11 — DSC scans at 2°C/min of magnesium stearate (A and B) and magnesium palmitate (C and D) dried at 90°C under vacuum (reproduced with permission from Miller & York, 1985).

endotherms were seen, with peaks at 96 and 123°C for magnesium stearate B and at 92 and 118°C for magnesium palmitate D. Each of the former endotherms were lost on drying. The initial two endotherms for samples A and C and the first endotherms for B and D were attributed to bound water since they corresponded to the temperatures of weight loss on TGA (Fig. 9.12). HSM revealed that considerable

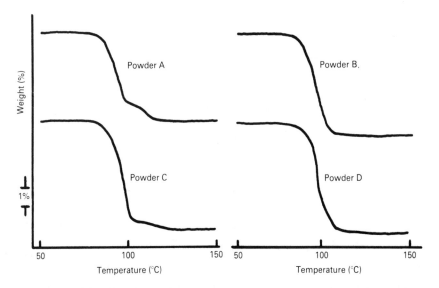

Fig. 9.12 — Thermogravimetric plots at 2°C/min of undried magnesium stearate (A and B) and magnesium palmitate (C and D) (reproduced with permission from Miller & York, 1985).

changes occurred in samples B and D corresponding to the development of diagonal striations across the particles and a loss of anisotropicity at 96°C, these changes not occuring for A and C to the same extent.

The remaining endotherms in Figs 9.10 and 9.11 are associated with melting which was a two-stage phenomenon. The first stage corresponds to the endotherms on Fig. 9.11 and the second, of such low energy transition as not to be detected by DSC, occurs a few degrees higher (Miller & York, 1985). Evidence for this was presented for magnesium stearate B whose final endotherm occurred at 126°C but which under HSM did not completely melt until 130°C. Similar phenomena occurred for all four powders. The additional small endotherms of the irregular forms were due to the existence within the sample of a different pseudopolymorph of the respective magnesium fatty acid salts which may have been responsible for the different binding of the water content which occurred within these samples (Miller & York, 1985).

The heats of activation for the major water loss and melting endotherms of the samples were determined (Miller & York, 1985) using the endotherm peak temperatures (T_m) derived at heating rates (ϕ) of 2, 5, 10, 20 and 50°C/min (see Chapter 4). The activation energies (E) were derived from plots of log (ϕ/T_m^2) against $1/T_m$. Values of E for the water loss endotherms and the melting endotherms were respectively 166 and 339 kJ/mol for magnesium stearate B and 153 and 371 kJ/mol for magnesium palmitate D.

9.2   SALT SELECTION

One of the most important criteria in the development of a product is selection of the correct salt form of a drug to provide maximum stability, formulation, solubility and dissolution characteristics concomitant with acceptable pharmacokinetics. Three examples of publications examining the thermal analysis of drug salt form prior to formulation show its use in this context. Serajuddin et al. (1986) examined the properties of a poorly water-soluble basic drug, α-pentyl-3-(2-quinolinylmethoxy)-benzenemethanol, and its hydrochloride salt. DSC was accomplished using a variety of conditions such as uncovered aluminium sample pans with and without a nitrogen flow rate and crimped sample pans or hermetically sealed aluminium sample pans. The base showed a sharp melting endotherm with T_0 at 63°C and T_m at 67°C with no further change. Crystallization did not occur on cooling. The hydrochloride salt existed in both anhydrous and monohydrate forms. Their DSC scans varied with the experimental conditions (Fig. 9.13). In closed pans the dehydration endotherm of the monohydrate increased in temperature and was followed by a degradation exotherm caused by a chemical reaction between released hydrochloric acid and the base powder wetted by the evolved moisture. The findings were confirmed using HSM and TG, the latter indicating that little weight loss occurred when the base was scanned up to 150°C. The monohydrate hydrochloride salt lost 6.5% of its weight in the range 25–64°C, 4.6% was loss of the monohydrate and the remaining 1.9% was unbound water. A similar weight loss (1.1%) was noted in the apparent anhydrous sample.

The changes observed were related by Serajuddin et al. (1986) to the van't Hoff equation. For solids reacting to form gas, the onset temperature, peak temperature

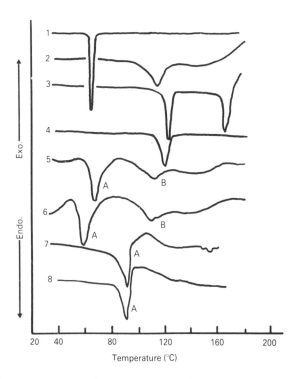

Fig. 9.13 — DSC scan of α-pentyl-3-(2-quinolinylmethoxy)benzenemethanol and its hydrochloride salt recorded by subjecting the samples to various atmospheric conditions. The scan of
α-pentyl-3-(2-quinolinylmethoxy)benzenemethanol (scan 1) did not change significantly with
the change in the atmospheric condition of the sample. Scans of the anhydrous salt were
recorded by placing the sample on an open pan without purging with nitrogen (scan 2), in a pan
closed by crimping (scan 3), and in an hermitically sealed pan (scan 4). Scans of the
monohydrate of the salt were recorded by placing the sample on an open pan without nitrogen
purging (scan 5) on an open pan with purging of nitrogen, 50 ml/min (scan 6), in a pan closed by
crimping (scan 7) and in a hermetically sealed pan (scan 8). The letters A and B represent
dehydration and melting endotherms of the monohydrate respectively. (Reproduced with
permission from Serajuddin et al., 1986.)

and peak shape of the endotherm will be modified by the gas pressure of the system
according to equation (9.1):

$$\ln\frac{(K_p)_2}{(K_p)_1} = \ln\frac{(P_c)_2}{(P_c)_1} = \frac{\Delta H(T_2 - T_1)}{R(T_1 \cdot T_2)} \tag{9.1}$$

where K_p, P_c and T are the equilibrium constant for the reaction, the partial pressure
of the produced gas and the temperature respectively (the subscripts represent two
different temperature conditions), H is the enthalpy and R is the gas constant. The
influence of sample preparation conditions can be predicted by equation (9.1). When
the partial pressure was reduced by nitrogen purging the onset of dehydration
temperature decreased; when the sample pans were sealed the partial pressure

increased and the onset temperature increased. Since no gas was produced from the anhydrous material the DSC scan was unaffected by sample preparation. Accelerated stability storage of the base produced no changes detectable by HPLC but the hydrochloride salt decomposed on storage. In addition to moisture further TG weight loss was apparent at above 90°C indicative of gas formation from the salt but not the free base. This decomposition on accelerated storage was due to hydrogen chloride liberation (Serajuddin *et al.*, 1986). Additionally since the exotherms were apparent in the monohydrate but not the anhydrous salt, degradation was possibly due also to moisture vapour which confirmed accelerated stability storage data.

Nyqvist & Graffner (1986) examined the selection of a precipitation medium as part of a preformulation program for amiflavine (+)-tartrate which was precipitated from ethanol or its blends with water. The product precipitated from ethanol:water blends was anhydrous but different forms were obtained. At a drying temperature of 20°C a monoclinic product A was formed containing 3.3% ethanol which was lost on prolonged storage and 1.3% water was absorbed. Drying at 110°C produced an orthorhombic product B_1 which did not contain ethanol. DSC and TG showed that A converted to B_1 with ethanol loss at 90°C, by melting/recrystallization processes. B_1 stored at 25°C and above 53% relative humidity produced another orthorhombic crystal B_2 which contained 2.6% water. Further storage under identical conditions produced yet another orthorhombic crystal B_3 which contained 5.2% water, equivalent to a monohydrate salt. TG and DSC scans of these salts indicated that each salt lost its respective solvent to form B_1, B_3 losing its water of hydration by a two-phase process. The compaction properties of the forms were different. The net work to form a compact of equivalent porosity decreased as B_1 hydrated to B_2 and similarly to B_3. The two anhydrates possessed different compaction properties. Products A and B_1 had high dissolution rates but moisture adsorption to form products B_2 or B_3 significantly lowered the rates.

Studies on fenoprofen (Hirsch *et al.*, 1978) utilized DTA and TG conjunction with solubility, humidity and photodegradation measurements to determine a suitable combination form with propoxyphene and codeine salts. The stability of several salts was determined by TG. Fenoprofen was unacceptable due to its low melting point (40°C). The sodium salt formed a crystalline dihydrate but an amorphous anhydrate. The potassium salt was crystalline but hygroscopic. The calcium salt was crystalline as its dihydrate but amorphous both as its monohydrate or anhydrate. Storage resulted in water loss from the sodium dihydrate salt but the calcium salt was stable at 25°C and 1% relative humidity. The amorphous calcium salt only absorbed enough water to form the dihydrate. TG of the preferred salts (Fig. 9.14) indicated that the calcium dihydrate lost its water of crystallization at 70°C but the amorphous calcium monohydrate and the sodium dihydrate salts lost their water of crystallization at room temperature. Incompatibilities between the sodium dihydrate salt and either propoxyphene or codeine salts ruled out its use in formulations. TG curves of 1:1 mixtures of fenoprofen calcium dihydrate with propoxyphene hydrochloride or propoxyphene napsylate monohydrate were unchanged following storage at 25°C for five years. Similar mixes with codeine sulphate displayed no changes on storage of eight months. DTA indicated that no changes occurred between fenoprofen calcium dihydrate and propoxyphene hydrochloride. Water loss was noted as an endotherm at 109°C and a eutectic reaction as an

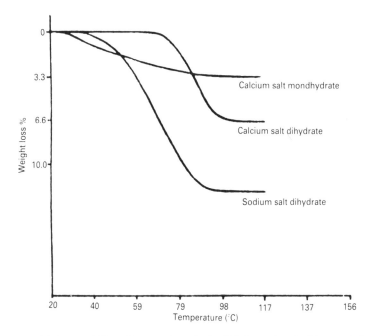

Fig. 9.14 — Hydrate stability of fenoprofen salts by TG (reproduced with permission from Hirsch *et al.*, 1978).

endotherm at 147°C. TG confirmed that the first endotherm corresponded to a weight loss due to moisture and that the second endotherm was not accompanied by weight loss. Fenoprofen sodium dihydrate was incompatible with propoxyphene salts as evidenced by the appearance of viscous oils and the loss of the dehydration endotherm following storage. The stability of fenoprofen calcium dihydrate with codeine sulphate pentahydrate was similarly predicted from DTA and TG results. Endotherms at 80 and 105°C were attributed to dehydration of the codeine and fenoprofen salts respectively. TG confirmed this weight loss and indicated that a melting endotherm at 190°C was not equivalent to a weight loss and was probably a eutectic melt. Storage of the mixture at 50°C resulted in the loss, within one week, of the endotherm due to dehydration of codeine sulphate. Fenoprofen sodium dihydrate interacted with the codeine sulphate pentahydrate and although it was suggested that an anhydrate might be more compatible, a crystalline non-hygroscopic form could not be prepared.

9.3   DISINTEGRATION AND DISSOLUTION STUDIES

Solution calorimetry has been used to determine the dissolution and disintegration properties of sugar-coated multivitamin tablets (Nakai *et al.*, 1974) in a study on disintegration of sugar coated tablets. One batch (A) possessed a water-protective film of cellulose acetate phthalate and acetylmonoglyceride, the other batch (B) did not. For tablet A (Fig. 9.15) the interval a–b endotherm represents the dissolution

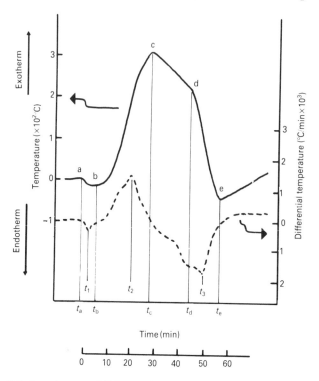

Fig. 9.15 — Calorimetric scan of disintegration of tablet A in acidic test solution. ———,
temperature; – – –, differential temperature. For explanation of symbols see text. (Reproduced from Nakai *et al.* (1974 with permission of the publishers.)

heat of the fine sugar particles coating the tablet surface, interval b–c is an
exothermic change due to the chemical reaction of calcium carbonate (present in the
sugar coat) with the acid in the buffer, interval c–d is a gradual temperature decrease
due to the cooling of the calorimeter cell indicating a period of little reaction due to
the protection of the internal core by the protective coat, and the endothermic
interval d–e is due to dissolution of the core. Finally interval e–f (exothermic) was
due to the decomposition of ascorbic acid in the core. The tablets without the
protective film (Fig. 9.16) displayed a small c–d interval. Physical dissolution and
disintegration were faster in batch B than batch A. The intervals indicated in the
figures as t_b, t_c, t_d and t_e were 5, 24, 38 and 56 min for Batch A and 5, 14 (t_c and t_b were
coincident) and 23 min for batch B. Ageing changes, following 18 months' storage,
were an increase in t_e values from 56 to 76 min.

9.4   FILM COATING

Film coating is the deposition of a thin film, usually 10–100 μm thick, on the surface
of a tablet or granule. The film is polymeric in nature and may contain plasticizers,
colorants or low levels of ingredients such as opacifiers. The film is prepared by
spraying a solution or suspension of the polymer and other additives onto the moving

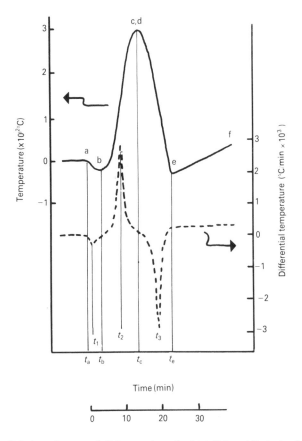

Fig. 9.16 — Calorimetric scan of disintegration of tablet B in acidic test solution. ————, temperature; – – –, differential temperature. For explanation of symbols see text. (Reproduced from Nakai *et al.* (1978) with permission of the publishers.)

mass of pellets, tablets or granules. The film is used to improve the acceptability of the appearance of the product, to reduce drug dissolution by retarding drug diffusion or allowing osmosis through its pores or, in enteric coating, to prevent dissolution by using a suitable pH-sensitive polymer which will dissolve at the pHs found away from the gastric contents. The polymer must coat the whole of the dosage form uniformly because thin regions or cracks in the film integrity will allow rapid release from the product. Water-insoluble polymers include ethylcellulose and methacrylate ester copolymers. The former is often used in conjunction with soluble polymers such as PEG and hydroxypropylmethylcellulose (HPMC) to provide water channels to modify the release rate of the drug. pH-Dependent polymers include cellulose acetate phthalate (CAP), hydroxypropylmethylcellulose phthalate (HPMCP) and poly(methacrylic acid–methylmethacrylate).

A knowledge of the glass transition temperature (T_g) of these coats is important. Attempting to cast and dry films at temperatures below their T_g results in cracking or flaws and a loss in film integrity. Drying must be accomplished at temperatures above the T_g so that the film is capable of flowing and uniformly covering the product. Since the T_g of most of the polymers used in film coating is above the coating temperature it is essential that a plasticizer be incorporated into the coating fluid formulation to produce an acceptable product. Whereas thermal analysis will give information regarding polymer interactions with plasticizers and other additives, it will not give information concerning the interactions of the film with the product that it coats. Values of T_gs of some polymers commonly used in film coating and the methods by which they were derived may be found in Table 9.1.

Table 9.1 — Some T_g values of polymers used in film coating

Polymer	Method T_g (°C)	Source
Cellulose acetate phthalate	DSC 18.5±0.5°	Porter & Ridgway (1983)
Cellulose acetate phthalate	TBA 163°	Sakellariou et al. (1985)
Ethylcellulose	DSC 129±1°	Entwistle & Rowe (1979)
Ethylcellulose	TBA 135°	Rowe et al. (1984)
Ethylcellulose	TBA 131.5°	Sakellariou et al. (1985)
Hydroxypropylcellulose	TBA 124°	Sakellariou et al. (1985)
Hydroxypropylmethylcellulose	DSC 177±1°	Entwistle & Rowe (1979)
Hydroxypropylmethylcellulose powder	DSC 180°	Sakellariou et al. (1985)
Hydroxypropylmethylcellulose film	DSC 157°	Sakellariou et al. (1985)
Hydroxypropylmethylcellulose	DTA 169–174°	Sakellariou et al. (1985)
Hydroxypropylmethylcellulose	TBA 158.5°	Sakellariou et al. (1985)
Hydroxypropylmethylcellulose phthalate[a]	TBA 133°	Sakellariou et al. (1985)
Hydroxypropylmethylcellulose phthalate[a]	TBA 137°	Sakellariou et al. (1985)
Polyvinylacetatephthalate	DSC 42.5±1°	Porter & Ridgway (1983)

[a]HPMCP HP 55® & HPMCP HP 50® are commercial brands of hydroxypropylmethylcellulose phthalate (Shin-Etsu Chemicals, Tokyo, Japan).

In mixes of polymers the T_g spread reflects the compatibility of polymers. An increase in spread is indicative that polymer blends are not compatible throughout their composition range. Once the compatibility limit is exceeded two glass transitions become apparent, originating from the compatible phase and the ingredient in excess. Film coats often contain more than one polymer and the type of dispersion around the tablet or granule may control drug release. When polymers are not fully compatible, e.g. there are islands of one polymer dispersed throughout the other, the drug may be released by erosion of the dispersed phase to leave pores or via diffusion through one of the phases. PEG, although used as a plasticizer, will, above its solubility within the carrier, be present as crystalline units. These may either retard drug diffusion by acting as a physical barrier or dissolve to leave a porous structure through which the drug may be transported. Many polymers are not fully compatible throughout their composition range, i.e. where they are not fully miscible the dispersed phases, rather than consisting solely of one polymer, may be a blend of both. TBA has been especially used in the quantification of such blends. Knowledge of T_g values helps to predict likely interactions. When two polymers are compatible only one T_g will be apparent throughout the composition range. When the polymers are only partially compatible one T_g will be apparent in the range of compatibility but where incompatible two T_g values will be noted corresponding to those of the compatible phase and the polymer present in excess. When the polymers are incompatible throughout the composition range both T_gs will be seen.

9.4.1 Uses of DSC

It is essential that a standard thermal history is provided for a sample prior to DSC analysis. Porter & Ridgeway (1979) prepared samples by adding the polymer/plasticizer solution in 50:50 dichloromethane:methanol dropwise to the sample crucible and allowing evaporation to proceed before addition of the next drop. The films were stored in a desiccator for two weeks prior to analysis. Many polymers, e.g. HPMC, HPC, HPMCP and CAP display water loss endotherms between 85 and 100°C without prior treatment (Sakellariou *et al.*, 1985). This occurred $\simeq$10°C lower for the more hydrophobic ethylcellulose (EC). Repeating the scan on the same sample allowed T_g determination due to a reduction in the moisture–loss endotherm. T_g values depended on the method of preparation and discrepancies with literature values were due to factors including sample size, preparation, heating rates or poor resolution from the baseline due to the small energy changes involved in the glass transition. DSC gave a value for the T_g of HPMC as 180°C in the powder form but 157°C when cast as a film from dichloromethane/methanol (Sakellariou *et al.*, 1985). Apparent values for HPC were 175–177°C by DTA but 177–183°C when obtained by DSC. The scan shapes were atypical of glass transitions and were caused by either viscous flow or molecular orientation (Sakellariou *et al.*, 1985).

9.4.1.1 *Plasticization*

Problems exist in deriving T_gs from DSC data. Entwistle & Rowe (1979) examined the T_gs of EC and HPMC plasticized by various compounds such as PEG and diethylphthalate. The DSC scans were complicated and the T_gs indistinct. Porter & Ridgway (1983) used DSC to investigate the interaction of the plasticizer diethyl phthalate with CAP and polyvinyl acetate phthalate (PVAP). Although X-ray diffraction indicated that the films were amorphous and T_gs should have been detected by DSC, their detection, especially in plasticized systems, was not easy. The transitions were obscured by baseline drift and low specific heat changes. Non-equilibrium conditions and too fast heating rates were considered as problems. A rate of 1°C/hour was considered acceptable but impractical to use. Impurities including unreacted monomer were a source of error. Porter & Ridgeway (1983) utilized equations (9.2) and (9.3) to predict the influence of plasticizer on the T_g of a polymer.

$$T_g^{(1,2)} = \frac{\phi_1 \Delta \alpha_1 T_g^{(1)} + \phi_2 \Delta \alpha_2 T_g^{(2)} + K\phi_1\phi_2}{\phi_1 \Delta \alpha_1 + \phi_2 \Delta \alpha_2} \qquad (9.2)$$

where ϕ_1 and ϕ_2 are the volume fractions of components 1 and 2, $\Delta\alpha_1$ and $\Delta\alpha_2$ are the differences in the thermal expansion of the components between the liquid and glass at T_g and K represents the interaction between components at or near the T_g.

Simplifying equation (9.2) gives equation (9.3),

$$T_g = \frac{(\alpha_p v_p T_{gp} + \alpha_d v_d T_{gd})}{\alpha_p v_p + \alpha_d v_p} \tag{9.3}$$

where T_g is the glass transition temperature of the mixture, T_{gp} and T_{gd} are those of the polymer and plasticizer respectively, α_p and α_d are their respective coefficients of volumetric expansion and v_p and v_d are the respective volume fractions in the mixture. Equation (9.3) was used by Porter & Ridgeway (1983) to compare their DSC-derived T_gs with computed values. Larger disparities were noted with PVAP and were probably a result of the problems mentioned above.

9.4.1.2 Crystallinity
York & Okhamafe (1985) utilized DSC to examine the T_g and crystallinity of free-cast films prepared from aqueous HPMC solutions containing PEG or PVA as plasticizer. For crystallinity measurements samples were held at 125°C for 5 min prior to analysis to remove water and quench-cooled using liquid nitrogen. Crystallinities were determined from the melting endotherms. Samples for T_g determination were held at 125°C for 10 min to avoid masking the T_g with moisture loss endotherms (Okhamafe & York, 1985), quench-cooled to and held at -40°C for 5 min prior to heating. T_gs were characterized by both the midpoint temperature of the heat capacity endothermic change and by the T_g spread, i.e. the temperature difference between the onset and end of transition (Fig. 9.17). HPMC gave an endotherm between 220 and 260°C but this disappeared when plasticizers were included. A melting point endotherm, equivalent to PVA, appeared in HPMC–PVA blends indicating partial crystallinity whereas the HPMC–PEG polymer blends were amorphous. Heats of fusion of the HPMC–PVA films allowed quantitative estimates of their crystallinities. Pure HPMC and PVA films displayed endotherms corresponding to transition heats of 32.31 and 33.6 J/g respectively. The latter was used to estimate film crystallinity. Films containing 10, 20 and 30% PVA gave values of 5.86, 6.31 and 5.99 J/g giving a crystallinity indices (their value expressed as a fraction of the value of the PVA film) of 0.175, 0.188 and 0.178 respectively. Films containing 40, 50 and 60% PVA had crystallinity indices of 0.216, 0.281 and 0.39 respectively indicating that once the limiting PVA content of 30% had been exceeded the films became progressively more crystalline. The increase in crystallinity indicates that the compatibility level of the two polymers is exceeded. The T_gs of PEG–HPMC films decreased with increasing PEG 1000 content indicative of plasticization but a more complex relationship occurred in HMPC–PVA films where T_gs were unchanged at PVA contents up to 20% but thereafter increased. Therefore the rigid crystalline PVA phase may have restricted molecular mobility and the compatible phase of the film via hydrogen bonding. The T_g spread reflects the compatibility of the HPMC and its plasticizer. Based on the appearance of a second T_g the compatibility limits were 20 and 40% for PEG 400 and PVA in HPMC (Okhamafe & York, 1985). A T_g was not obtained for PEG 1000 but the appearance of an endotherm due to PEG 1000 melting indicated that the limit of compatibility was 15% PEG 1000.

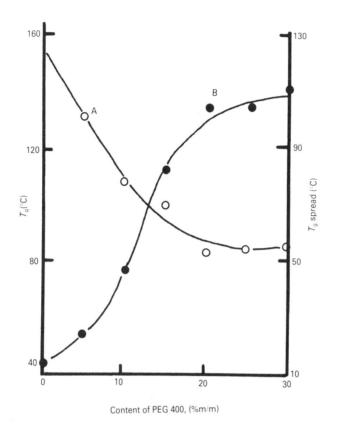

Fig. 9.17 — Glass transition plots for HPMC — PEG400 films: A, glass transition temperature
(T_g); B, glass transition spread (T_g spread) (reproduced with permission from York &
Okhamafe, 1985).

9.4.2   Use of TBA

The problems of accurate determination of T_gs by DSC and DTA can be overcome
by using a torsional braid pendulum (Rowe *et al.*, 1984). Sakellariou *et al.* (1985)
examined the properties of cellulose-based film coatings by DSC, DTA and TBA.
DTA and DSC gave discrepancies with literature values, due partly to the poor
resolution of these techniques. Glass transition measurements are sensitive to
molecular structure, presence of crystallinity, rate effects, residual solvent, diluents
and molecular weight variation. TBA should give higher values than static methods
such as DSC and DTA because it is a dynamic method. Sakellariou *et al.* (1985),
using heating rates of 20°C/min for DSC and DTA but only 1°C/min for TBA,
predicted that this rate effect should produce values about 9°C higher when using
DSC or DTA. The values found were in this range with the exception of HPMC when
there was a 16–26.5°C difference. This reflected the problem of solvent removal and
the need for careful sample preparation since TBA examined cast films whereas
powdered polymer without casting was used for DSC and DTA. Without complete

certainty that residual solvent has been removed it is very difficult to rely on T_g values quoted by different authors.

As an example of braid preparation Sakellariou *et al.* (1985) made the braids by desizing at 500°C for one hour and then these were impregnated by the polymer by immersion into solvent and drying to constant weight at 85°C. Samples were examined at a resonant oscillation frequency of 1 Hz at a heating rate of 1°C/min. Peak fitting was used to determine the period of oscillation (P). The data treatment used in TBA to predict T_g involves plotting either relative rigidity ($1/P^2$) or logarithmic decrement (Δ) as a function of temperature. The T_g is represented for most polymers by a simple peak in the log decrement curve and a decrease in the relative rigidity curve (Sakellariou *et al.*, (1985). Fig. 9.18 indicates a typical scan for

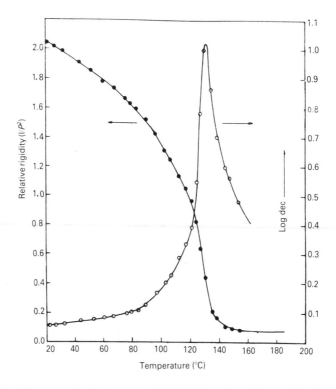

Fig. 9.18 — Thermomechanical spectra of ethylcellulose: ●, relative rigidity; ○, logarithmic decrement (reproduced with permission from Sakellariou *et al.*, 1985).

EC. No changes were noted in the range −20 to 40°C. A single transition in the log decrement curve at 131.5°C is concomitant with a decrease in the relative rigidity curve. Molecular movement commenced at 60–70°C with a maximum energy dissipation at 131.5°C. The transition was complete by 150°C. Other values of T_gs for polymers obtained by Sakellariou *et al.* (1985) are included in Table 9.1. The two values for HPMC were caused by one batch containing a larger portion of lower molecular weight material giving additional free volume, increased mobility and an increased number of end-chains. Two batches of HPMCP (Shin-Etsu Chemicals, Tokyo, Japan) produced different T_g values because HP 55® had a higher content of σ-carboxybenzoyl units than HP 50®. The bulky phthalyl groups increased the free volume and thereby increased chain mobility (Sakellariou *et al.* 1985).

In contrast HPC provides an example of a polymer having a complex thermal spectrum (Fig. 9.19). The relative rigidity curve showed a gradual decrease from −20

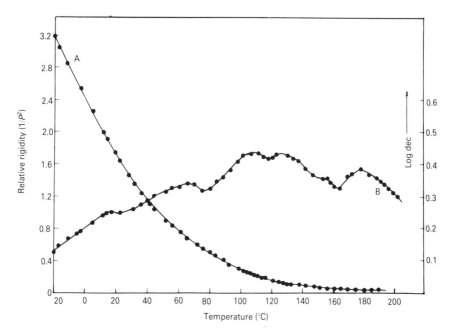

Fig. 9.19 — The thermomechanical spectra of hydroxypropylcellulose: A, relative rigidity; B, logarithmic decrement (reproduced with permission from Sakellariou *et al.*, 1985).

to 180°C before levelling off with slope changes apparent at 65, 100 and 124°C. The log decrement curve displayed several peaks and plateaus. The plateau at 17°C corresponded to a secondary relaxation of weak intermolecular hydrogen bonds and the peak at 63–67°C was methanol evaporation (Sakellariou *et al.*, 1985). The curve displayed two almost merged peaks at 105 and 124°C, a plateau at 145–150°C leading to a minimum at 162°C and a final peak at 178°C. A DTA endotherm at 105°C was due to water evaporation leaving the peaks at either 124 or 178°C to represent the T_g. Neither DSC or DTA gave evidence for a T_g at 178°C and since DSC of a water-cast film did not show a T_g at 140–170°C, it seemed that the T_g was at 124°C. The peak at 178°C corresponded to the start of viscous flow (Sakellariou *et al.*, 1985).

9.4.2.1 Plasticization
Rowe *et al.* (1984) determined the influence of various substituted phthalates on the T_g of EC by TBA comparing with theoretical values calculated from equation (9.3).

Plasticizing efficiency ranked as diethyl phthalate > dimethyl phthalate > dibutyl phthalate > dioctyl phthalate, the latter showing little plasticization even at 40% concentration. Peak shape was used as an index of the compatibility of the plasticizer and the polymer. Diethyl and dimethyl phthalates showed little peak broadening indicative of good compatibility. Broadening occurred at 30% dibutyl phthalate but dioctyl phthalate induced broadening at all concentrations and was coincident with a pronounced phase separation of the two ingredients. Deviations of actual values from theroretical values were connected with the fact that equation (9.3) was derived for systems that are compatible throughout their composition range and is not applicable where phase separation occurs. Theoretical and actual values most acceptably correlated for diethyl phthalate at low levels. Generally equation (9.3) was invalid for the alkyl phthalates. Sakellariou *et al.* (1986a) examined HPMC and EC films containing PEG of different molecular weights as plasticizer. The peak of the EC log decrement curve decreased with increasing levels of PEG. Additionally in the films containing plasticizer a broad plateau occurred in log decrement curves (Fig. 9.20) in the range 50–80°C. This may indicate melting of the PEG fraction but, of the molecular weights of PEG used, only PEG 6000 melted in this range. The peak widths are representative of the polymer-PEG interaction. This values were expressed as the ratio of the area of the log decrement peak at half height divided by the peak height. PEG 6000 gave a narrow transition and lower molecular weights gave wider peaks which were broader at increasing PEG content with the peak of the log decrement curves also occurring at lower heights. Plasticization efficiency decreased with an increase in molecular weights, due to a decrease in the number of terminal hydroxy groups. Similar results were found for the plasticization of HPMC films by PEG (Sakellariou *et al.*, 1986a).

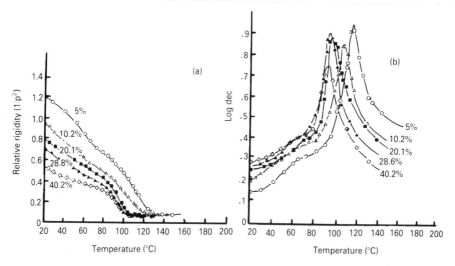

Fig. 9.20 — The thermomechanical spectra of ethylcellulose plasticized with PEG 200: (a) relative rigidity; (b) logarithmic decrement (reproduced with permission from Sakellariou *et al.*, 1986a).

Sakellariou *et al.* (1986a) used equation (9.2) to compare theoretical results with experimentally determined T_g values. Close agreement for HPMC plasticized with

PEG was obtained for PEG concentrations up to 15% PEG except for PEG 1000 (10%) and for PEG 6000 when there was poor agreement. For EC there was only good agreement for PEG 200 and then for concentrations only up to 10% PEG 200. The experimental results always deviated towards higher values and poor agreement was indicative of phase separation. Consequently thermal analysis can be used to determine compatibility and phase separation. The latter results in a concentration of PEG acting as plasticizer less than the nominal value incorporated in the films.

9.4.2.2 Polymer compatibility
TBA has been especially used in the quantification of polymer compatibility in blends. In the log decrement curves for EC/HPMC blends (Fig. 9.21) transitions

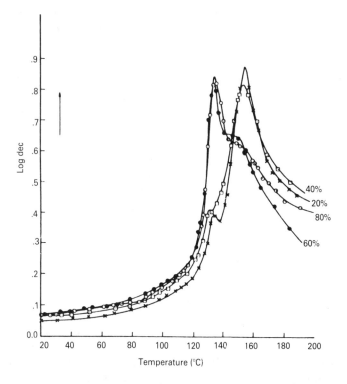

Fig. 9.21 — The thermomechanical spectra (logarithmic decrement curves) for ethylcellulose/ hydroxypropylmethylcellulose blends. Percentages are of ethylcellulose by weight. (Reproduced with permission from Sakellariou *et al.*, 1986b.)

were noted at 132.5–133.5°C corresponding to EC and a plateau attributable to HPMC at 153°C for compositions where EC was the major component (Sakellariou *et al.* 1986b). In compositions containing HPMC as the major component peaks were observed at 130–133 and 153–154°C. The fact that the peaks were broadened and that incomplete separation of the transition occurred indicated that the polymers had a limited mutual solubility.

Sakellariou *et al.* (1986b) examined several other polymer–polymer systems. The results for EC and CAP blends indicated phase separation at all compositions. The fact that the transitions corresponding to EC and CAP were decreased by 1.5–4.5 and 9–11°C respectively indicated a limited amount of mixing and interaction between the two polymers. The dispersion of EC in CAP was favoured since the latter values showed the greater depression. Blends of HPMCP with EC presented problems in analysis because of the closeness of their T_g values. In EC-HP55® blends only one relatively sharp peak was seen at 128–129°C and since its T_g was lower than the T_g of either individual component Sakellariou et al. (1986b) considered that plasticization occurred. HP55® was a better plasticizer than HP50®. Blends of HPC with EC gave complicated spectra due to the complex nature of the transitions of HPC (Sakellariou et al., 1985). The log decrement curves of PEG 6000 and EC blends (Sakellariou et al., 1986b) were the result of a combination of an amorphous polymer (EC) and a crystalline polymer (PEG 6000). Three major transitions were apparent. The transition at −47 to −40°C corresponded to the T_g of PEG 6000 and increased in height but decreased in breadth with an increase in PEG concentration. Increase in EC concentration increasingly restricted movement of the PEG chains and was associated with a decrease in the degree of crystallinity. This was reflected in a decrease in temperature of the peak at 60°C which was due to the melting of the PEG fraction. This confirmed the presence of either more amorphous PEG or smaller crystallites. A third transition at 118–126°C was due to the T_g of EC; its height increased and its breadth decreased with increase in EC concentration.

The data of Sakellariou et al. (1986b) confirmed that although the polymers were incompatible with EC there was some interaction between their chains. In the blends where there were sharp T_g transitions it was possible to determine the composition of the two incompatible phases using either equation (9.4) (Fox, 1956) or equation (9.5) (Stoelting et al., 1970).

$$1/T_g = W_1/T_{g1} + W_2/T_{g2} \tag{9.4}$$

$$T_g = W_1 . T_{g1} + W_2 . T_{g2} \tag{9.5}$$

Equation (9.4) showed that for the EC:CAP system containing 80% EC, the EC-rich phase contained 14% CAP whilst the CAP-rich phase contained 31% EC. The values from equation (9.5) were 14 and 32% respectively. Similar values calculated according to equation (9.4) for the PEG 6000/EC system indicated that for the system nominally containing 80% EC the PEG phase contained 21% EC and the EC phase contained 2% PEG 6000. This indicated a lower compatibility and that the dispersion of EC in the PEG 6000 phase is favoured.

The area under the log decrement curve is associated with energy dissipation (Sakellariou et al., 1986b) and is a measure of the overall toughness of the film. When a second component caused a broadening of the peak or formation of a second peak or shoulder, film toughening would be expected.

9.4.3   Use of TMA

In the penetration mode a film under examination will be initially resistant to penetration because the movement of individual atoms is limited as there is no thermal energy. With increase in temperature immobilized segments become freer and the film more flexible. Near the T_g an increase in the void volume occurs and the polymer becomes more penetrable. Masilungan & Lordi (1984), using TMA, referred to a softening temperature equivalent to the intersection of the extrapolations of the baseline and penetration line. Typical results (Fig. 9.22) illustrating the

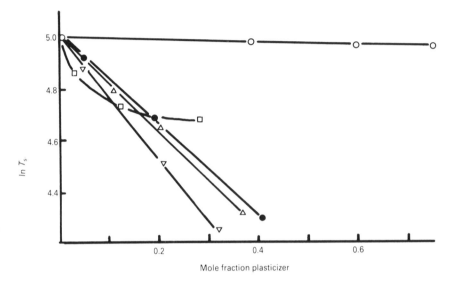

Fig. 9.22 — Effect of plasticizers on the softening temperature of ethylcellulose: ○, propylene glycol; △, glyceryl acetate; ▽, diethyl phthalate; □, PEG 400; ●, castor oil (reproduced with permission from Masilungan & Lordi 1984).

effects of plasticizers on the softening temperatures of EC films. Masilungan & Lordi (1984) used equation (9.6) (Moelter & Schweizer, 1949) to describe the influence of plasticizers on the softening temperature (T_s).

$$T_s = T_0\, e^{-kn} \tag{9.6}$$

T_0 is the softening temperature of the unplasticized film, n is the mole fraction of the plasticizer and k is the softening point depression coefficient. k is a measure of the plasticizing efficiency, high values representing high plasticizing efficiency and low values a low efficiency. Fig. 9.22 indicates the general straight-line relationship predicted by equation (9.6) with a slope equivalent to k. The curvature for PEG 400 may have been due to loss of the plasticizer since the high temperature of testing resulted in migration of the PEG 400. On the basis of Fig. 9.22 diethyl phthalate was considered the best plasticizer for EC (Masilungan & Lordi, 1984). Similar data indicated that PEG 400 was the best plasticizer for both HPMC and HPMCP. Solvent retention in the film also depressed the softening temperature as did moisture adsorption when the films were stored under controlled relative humidities.

9.5   THERMAL ANALYSIS OF FORMULATED PRODUCTS

The thermal analysis of entire formulations is very complicated because of the possibility of eutectic interactions, solid solution formation, polymorphic and polymeric transitions, melting or decomposition of the active or any added excipients and the possibility of interactions and instability between ingredients at high temperatures. Much of the problems of stability prediction by thermal analysis is covered in Chapter 10. Nevertheless papers detailing the complex DSC, DTA or TG scans of fully formulated, commercial preparations have appeared in the literature. Those preparations examined include antacids (Wendlandt, 1974), analgesics (Wendlandt & Collins, 1974), vitamins (Collins & Wendlandt, 1975), chemotherapeutic preparations and expectorants (Radecki & Wesolowski, 1980), neuroleptic drugs and vitamins (Wesolowski, 1981) and there was also a study on DTA and TGA techniques for checking the composition of 117 pharmaceutical preparations (Radecki & Wesolowski, 1979). The major applications are identification and assay although the difficulty in assigning peaks to a particular transition has been noted (Wendlandt, 1974).

9.5.1   Identification
Thermal analysis by TG, DSC or DTA has, in certain circumstances, the ability to aid identification of tablets. Scans can be compared with authentic scans to facilitate comparison. The method is not without its problems. The DTA scans of aspirin tablets, which contain mostly drug, resembled those of the pure drug (Wendlandt & Collins, 1974). This meant that several commercial brands of aspirin tablets could not be differentiated on the basis of DTA or TG alone because of the similarity of their curves. Differences in the formulated products above 200°C were due to the binder. In compound preparations containing more than one active agent the DTA peak due to aspirin was shifted to lower temperatures and the scans were more complicated. It was possible to identify on the basis of TG and DTA data the preparations, with the exception of those containing only aspirin.

The problem of interference by binders and other excipients is highlighted by vitamin preparations. These often contain many vitamins and large amounts of non-active agents. The complexity of these formulations made it difficult to assign specific curve peaks to a given vitamin (Collins & Wendlandt, 1975). Evolution of large portions of gas caused inconsistent DTA scans which were partially overcome by pelleting the sample prior to analysis. Although the individual preparations gave highly individualistic scans and made comparison between products simple the technique proved useless in the identification of the incorporated vitamins.

9.5.2   Assay
Radecki & Wesolowski (1980) considered that a combination of DTG and TG could be used to assay preparations on the basis of their decomposition profiles provided that they permit discrimination of individual steps of thermal decomposition and precise determination of the weight loss. The types of reaction that could be suitable for assaying pharmaceutical components (Radecki & Wesolowski, 1979) are dehydration, decarboxylation, weight loss due to reactions between effervescent components, weight losses due to the formation of intermediates in reactions and weight

losses due to decomposition and combustion. It must be stated that such results would be nowhere near as reliable as conventional assaying procedures and thermal analysis should not be used in this manner because of uncertainties concerning interpretation of curves. Radecki & Wesolowski (1980) differentiated and assayed 18 out of 31 drugs without the need to separate them from the remaining ingredients of the formulation. However it was not possible to assay small levels of ingredients due to interference by other excipients.

The problem with the thermal analysis of whole tablets is that the precise composition of the tablet under test, whether entire or previously triturated (which may lead to changes in the thermal scans), is often unknown. Wesolowski (1980) considered that best results were obtained when (a) the excipients were stable over the range where the drug decomposed, (b) the decomposition characteristics of the pure drugs and excipients were known, (c) the drug constituted more than 10% of the tablet mass, (d) the drug's decomposition occurred over a narrow range, (e) the decomposition equated to a large mass loss, (f) intermediate decomposition products were stable over a broad temperature range and (g) the residue weight following combustion was used only if the other components did not produce a residue.

9.6 MISCELLANEOUS USES

Thermal analysis has been used to examine crystal growth on tablets during storage and to assess the loss of volatile components. TG was used in the assessment of a stable formulation of the volatile drug glyceryl trinatrate. Historically this drug has created many problems due to its high volatility and component loss on even short storage periods. Since these tablets are manufactured by moulding techniques (Goodhart et al., 1976) problems occur when the damp triturate is dried to remove the residual granulating solvent. This resulted in a concentration of the drug at the surface which enhanced the drug loss. Goodhart et al. (1976) assessed microcrystalline cellulose and povidone as potential stabilizers by placing tablets on a TG balance and holding isothermally at 80°C for 90 minutes. Tablets prepared by compression and containing the stabilising agents lost only 4% of their drug during TG whereas two brands of commercial moulded tablets lost 44 and 78%. Extending the treatment up to 4 hours may be useful in preparations whose loss is similar in the 90-minute period (Gucluyildiz et al., 1977). Fig. 9.23 indicates data for seven formulations containing 0.6 mg glyceryl trinatrate and a placebo. Two trends may be observed. Batches A and B were made from anhydrous materials and their weight loss was attributed to loss of the volatile drug only. Weight loss for the remaining formulations, including the placebo, was biphasic. The initial 10-minute loss was due to the removal of water from excipients such as starch but the remaining loss was attributed to drug. The volatility for these formulations was represented by nitroglycerine loss subsequent to this initial 10-minute period (Gucluyildiz et al., 1977). Povidone provided an acceptable means of stabilizing the drug and gave superior results to EC and HPMC. TG results were directly comparable to those obtained by chemical assay.

Ando et al. (1985, 1986) used DSC to identify the nature of substances which recrystallized on the surface of tablets during storage under relatively humid conditions. Crystals growing on tablets stored at 37°C at 59, 75, or 90% relative

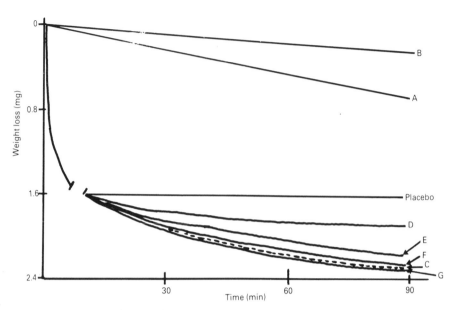

Fig. 9.23 — Typical TG scans of nitroglycerine tablets containing various binders: (A) lactose; (B) lactose, povidone; (C) lactose, starch; (D) lactose, starch, povidone; (E) lactose, starch, HPMC; (F) lactose, starch, ethylcellulose; (G) lactose starch gelatine (reproduced with permission from Gucluyildiz et al., 1977).

humidity were examined by DSC, thin layer chromatography (TLC) and scanning electron microscopy. Samples in the range 0.4 to 2.0 mg were used for DSC. Under humid conditions water would be absorbed by the more hygroscopic substance, diffuse to the less hygroscopic substance and evaporate from the latter, which if soluble in water would recrystallize out. Crystal growth of lactose occurred in tablets containing the hygroscopic materials docusate sodium (sample A, Fig. 9.24) or magnesium chloride (sample B, Fig. 9.24). The similarities of the DSC scans to that of lactose (dehydration endotherm at 140° and fusion endotherm at about 210°C) coupled with TLC similarities indicated that lactose had recrystallized on the surface of the tablets. Mannitol similarly recrystallized in the presence of docusate sodium, magnesium chloride and potassium acetate and was partially identified by an endothermic melting transition at about 166°C. Similar recrystallization of theophylline monohydrate occurred in tablets contining anhydrous theophylline and magnesium chloride or potassium acetate (Ando et al., 1986).

REFERENCES

Ando, H., Watanabe, S., Ohwaki, T. & Miyake, Y. (1985) *J. Pharm. Sci.,* **74,** 128–131.
Ando, H., Ohwaki, T., Ishii, M., Watanabe, S. & Miyake, Y. (1986) *Int. J. Pharm.,* **34,** 153–156.
Anon. (1984) The significance of differential scanning colrimetry in the characterisa-

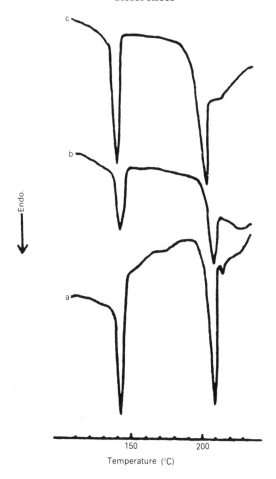

Fig. 9.24 — DSC scans of lactose (a), sample A (b) and sample B (c). For key to lettering see text (reproduced with permission from Ando *et al.*, 1985).

tion of lactose products for direct compression, *Proceedings of PharmaTech,* Fulmer Grange, London, 14–21.

Bechtel, W. G., Geddes, W. F. & Gilles, K. A. (1964) In *Wheat: chemistry and technology,* American Association of Cereal Chemists, 277–352.

Berlin, E., Kliman, P. G., Anderson, B. A. & Pallansch, M. J. (1971) *Thermochim. Acta.,* **2**, 143–152.

Biliaderis, C. G., Maurice, T. J. & Vose, J. R. (1980) *J. Food Sci.,* **45**, 1669–1674.

Collins, L. W. & Wendlandt, W. W. (1975) *Thermochim. Acta.,* **11**, 253–260.

Dollimore, D. & Hoath, J. M. (1981), In: *Proceedings 2nd European Symposium on Thermal Analysis.* Dollimore, D. (ed.) Heyden, London. 576–579.

Donovan, J. W., Lorenz, K. & Kulp, K. (1983) *Cereal Chem.,* **60**, 381–387.

Entwhistle, C. A. & Rowe, R. C. (1979) *J. Pharm. Pharmacol.,* **31**, 269–272.

Fox, T. G. (1956) *Bull. Am. Phys. Soc.,* **1**, 123.

Goodhart, F. W., Gucluyildiz, H., Daly, R. E., Chafetz, L. & Ninger, F. C. (1976) *J. Pharm. Sci.,* **65**, 1466–1471.

Gucluyildiz, H., Goodhart, F. W. & Ninger, F. C. (1977) *J. Pharm. Sci.,* **66**, 265–266.

Hirsch, C. A., Messenger, R. J. & Brannon, J. L. (1978) *J. Pharm. Sci.,* **67**, 231–236.

Itoh, T., Satoh, M. & Adachi, S. (1977) *J. Dairy Sci.,* **60**, 1230–1235.

Itoh, T., Katoh, M. & Adachi, S. (1978) *J. Dairy Res.,* **45**, 363–371.

Kilzer, F. J. (1971) *High Polymer,* **5**, 1015–1046.

Kilzer, F. J. & Broido, A. (1965) *Pyrodynamics,* **2**, 151–163.

Krycer, I. & Hersey, J. A. (1981) *Int. J. Pharm. Tech. & Prod. Mfr.,* **2**, 55–56.

Lerk, C. F., Andreae, A. C., de Boer, A. H., de Hoog, P., Kussendrager, K. & van Leverink, J. (1984a) *J. Pharm. Sci.,* **73**, 856–857.

Lerk, C. F., Andreae, A. C., de Boer, A. H., de Hoog, P., Kussendrager, K. & van Leverink, J. (1984b) *J. Pharm. Sci.,* **73**, 857–859.

Lerk, C. F., Buma, T. J. & Andreae, A. C. (1980) *Neth. Milk Dairy J.,* **34**, 69–73.

Lund, D. (1984) *C.R.C. critical reviews in food science and nutrition,* **20**, 249–273

Marshall, A. S. & Petrie, S. E. B. (1980) *J. Photographic Science,* **28**, 128–134.

Masilungan, F. C. & Lordi, N. G. (1984) *Int. J. Pharm.,* **20**, 295–305.

Miller, T. A. & York, P. (1985) *Int. J. Pharm.,* **23**, 55–67.

Moelter, G. M. & Schweizer, E. (1949) *Ind. Eng. Chem.,* **41**, 684–689.

Morita, H. (1956) *Anal. Chem.,* **28**, 64–67.

Morita, M., Nakai, Y., Fukuoka, E. & Nakajima, S-I. (1984) *Chem. Pharm. Bull.,* **32**, 4076–4083.

Muhr, A. H., Wetton, R. E. & Blanshard, J. M. V. (1982) *Carbohyd. Polym.,* **2**, 91–102.

Muller, B. W. (1977) *Arch. Pharm.,* **310**, 693–704.

Nakai, Y., Nakajima, S. & Kakizawa, H. (1974) *Chem. Pharm. Bull.,* **22**, 2910–2915.

Nyqvist, H. & Graffner, C. (1986) *Acta Pharm. Suec.,* **23**, 257–270.

Nyqvist, H., Lundgren, P. & Nystrom, C. (1978) *Acta Pharm. Suec.,* **15**, 150–159.

Okhamafe, A. O. & York, P. (1985) *Pharm. Res.,* **2**, 19–23.

Olkku, J. & Rha, C. (1978) *Food Chem.,* **3**, 293–317.

Porter, S. C. & Ridgway, K. (1983) *J. Pharm. Pharmacol.,* **35**, 341–344.

Radecki, A. & Wesolowski, M. (1979) *J. Therm. Anal.,* **17**, 73–80.

Radecki, A. & Wesolowski, M. (1980) *Talanta,* **27**, 507–512.

Rowe, R. C., Kotaras, A. D. & White, E. F. T. (1984) *Int. J. Pharm.,* **22**, 57–62.

Sakellariou, P., Rowe, R. C. & White, E. F. T. (1985) *Int. J. Pharm.,* **27**, 267–277.

Sakellariou, P., Rowe, R. C. & White, E. F. T. (1986a) *Int. J. Pharm.,* **31**, 55–64.

Sakellariou, P., Rowe, R. C. & White, E. F. T. (1986b) *Int. J. Pharm.,* **34**, 93–103.

Serajuddin, A. T. M., Sheen, P-C., Mufson, D., Bernstein, D. F. & Augustine, M. A. (1986) *J. Pharm. Sci.,* **75**, 492–496.

Stoelting, J., Karasz, F. E. & MacKnight, W. J. (1970) *Polym. Eng. Sci.,* **10**, 133–138.

Vromans, H., de Boer, A. H., Bolhuis, G. K., Lerk, C. F. & Kussendrager, K. D. (1985) *Acta Pharm. Suec.,* **22**, 163–172.

Wendlandt, W. W. (1974) *Thermochim. Acta,* **10**, 93–99.

Wendlandt, W. W. & Collins, L. W. (1974) *Anal. Chim. Acta,* **71**, 411–417.

Wesolowski, M. (1980) *Mikrochim. Acta,* **1**, 199–213.
Wesolowski, M. (1981) *Microchim. J.,* **26**, 105–119.
York, P. & Okhamafe, A. O. (1985) *Anal. Proc.,* **22**, 40–41.

10

Application of thermal analysis to compatibility studies for solid dosage forms

10.1 INTRODUCTION

In the early preformulation studies of solid dosage forms (tablets, capsules, granules or powders) one major concern will be whether the active ingredient will be compatible with any of the included inactive ingredients, or in the case of combination products whether active ingredients are compatible with each other. Incompatibility might lead to accelerated potency loss, complex formation, acid/base interactions or eutectic formation. These different modes of incompatibility might have differing significance resulting in products of poor stability or altered bioavailability and it is essential to avoid incompatibilities. Classically formulators would blend various binary mixes of active and inactive materials and store these mixtures under various conditions of accelerated temperature and humidity. Akers (1975) has suggested suitable ratios in mixtures for study. Aliquots would be analysed at intervals for intact drug to identify those excipients promoting instability of the active ingredient. Analysis might be by TLC or HPLC. In the former case a larger number of samples can be run simultaneously. In the latter case analysis times might be short allowing many samples to be run per day and automation is possible allowing continuous operation. In either case the method should ideally detect decomposition of the drug at an early stage to shorten the time required to identify excipients promoting degradation. In many cases it takes from several weeks to several months to generate adequate data to classify acceptable and unacceptable excipients. Whilst this methodology will detect chemical interaction leading to instability it might be less sensitive in detecting physical interactions leading to, for example, eutectic formation or adsorption.

Thermal analytical techniques are, in many situations, able to detect both chemical and physical interactions and allow rapid sample turnaround as several mixtures can be evaluated in one day. Perhaps the first reference to the application of thermal analysis to pharmaceutical compatibility testing using DTA was by Simon in 1967. A large number of variations have appeared since and are discussed below.

10.2   PRACTICAL ASPECTS OF COMPATIBILITY STUDIES BY THERMAL ANALYSIS

10.2.1   Simple evaluations of peak characteristics

One of the first references to the application of thermal analysis to compatibility studies of pharmaceutical materials was a paper indicating the incompatibility between triampyzine sulphate and magnesium stearate as determined by DTA (Simon, 1967). Several workers in the pharmaceutical industry took up the approach and one early publication from workers at Squibb indicated compatibility between penicillins and stearic acid lubricants (Jacobson & Reier, 1969). The technique was to mix the components under study, compress them and size reduce for the thermal analysis. Binary and some tertiary or quaternary mixtures simulating complete capsule formulations were studied. In the case of all DTA scans where stearic acid was introduced into the mixture with the active material dicloxacillin, the features due to dicloxacillin were eliminated from the DTA scans. This feature correlated well with chemical stability and DTA was able to predict interactions between stearic acid and sodium oxacillin monohydrate or potassium penicillin G and predict lack of interaction between ampicillin trihydrate and stearic acid (Fig. 10.1).

This early example of using thermal analysis for compatibility testing demonstrates a common approach and includes correlation with chemical data. Of course concerns for economy of time and scale might demand that only the thermal analysis approach is used routinely. Smith (1982) has described a broadly applicable approach where a few milligrams of each excipient and the drug candidate are separately studied first by DSC using a standard scanning rate, e.g. 5°C/min, 10°C/min, and usually under a nitrogen atmosphere. Binary mixtures of the drug with each of the excipients in the formulation are then scanned. To maximize the likelihood of an interaction 1:1 mixtures of drug and excipient are employed rather than the 'realistic' ratios proposed by Akers (1975) for classical compatibility studies. The temperature range used was selected to encompass all thermal features of excipients and drug (e.g. desolvation, melting, decomposition).

Interactions in the samples are derived or deduced from DSC by changes in thermal events such as elimination of an endothermic or exothermic peak or appearance of a new endothermic or exothermic peak. Changes in peak shape, peak onset or peak maximum temperature and relative peak heights are changes which might also be considered. However, it should be cautioned that some broadening of peaks leading to changes in area, onset or peak temperatures are simply due to mixing the components without indicating an interaction. Provided that all thermal features more or less remain in the sample, compatibility can be accepted.

Should an interaction be seen to occur then either the implicated excipient is avoided or the significance of the interaction needs to be evaluated. This would involve classic accelerated storage tests but, by prior use of thermal analysis screening, the number of these tests carried out might be reduced. Subsequent to the work of Simon (1967) and Jacobson & Reier (1969) a number of workers have

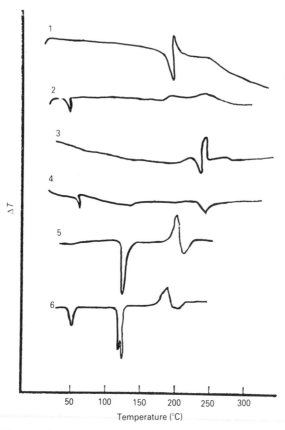

Fig. 10.1 — DTA scans of penicillins and mixtures thereof with 5% stearic acid. Scans 1 and 2 refer to sodium oxacillin monohydrate and its mixture with stearic acid respectively; scans 3 and 4 to penicillin G and its mixture with stearic acid, respectively; curves 5 and 6 to ampicillin trihydrate and its mixture with stearic acid, respectively. (Reproduced with permission from Jacobson & Reier, 1969.)

indicated the use of DTA or DSC in drug–drug or drug–excipient compatibility testing for solid oral dose forms by the simple method of peak evaluation as described above (Table 10.1). Modifications to the straightforward technique including use of two heating cycles (Hentze & Voege, 1971a), study of aged formulations by DTA (Geneidi & El-Sayed, 1978), evaulation of fresh samples along with those subjected to a short-term accelerated storage test (van Dooren, 1983) and construction of phase diagrams where interactions occur should be considered (Gordon et al., 1984).

A typical DSC trace showing incompatibility between a research compound and magnesium stearate is shown in Fig. 10.2).

10.2.2 Comparison of enthalpies of fusion

DSC studies can yield more data than studies on compatibility undertaken by DTA, as the quantity of heat required to complete a transition can be measured accurately. Thus changes in enthalpy of fusion as a result of incompatibility on mixing a drug with

Table 10.1 — Inspection technique for DSC/DTA compatibility studies

Material studied	Reference
Tetracycline/oxytetracycline with	Geneidi *et al.* (1981)
excipients/vitamins	Geneidi & El-Sayed (1978)
Penicillin, phenazone, novalgin, benzbromerone	Muller (1977)
isoxsuprine with polyethylene glycol	
Dexchlorpheniramine with excipients	Graf *et al.* (1985)
Binary component of cold relief product	Botha *et al.* (1986, 1987a,b)
ingredients	
Bromazepam with PVP	Fassihi & Persicaner (1987)
Cephradine with excipients	Jacobson & Gibbs (1973)
Oxytetracycline with excipients	Lee & Hersey (1977)
Ibuprofen with lubricants	Gordon *et al.* (1984)
Aspirin with excipients	Ager *et al.* (1986)
Mepyramine maleate with aspirin	Li Wan Po & Mroso (1984)
Penicillins with excipients	Jacobson & Reier (1969)
Aspirin, penicillins, tetracycline, vitamins,	Hentze & Voege (1971a)
chloramphenicol, isoniazid with excipients	
Chlorpropamide with excipients	Ford & Rubinstein (1981)
Oxytetracycline with excipients	Remon *et al.* (1978)

an excipient can be measured. Incompatibilities which manifest themselves only as changes in enthalpy of fusion can be detected in this way and might be overlooked by the inspection techniques detailed in section 10.2.1.

A series of papers by El-Shattawy and various coworkers have employed this technique to look at a range of drug–excipient compatibilities including those in direct compression formulations for aspartamine, ampicillin, cephalexin and erythromycin as well as studies on other powder mixes and blends forming complexes. These are detailed in Table 10.2.

El-Shattawy and his coworkers used the enthalpy approach in addition to inspection. Thus incompatibility between nalidixic acid and magnesium stearate (El-Shattawy *et al.*, 1984) was judged solely by inspection (Fig. 10.3) whereas an interaction between cephalexin and stearic acid (El-Shattawy *et al.*, 1982d) was indicated by inspection but not supported by enthalpy measurement up to the point of decomposition. This might indicate that the incompatibility is not significant at room temperature (Fig. 10.4). Incompatibility between erythromycin and magnesium stearate (El-Shattawy *et al.*, 1982e) may not readily be deduced by inspection (Fig. 10.5) but was suggested by the measured enthalpy change from the DSC trace of the mixture, which was 66% of that predicted from enthalpies determined from running individual components of the mixture.

10.2.3   Other approaches
It has been indicated above that simple binary mixtures should be evaluated for compatibility. Their preparation may or may not bring materials into intimate

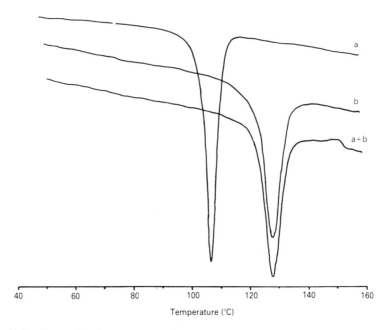

Fig. 10.2 — Interactions between magnesium stearate and research compound demonstrating incompatibility. Scans: a, research compound, b, magnesium stearate, a+b, 1:1 blend of research compound and magnesium stearate.

Table 10.2 — Compatibility studies of thermal analysis utilizing enthalpy of fusion evaluations

Materials studied	References
Aspartame with excipients	El-Shattawy *et al.* (1981)
Aspartame with mannitol	El-Shattawy *et al.* (1982a, 1984)
Cimetidine with caffeine	El-Ridy *et al.* (1982)
Aspartame with caffeine	El-Shattawy *et al.* (1982b)
Ampicillin with dextrose	El-Shattawy *et al.* (1982c)
Ampicillin with excipients	El-Shattawy (1982)
Cephalexin with excipients	El-Shattawy *et al.* (1982d)
Erythromycin with excipients	El-Shattawy *et al.* (1982e)
Combined digestive enzymes	El-Shattawy (1983)
Naladixic acid with excipients	El-Shattawy (1984)
Tetracaine with dextrose	El-Ridy & Kildsig (1985)

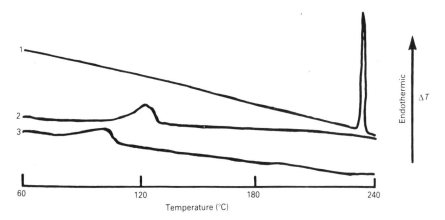

Fig. 10.3 — DSC scans of nalidixic acid (1), magnesium stearate (2) and 1:1 nalidixic acid–magnesium stearate (3). (Reprinted from El-Shattawy (1984) p. 502, by courtesy of Marcel Dekker, Inc.)

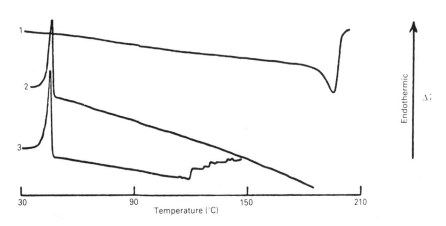

Fig. 10.4 — DSC scans of cephalexin (1), stearic acid (2) and 1:1 cephalexin–stearic acid (3). (Reprinted from El-Shattawy *et al.* (1982d) p. 906, by courtesy of Marcel Dekker, Inc.)

contact for interaction as would occur, for example, in a tablet formulation. Jacobson & Reier (1969) slugged materials they examined by DTA to simulate conditions in unstable capsule formulations. A recent publication by Cotton *et al.* (1987) evaluated the increase in sensitivity to incompatability between enalapril maleate and microcrystalline cellulose by comparing tumble-mixed compressed blends, ground mixtures and ground compressed mixtures. Reduction in apparent heat of fusion, indicative of incompatibility, was greater in compressed mixtures

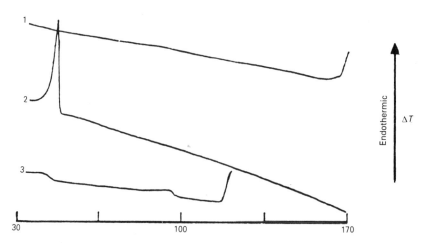

Fig. 10.5 — DSC scans of erythromycin (1), stearic acid (2) and 1:1 erythromycin–stearic acid (3). (Reprinted from El-Shattawy *et al.* (1982e) p. 944, by courtesy of Marcel Dekker, Inc.)

where there was better surface contact between drug and excipient particles. Cotton *et al.* (1987) utilized TG, HSM, scanning electron microscopy, IR spectroscopy and HPLC to evaluate the nature of the changes occurring on mixing with microcrystalline cellulose. A weak interaction with the cellulose appeared to alter crystal structure and thus reduce the heat of fusion. HSM was applied in conjunction with X-ray diffraction by El-Shattawy *et al.* (1984) to characterize aspartame–mannitol fused mixtures as eutectics with some solid–solid solubility, giving advantages in chewable tablets.

To enhance contact between drug and excipient in studies on aspirin by DSC, Ager *et al.* (1986) studied not only mixtures and tabletted mixtures but mixtures slurried in chloroform followed by removal of chloroform as demonstrated by NMR analysis. The last method of mixture formation of course might have the disadvantage of altering the crystallinity or crystal form of the mixtures and lead to erroneous interpretation of traces as indicating incompatibility. Its general application should be treated with caution.

Mass spectroscopy was applied by Signoretti *et al.* (1986) to the contents of DSC pans in which scans of clenbuterol/excipient mixtures had been made, in order to confirm compatibility where inspection did not clarify the situation.

10.3 LIMITATIONS

Most users of thermal analysis in compatibility testing would agree there are a number of difficulties in its general application. Evaluation of the curves via inspection can be difficult due to peak broadening etc. and this has led van Dooren (1983) to propose the use of comparing fresh samples with those aged for three weeks at 55°C. Not only van Dooren (1983) but Li Wan Po & Mroso (1984) and Nyqvist (1986) have recommended routine supplementing of DSC data with those from conventional accelerated storage tests. This would negate some of the advantage of using DSC in terms of speed but its utility as a screening tool should not be dismissed. It should be possible, as Smith (1982) indicated, to screen out those drug–excipient mixtures where interactions are probable and press on with formulating those mixtures where interactions appear not to occur. As back-up those mixtures could be placed on accelerated storage testing while formulation studies commence. If these studies confirm lack of interactions much time has been saved whereas if they indicate an interaction not clearly indicated by DSC only a few weeks of time might have been lost in beginning formulation studies indicating the suspect excipient.

Some problems in the application of DSC alone are indicated by Chzanowski *et al.* (1986) who pointed out that DSC indicated no incompatibilities in fenretinide–excipient or mefindil–excipient mixtures, whereas classic accelerated storage tests showed some incompatibilites. However, samples were not prepared identically for DSC and accelerated storage testing, which may contribute to the differences seen. Such reports nevertheless point to caution in relying solely on DSC data.

Other problems in the application of DSC to compatibility testing occur when materials being tested have close or coincident melting or if other thermal events which would make data interpretation difficult, e.g. eutectic formation, occur. The formation of eutectic, where a defined mixture of two materials melts as if a single entity and at a temperature lower than that of either component, may or may not be a problem; the eutectic may not present any chemical stability problems but could possibly present physical problems. Examples of the latter may be a eutectic in a tablet formulation and melting a little above room temperature. Such a eutectic could lead to sticking and filming problems on tablet punches. A further problem in interpretation of DSC or DTA curves of a mixture of two components is the possibility of dissolution of the second component in the melt of the first. Again this may not present itself as any real problem under normal storage conditions.

10.4   CONCLUSIONS

Thermal analytical techniques, notably DSC and to a lesser extent DTA and TG, are useful in the evaluation of incompatibilities in binary mixtures of ingredients of solid pharmaceutical dose forms. The value of hot stage microscopy as a back-up technique should not be overlooked. However, thermal analytical techniques may not provide unequivocal evidence of incompatibility and should be utilized as a primary technique supplementary to or supplemented by conventional accelerated isothermal stability testing.

REFERENCES

Akers, M. J. (1975) *Can. J. Pharm.*, **11**, 1–10.

Ager, D. J., Alexander, K. S., Bhatti, A. S., Blackburn, J. S., Dollimore, D., Koogan, T. S., Mooseman, K. A., Muhvic, G. M., Sims, B. & Webb, V. J. (1986) *J. Pharm. Sci.*, **75**, 97–101.

Botha, S. A., duPreez, J. L. & Lotter, A. P. (1986) *Drug Dev. Ind. Pharm.*, **12**, 811–827.

Botha, S. A., Lotter, A. P., & duPreez, J. L. (1987a) *Drug Dev. Ind. Pharm.*, **13**, 345–354.

Botha, S. A., Lotter, A. P., & duPreez, J. L. (1987b) *Drug Dev. Ind. Pharm.*, **13**, 1197–1215.

Chzanowski, F. A., Ulissi, L. A. Fegely, B. J. & Newman, A. C. (1986) *Drug Dev. Ind. Pharm.*, **12**, 783–800.

Cotton, M. L., Wu, D. W. & Vadas, E. B. (1987) *Int. J. Pharm.*, **40**, 129–142.

El-Ridy, M. S. & Kildsig, D. O. (1985) *Pharm. Ind.*, **47**, 330–331.

El-Ridy, M. S., Mayer, P. R., Peck, G. E. & Kildsig, D. O. (1982) *Drug Dev. Ind. Pharm.*, **8**, 595–603.

El-Shattawy, H. H. (1982) *Drug Dev. Ind. Pharm.*, **8**, 819–831.

El-Shattawy, H. H. (1983) *Drug Dev. Ind. Pharm.*, **9**, 1435–1443.

El-Shattawy, H. H. (1984) *Drug Dev. Ind. Pharm.*, **10**, 491–504.

El-Shattawy, H. H., Peck, G. E., & Kildsig, D. O. (1981) *Drug Dev. Ind. Pharm.*, **7**, 605–619.

El-Shattawy, H. H., Kildsig, D. O. & Peck, G. E. (1982a) *Drug Dev. Ind. Pharm.*, **8**, 429–443.

El-Shattawy, H. H., Kildsig, D. O. & Peck, G. E. (1982b) *Drug Dev. Ind. Pharm.*, **8**, 651–662.

El-Shattawy, H. H., Kildsig, D. O. & Peck, G. E. (1982c) *Drug Dev. Ind. Pharm.*, **8**, 739–749.

El-Shattawy, H. H., Kildsig, D. O. & Peck, G. E. (1982d) *Drug Dev. Ind. Pharm.*, **8**, 897–909.

El-Shattawy, H. H., Kildsig, D. O. & Peck, G. E. (1982e) *Drug Dev. Ind. Pharm.*, **8**, 937–947.

El-Shattawy, H. H., Kildsig, D. O. & Peck, G. E. (1984) *Drug Dev. Ind. Pharm.*, **10**, 1–17.

Fassihi, A. R. & Persicaner, P. H. R. (1987) *Int. J. Pharm.*, **37**, 167.

Ford, J. L. & Rubinstein, M. H. (1981) *Drug Dev. Ind. Pharm.*, **7**, 675–682.

Geneidi, A. S. & El-Sayed, L. (1978). *Pharm. Ind.*, **40**, 1074–1076.

Geneidi, G. S., El-Shattawy, H. H. & Geneidi, A. S. (1981) *Sci. Pharm.*, **49**, 172–179.

Gordon, R. G., Van Koevering, C. L. & Reits, D. J. (1984) *Int. J. Pharm.*, **21**, 99–105.

Graf, E., Fawzy, A. A. & Tsaktanis, I. (1984) *Acta Pharm. Tech.*, **30**, 25–29.

Hentze, G. & Voege, H. (1971a) *Drugs Made Ger.*, **14**, 142–146.

Hentze, G. & Voege, H. (1971b) *Pharm. Ind.*, **33**, 519–522.

Jacobson, H. & Gibbs, I. (1973) *J. Pharm. Sci.*, **62**, 1543–1545.

Jacobson, H. & Reier, G. (1969) *J. Pharm. Sci.*, **58**, 631–633.

Lee, K. C. & Hersey, J. A. (1977) *J. Pharm. Pharmacol.*, **29**, 515–516.

Li Wan Po, A. & Mroso, P. V. (1984) *Int. J. Pharm.*, **18**, 287–298.

Muller, B. W. (1977) *Acta Pharm. Tech.*, **23**, 257–266.

Nyqvist, H. (1986) *Drug Dev. Ind. Pharm.*, **12**, 953–968.

Remon, J. P., Van Severen, R. V., Braeckman, P. (1978) *J. Pharm. Pharmac.*, **30**, 204.

Signoretti, E., Dell'Urti, A., DeSalvo, A. & Donini, A. (1986) *Drug Dev. Ind. Pharm.*, **12**, 603–620.

Simon, T. H. (1967) Paper presented at American Pharmaceutical Association
    Meeting, Las Vegas.
Smith, A. (1982) *Anal. Proc.*, **19**, 559–561.
van Dooren, A. A. (1983) *Drug Dev. Ind. Pharm.*, **9**, 43–55.

11

Thermal analysis and semi-solid pharmaceutical systems

11.1 INTRODUCTION

Semi-solid pharmaceutical systems covered in this section include lotions, creams and ointments applied for the treatment of dermatological conditions, and suppositories and pessaries inserted into the rectum or vagina respectively for local or systemic drug delivery. The applications of thermal analysis to the component fats and waxes of such systems and to emulsions, being a dispersion of one immiscible phase within another, are included. Note that emulsions should be regarded as very simple models of topical pharmaceutical products somewhat less complex than typical oil-in-water or water-in-oil creams.

In the case of creams, thermal analysis, including DSC and TG, has been utilized along with other techniques in determining the microstructure of these semi-solids.

11.2 FATS AND WAXES

Fatty materials and waxes constitute some of the principal ingredients of semi-solid pharmaceutical products. The characterization of these materials could include thermal analytical evaluation. Apart from work on triglyceride suppository bases (see section 11.3) there would appear to be little pharmaceutical literature, however. Nevertheless data recorded in references culled from food chemistry, oil chemistry, petroleum chemistry and dental science (dental impression waxes) sources provide information and direction for the pharmaceutical scientist wishing to explore this area.

Vegetable oils such as palm kernel oil and cocoa butter may be utilized in pharmaceuticals. The polymorphic changes in palm kernel oil crystallized under varied conditions was examined by Kawamura (1979). The retardant effect of sorbitan tristearate on the transformation of unstable polymorphs to a stable one was also characterized in this way (Kawamura, 1980). Analytical conditions, heating and cooling rate, had marked effects on the observed retardation. Polymorphism in cocoa butter is well known in the food industry with regard to processing (including tempering and cooling) and 'bloom' in chocolate. It can create problems of poor quality in cocoa butter suppositories, cocoa butter, however, having been largely replaced by other triglycerides. The processing effects on formation of polymorphs and the correlation with process effects such as fat bloom have been described by Chapman *et al.* (1971) and Lovegren *et al.* (1976a,b).

The oxidative stability of vegetable oils can be established by DSC and TG methods. In the DSC method a pressure DSC cell, such as the DuPont Cell, is charged with oxygen under pressure after loading with sample. The instrument is then run under elevated temperature isothermal conditions and the time required for deviation from the baseline to occur measured. This indicates onset of oxidation, and comparing times for different oil sample allows comparison of their oxidative stability (Hassel, 1976). Normal-pressure DSC can be used but run times can be several hours rather than one to two hours (Cross, 1970). Hassel (1976) also describes a TG method for measuring the first deviation of the baseline under isothermal conditions in an atmosphere of oxygen. The gain of weight indicates onset of oxidation and comparative data for various oils can be generated as for DSC.

Triglycerides show polymorphism which may be important in the development of suppository dosage forms (see section 11.3) as lower melting, less stable, polymorphic forms of the triglyceride may convert to higher melting, more stable, forms on storage and alter dosage-form performance characteristics.

Heat of fusion data for various single fatty acid triglycerides was determined by DSC by Hampson and Rothbart (1969). DTA of pure and mixed triglycerides derived from palmitic and stearic acids illustrated polymorphism and that the form yielded depended on the rate of cooling of a melt (Perron *et al.* 1969).

The effect of chain length and unsaturation on the DSC behaviour of 13 single-acid triglycerides was studied by Hageman *et al.* (1972). Although only a single (β') form is usually exhibited between the least stable (α) and most stable (β) polymorph for all triglycerides, all samples studied showed evidence of at least two intermediate endotherms suggestive of additional forms. For some unsaturated triglycerides, e.g. triolein, further intermediate endotherms were observed between the endotherms for the α and β polymorphs. These were attributed to the possible alternative arrangements of the polymethylene chains on either side of the double bonds.

HSM supplemented by DSC, X-ray diffraction, NMR and IR data was utilized by Whittam and Rosano (1975) to study the physical ageing of triglycerides on storage (i.e. α to β transformation under normal storage conditions). Some waxes such as carnauba wax, microcrystalline wax, polyethylene wax and paraffin wax are used in pharmaceuticals, but their use in compounding dental impression waxes has prompted dental scientists to examine their properties by thermal analysis. Craig and various co-workers have used DTA (Craig *et al.*, 1967), DSC (Craig *et al.*, 1969), TG (Craig *et al.*, 1971) and TMA (Powers and Craig, 1978). DTA was of little value for qualitative or quantitative work, TG gave some indication of molecular weight distributions in the waxes, DSC allowed detection of interaction between wax components, whilst, most usefully, TMA indicated those waxes that might produce distributed impressions.

Both Flaherty (1971) and Miller and Dawson (1980) indicated that the DSC curve may be useful in characterizing the type of wax under study, whether microcrystalline, polyethylene, paraffin or synthetic. Thus it can form part of a quality control system for identifying and even 'finger printing' such waxes. Currell and Robinson (1967) had previously demonstrated that DTA could be used for determining the content of microcrystalline or polyethylene wax in mixtures with paraffin wax by measurement of an endotherm above 460°C, and not found in the paraffin waxes. One pharmaceutical material, anhydrous lanolin USP, has had its performance characteristics correlated with thermal analysis data. Radebaugh and Simonelli (1983) compared the viscoelastic properties of this material at various temperatures. There was some evidence of structural changes in DTA traces in the form of a peak maximum that could be correlated with the viscoelastic properties at that temperature.

Fats and waxes, especially triglycerides, suppository bases and modified triglycerides (e.g. ethoxylated triglycerides used in topical formulations) are of course complex mixtures and thermal analysis such as DSC often provides broad melting endotherms over a range of temperatures. DSC lends itself to better characterizing these materials by rapid determination of the solid fat index (SFI). This is classically determined by dilatometry (AOCS, tentative method, Official and Tentative Methods of the Am. Oil Chem. Soc., Solid Fats Index Cd10–57) and essentially measures the amount of melted solid in the sample at various temperatures (usually 10, 21.1, 26.7, 33.3 and 37.8°C).

Prendergast (1969) used specially designed sample cells in a DuPont DSC calorimeter to ensure good thermal conduct for the melting sample. The sample is melted in the DSC and allowed to cool and solidify, thus avoiding effects due to previous thermal history. The sample is run through a heating cycle until completely melted and the DSC curve recorded. The baseline is drawn in and perpendiculars dropped from the curve to the baseline at each SFI temperature. The areas under the curves up to each temperature are determined and converted to SFI values by correlating with a calibrant with known SFI values run under equivalent conditions. Walker and Bosin (1971) showed the DSC method to compare favourably with NMR and dilatometric methods of determining SFI.

11.3 SUPPOSITORIES AND PESSARIES

Suppositories are intended for rectal delivery of drugs and are usually prepared either from bases that are complex mixtures of triglycerides or from polyethylene glycols. Pessaries are similar products intended for vaginal delivery of drugs. Some pessaries are prepared by compression of non-fatty excipients on tabletting machines but this section concentrates solely on those dosage forms prepared from triglycerides or polyethylene glycols. The term suppository is applied throughout but implies either suppositories or pessaries.

DTA and DSC have been applied to the characterization of interactions between excipient components in triglyceride suppositories and also to drug excipient interactions. Liversidge et al. (1981, 1982) evaluated pure triglycerides singly and in binary mixtures to characterize the melting behaviour and polymorphic changes, especially on storage. Such model systems were deemed essential due to the extreme complexity of commercial suppository bases. The example of the Gattefosse

Suppocire A, was cited as an instance of this by Liversidge *et al.* (1981) where the varied fatty acid content, structural isomerism, polymorphism and possibility of solid solutions leads to thousands of potential individual transitions.

Care in the choice of experimental conditions was emphasized in these DTA studies where a heating rate of 2°C/min was found optimum, faster rates resulting in the obscuring of some endotherms. Furthermore, in the case of these triglycerides, sample weight needed to be optimized to provide for maximum sensitivity (large sample size) without compromising detection of transitions due to excessive sample size, this latter being a result of poor thermal conductivity of the samples. As HSM showed samples melted over a broad range, Liversidge *et al.* (1981) recorded the melting point 'as the temperature of the last peak maximum under DTA or the melting point of the highest melting component under HSM'. These melting points agreed within 1°C.

In their studies Liversidge *et al.* (1982) showed that the two unstable polymorphs (α and β') of trilaurin and tripalmitin slowly converted to the stable β form on long-term room-temperature storage. The relative amount of each polymorph was estimated by measurement of the peak areas of the melting endotherms (Fig. 11.1). These observations explain the increase in melting temperatures of commercial suppository bases on room-temperature storage (Fig. 11.2).

Giron *et al.* (1985) examined the effect of storage temperature on the melting behaviour of triglyceride suppository bases using DSC. Characteristically most suppository bases showed melting endotherms as up to three peaks in low, middle and high ranges. Although newly prepared suppositories showed only the two higher endotherms, aged suppositories stored below 25°C showed transition to the higher melting forms. Suppositories stored at 27.5–30°C, due to partial melting, showed appearance of the lower melting peak associated with the less stable form. Such data are useful in determining processing and storage conditions to enable better control quality of the finished product.

Similar applications of DSC to commercial suppository bases have been reported by Coben and Lordi (1980). Optimization of sample heating rate again indicated rates above 2°C/min provided limited data due to poor heat transfer. During long-term storage low-melting endotherms are lost with an increase in height and sharpness of higher-melting endotherms due to polymorphic transitions (Fig. 11.3). The DSC changes correlated with increases in suppository hardness and increases in fat index (i.e. crystalline fat level versus amorphous fat level is increased). Liversidge *et al.* (1982) indicated that two model drugs, ketoprofen and metronidazole, had little effect on phase diagrams of binary mixtures of triglycerides of the two model compounds in the triglycerides. Presumably more soluble compounds would have had more significant effects.

Srcic *et al.* (1985) have suggested the application of DSC to process control of manufacture of polyethylene glycol suppositories. Different cooling rates on casting molten suppository mixtures into moulds was found to affect critical properties of finished suppositories such as hardness, density and drug release rate. DSC of the surface scrapings of such PEG 4000 suppositories showed three endotherms (Fig. 11.4). The first is attributed to density change prior to melting, the second to melting itself and the third to a viscosity change of the molten polyethylene glycol. The scans obtained by running scrapings from suppositories cooled slowly in the sample cell

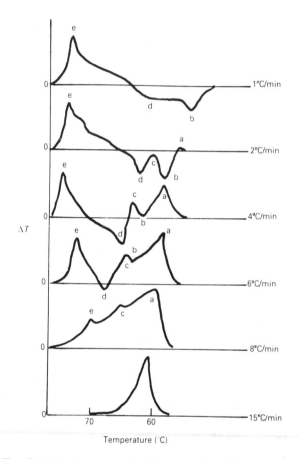

Fig. 11.1 — The effect of heating rate on the DTA curve of tristearin (6 mg). The assignment of the peaks is as follows: a, melting of α-polymorph; b, crystallization of β′-polymorph; c, melting of β′-polymorph; d, crystallization of β-polymorph; e, melting of β-polymorph. For each trace, endotherms are above and exotherms below the appropriate $\Delta T = 0$ line. (Reproduced with permission from Liversidge *et al.*, 1981.)

against those from fast-cooled suppositories in the reference cell indicated greater enthalpy of melting for the slowly cooled product, correlating with increased crystallinity. Increased crystallinity in the core versus the crust of PEG suppositories was confirmed similarly by this differential DSC technique.

The use of DSC in characterizing changes in suppositories bases on storage has also been evaluated by Bornschein *et al.* (1980) who used DTA to follow changes in melting behaviour of suppositories and correlated these with changes in drug release rates.

As a quality-control tool, therefore, thermal analysis, particularly DSC and DTA, has a role in the evaluation of suppositories in addition to a role in the pharmaceutical development of such products. Wesolowski (1982) extended the

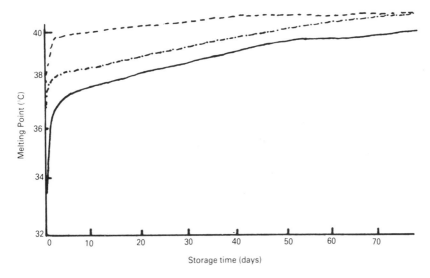

Fig. 11.2 — The effect of storage time at 22°C on the DTA melting point of the following commercial suppository bases: Suppocire A ——; Witepsol E75 – – –; Witepsol W35 – · – · –. (Reproduced with permission from Liversidge *et al.*, 1981.)

quality-control role of DTA and included TG for the monitoring of the composition (i.e. active-ingredient control) of suppositories.

11.4 EMULSIONS, CREAMS AND OINTMENTS

Thermal analysis has made several contributions to our understanding of the microstructure of topical formulations in the form of emulsions, creams and ointments. It is first intended to concentrate on simple systems and carry on into creams. This is a logical development as the simple topical systems studied by thermal analysis consist of water, fatty alcohols and emulsifier, and the cream develops from the system by the inclusion of a dispersed oil phase.

Fukushima *et al.* (1976) investigated the stability of model formulations containing liquid paraffin, soft paraffin, polyoxyethylene (15) oleyl ether (emulsifier) and water, and stearyl alcohol, cetostearyl alcohol or cetyl alcohol as the fatty alcohol. DSC was utilized along the rheological evaluation, X-ray diffraction and light microscopy. The emulsion containing pure fatty alcohols showed two endotherms, one at about 38 and 48°C (cetyl and stearyl alcohols respectively) and a higher one about 53 and 62°C (cetyl and stearyl alcohols respectively). In mixtures in the ratios cetyl:stearyl 9:1 to 3:7 only one distinct endotherm at approximately 55–57°C was apparent. The higher peak temperature in the pure alcohol systems is a melting point whereas the lower one is due to polymorphic transition of the alcohol. This effect is therefore lost in the admixtures and is responsible for increased physical stability of the emulsions (Fukushima *et al.* 1977). Eccleston (1985) found similar results. Both DTA and TG have been applied in the study of the phase inversion of oil in water emulsions (Matsumoto and Sherman, 1970; Frenkel and Garti, 1980). It has been

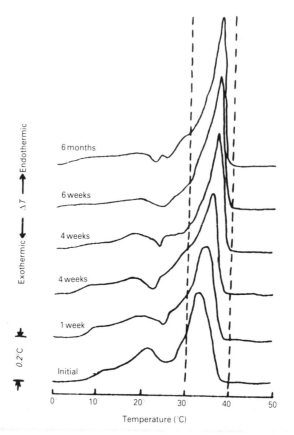

Fig. 11.3 — DSC endotherm stability scan of Suppocire BCM stored at 22°C. The heating rate was 2°C/minute. (Reproduced with permission from Coben & Lordi, 1980.)

claimed that TG can provide a more accurate measure of the phase inversion temperature of an emulsion (Frenkel and Garti, 1980).

Junginger and co-workers have applied thermal analysis, notably DSC, along with other physical techniques in the characterization of the microstructure of creams. Early work (Junginger *et al.*, 1979) looked at the effect of adding water to the hydrophilic ointment, DAB 7, which contains cetostearyl alcohol, self-emulsifying cetostearyl alcohol, white soft paraffin and liquid paraffin. DSC curves indicated polymorphic transitions and melting behaviour of water-containing emulsifier components when examined in isolation. In the formulated cream (ointment plus water) the same endotherms are seen but the peak temperature is lowered slightly and the peak broadened.

By combining them with other data these results were explained in terms of the swelling of a lamellar lattice layer derived from the cetostearyl alcohol and emulsifying cetostearyl alcohol on addition of water. The lattice yields the endotherm, due to polymorphic transition and melting. The melting-point depression and peak-broadening in the presence of the paraffins is attributed to interaction with the lattice. This interaction is weak and is essentially mechanical immobilization of the oils, thus

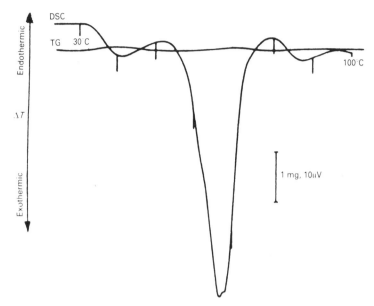

Fig. 11.4 — DSC of PEG 4000 suppository crust. (Reproduced with permission from Srcic *et al.*, 1985).

there are only slight changes in the endotherms between single components and the finished cream.

Further work (Junginger *et al.*, 1984) included the use of TG and differential TG to characterize water distribution in the cream. Starting at 20°C and heating at 2°C/min the rate of water loss was monitored (Fig. 11.5). The rate increases at first, reaching a maximum at about 46°C, then declining with increasing temperature to about 56°C, where the rate levels off to about 65°C before again it starts to decline. The rate of water loss increases sharply at 72°C, reaching a maximum at 78°C.

Up to 56°C the loss of bulk water is seen and at 56°C loss of water of hydration and mechanically entrapped water from the gel network results in the slight hump on the declining rate of water loss. The gel phase melts at 72°C, liberating the lamellarly fixed water. These observations, earlier DSC observations and other physical data (e.g. from X-ray diffraction) resulted in the proposed microscopic structure for this cream (Fig. 11.6).

DSC on single components, simple mixtures and complete formulations of the non-ionic emulsifying cream DAC helped understand the gel structure in this product (de Vringer *et al.*, 1986). Its formula is as follows:

Polyoxyethylene-20-glycerol monostearate	5.0%
Liquid paraffin	7.5%
Cetyl alcohol	5.0%
Stearyl alcohol	5.0%
Glycerol	8.5%
White soft paraffin	17.5%
Water	51.5%

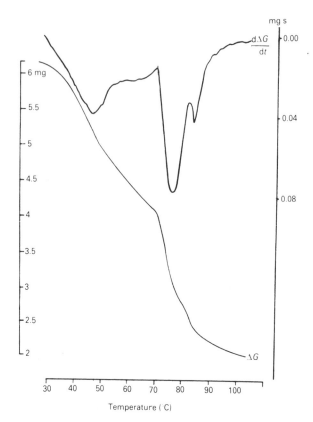

Fig. 11.5 — TG-curve (ΔG-curve) and DTG-curve ($d\Delta G$-/dt-curve) of water-containing hydrophilic ointment DAB 8 (reproduced with permission from Junginger *et al.*, 1984).

Studies on the polyoxyethylene-20-glycerol monostearate suggested this bound two molecules of water per polymer unit (as determined by quantifying freezable water).

In finished creams the endotherm due to the melting of cetyl/stearyl alcohol only is seen at 59°C, with some contribution as a shoulder at 56°C from melting of cetyl/stearyl alcohol not interacted with polyoxyethylene-20-glycol monostearate. The characteristic of the finished cream determined by thermal analysis could thus be explained from the analogous behaviour of the single components.

There is little literature on the thermal analysis of ointments. Campigli *et al.* (1980) suggested interactions occur between isopropyl myristate and hydrocarbons on the basis of DSC in a white soft paraffin–isopropyl myristate–hard paraffin ointment. These interactions were not further investigated however.

Jackson and Wood (1974) used DTA as a tool to assay quantitatively the amount of white soft paraffin in white soft paraffin–liquid paraffin mixtures extracted from dermatological formulations. The area of the melting endotherm recorded was proportional to the amount of white soft paraffin present. The method obtained results within 10% of theoretical and showed a typical coefficient of variation of 5%.

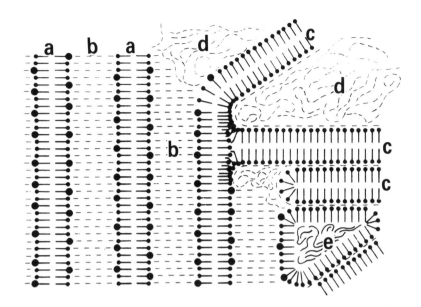

Fig. 11.6 — Gel structures of the water-containing hydrophilic ointment DAB 8: a, mixed crystal bilayer of cetostearyl alcohol and cetostearyl alcohol sulphate; b, interlamellarly fixed water layer; a+b, hydrophilic gel phase; c, cetostearyl alcohol semihydrate lipophilic gel phase; d, bulk water phase; e, lipophilic components (dispersed phase) (reproduced with permission from Junginger *et al.*, 1984).

12.5   OTHER SEMI-SOLID APPLICATIONS

Thermal analysis has been much applied to the components of semi-solid formulations and is beginning to demonstrate its value as a research tool in the characterization of finished products. The evaluation of newer semi-solid systems such as paste-filled hard gelatin capsules will no doubt provide further applications for thermal analysis.

REFERENCES

Bornschein, M., Grohmann, A. & Voight, R. (1980) *Pharmazie*, **35**, 40–42.
Campigli, V., Carelli, V., DiColo, G., Nannipieri, E., Serafini, M. F. & Vitale, D. (1980) *Pharm. Acta Helv.*, **61**, 198–204.
Chapman, G. M., Akehurst, E. E. & Wright, W. B. (1971) *J. Am. Oil. Chem. Soc.*, **48**, 824–830.
Coben, L. J. & Lordi, N. G. (1980) *J. Pharm. Sci.*, **69**, 955–960.
Craig, R. G., Powers, J. M. & Peyton, F. A. (1967) *J. Dent. Res.*, **46**, 1090–1097.
Craig, R. G., Powers, J. M. & Peyton, F. A. (1969) In *Analytical Calorimetry*, eds. Porter, S. & Johnson, J. E., Plenum Press, New York, 157–166.
Craig, R. G., Powers, J. M. & Peyton, F. A. (1971) *J. Dent. Res.*, **50**, 450–454.
Cross, C. K. (1970) *J. Am. Oil. Chem. Soc.*, **37**, 229–230.
Currell, B. R. & Robinson, B. (1967) *Talanta*, **14**, 421–424.

de Vringer, T., Joosten, J. G. H. & Junginger, H. E. (1986) *Coll. Polym. Sci.*, **264**, 691–700.

Eccleston, G. M. (1985) *Int. J. Pharm.*, **27**, 311–323.

Flaherty, B. (1971) *J. Appl. Chem. Biotechnol.*, **21**, 144–148.

Frenkel, M. & Garti, N. (1980) *Thermochim. Acta*, **42**, 265–272.

Fukushima, S., Takahashi, M. & Yamaguchi, M. (1976) *J. Coll. Int. Sci.*, **57**, 201–206.

Fukushima, S., Yamaguchi, M. & Harusawa, F. (1977) *J. Coll. Int. Sci.*, **59**, 159–165.

Giron, D., Riva, A. & Steiger, M. (1985) *Thermochim. Acta.*, **85**, 509–512.

Hagemann, J. W., Tallent, W. H. & Kolb, K. E. (1972) *J. Am. Oil Chem. Soc.*, **49**, 118–123.

Hampson, J. W. & Rothbart, H. L. (1969) *J. Am. Oil Chem. Soc.*, **46**, 143–144.

Hassel, R. L. (1976) *J. Am. Oil Chem. Soc.*, **53**, 179–181.

Jackson, I. M. & Wood, P. R. (1974) *Proc. Soc. Analyt. Chem.*, **11**, 96–97.

Junginger, H., Fuhrer, C., Ziegenmeyer, J. & Friberg, S. (1979) *J. Soc. Cosmet. Chem.*, **30**, 9–23.

Junginger, H., Akkermans, A. A. M. D. & Heering, W. (1984) *J. Soc. Cosmet. Chem.*, **35**, 45–57.

Kawamura, K. (1979) *J. Am. Oil Chem. Soc.*, **56**, 753–758.

Kawamura, K. (1980) *J. Am. Oil Chem. Soc.*, **57**, 48–52.

Liversidge, G. G., Grant, D. J. W. & Padfield, J. M. (1981) *Int. J. Pharm.*, **7**, 211–223.

Liversidge, G. G., Grant, D. J. W. & Padfield, J. M. (1982) *Anal. Proc.*, **19**, 549–553.

Lovegren, N. V., Gray, M. S. & Feuge, R. O. (1976a) *J. Am. Oil Chem. Soc.*, **53**, 83–88.

Lovegren, N. V., Gray, M. S. & Feuge, R. O. (1976b) *J. Am. Oil Chem. Soc.*, **53**, 108–112.

Matsumoto, S. & Sherman, P. (1970) *J. Col. Int. Sci.*, **33**, 294–298.

Miller, R. & Dawson, G. (1980) *Thermochim. Acta*, **41**, 93–105.

Perron, R., Petit, J. & Mathieu, A. (1969) *Chem. Phys. Lipids*, **3**, 11–28.

Prendergast, J. A. (1969) In *Thermal Analysis Vol. 2, Inorganic materials and physical chemistry*, eds Schwenker, R. F. & Garn, P. D. Academic Press, New York, 1317–1327.

Powers, J. M. & Craig, R. G. (1978) *J. Dent. Res.*, **57**, 37–41.

Radebaugh, G. W. & Simonelli, A. P. (1983) *J. Pharm. Sci.*, **72**, 415–421.

Srcic, S., Bukovec, P. & Smid-Korbar, J. (1985) *Thermochim. Acta.*, **92**, 333–336.

Walker, R. C. & Bosin, W. A. (1971) *J. Am. Oil Chem. Soc.*, **48**, 50–53.

Wesolowski, M. (1982) *Int. J. Pharm.*, **11**, 35–44.

Whittam, J. H. & Rosano, H. L. (1975) *J. Am. Oil Chem. Soc.*, **52**, 128–133.

12

The use of thermal analysis in the study of liposomes

12.1 INTRODUCTION

The membranes of living cells are composed of many different lipids in conjunction with proteins and carbohydrates. Important lipids in these membranes are the phospholipids, sphingolipids and sterols, the latter including cholesterol (Fig. 12.1). When purified phospholipids are dispersed in water, liposomes are formed. Liposomes are multi-layered vesicles consisting of concentric structural bilayers of phospholipid separating discrete aqueous compartments (Fig. 12.2).

The pharmaceutical applications of liposomes arise from a two-fold interest. One concerns their potential use as drug delivery systems. Their ability to passively target to the reticulo-endothelial system has resulted in much research into their use to deliver anti-infective and anti-parasitic drugs. Furthermore researchers have evaluated approaches to targetting to other areas, e.g. with anti-tumour drugs. The other application involves their use as model membrane systems to provide a simple model of drug activity. The latter part of this chapter will concentrate on this particular aspect with emphasis on the mode of action of drugs with antimicrobial activity.

The characterization of the phospholipid materials, particularly in their hydrated bilayer liposomal state, and the interaction of those bilayers with drug substances and non-drug bilayer modifying materials (e.g. sterols) utilizes thermal analysis among the techniques suitable for their study.

Purified phospholipids can be prepared by appropriate solvent extraction of suitable phospholipid sources, eggs being widely utilized for this purpose. Column chromatography is extensively used for the final purification steps. In addition to purified egg lecithin, semisynthetic phospholipids, such as dimyristoylphosphatidyl-choline (DMPC) (Fig. 12.1), can be prepared by substitution of fatty acid groups which may thus modify the physico-chemical characteristics of the material.

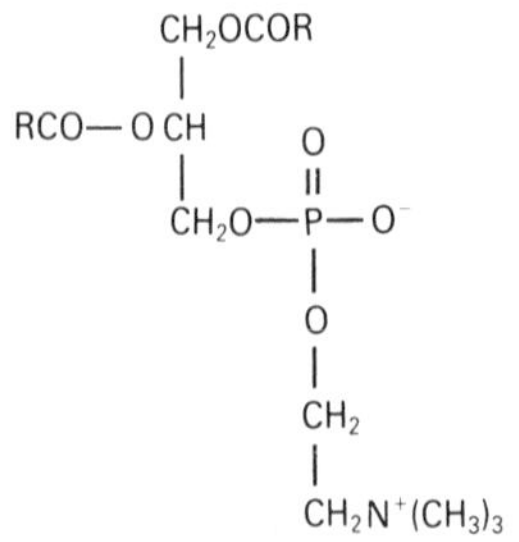

Fig. 12.1 — Formulae of some representative phospholipids, sphingolipids and sterols: (a) phosphatidylcholine (R', R" = acyl); (b), sphingomyelin (R = acyl); (c) cholesterol.

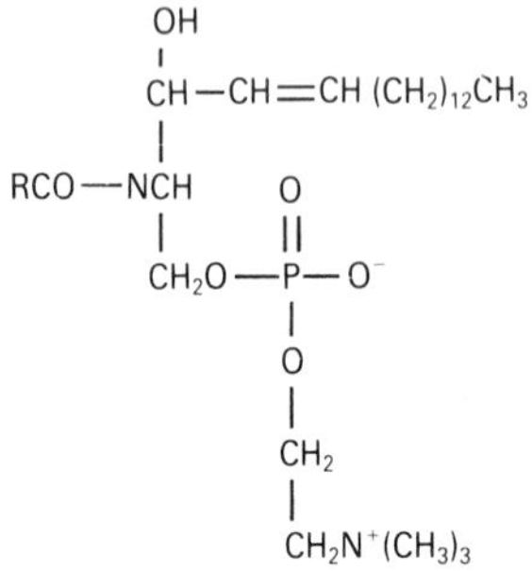

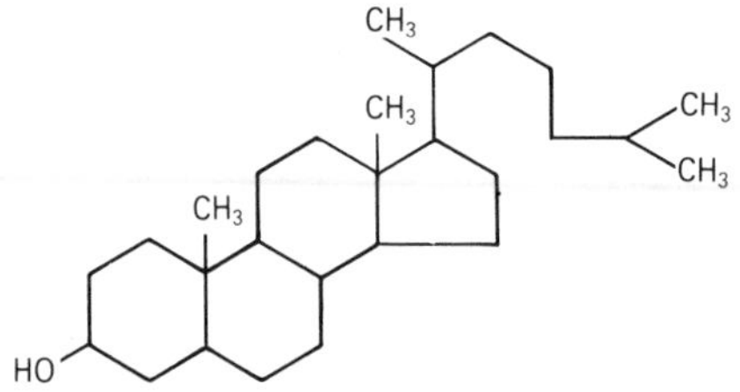

12.2 INSTRUMENTS AND MEASURED PARAMETERS

Until recently commercial instruments such as required for DTA and DSC suffered from the limitation of poor sensitivity and small sample sizes (below 100 μl) which, coupled with relatively fast scanning rates meant that low energy transitions were often lost during analysis of liposomal suspensions. The low concentration of lipids and their low energies of transition necessitate that instruments should be sensitive or capable of dealing with high volumes. This mitigated against the use of old DSC or DTA instrumentation. Modern instruments are capable of greater accuracy and sensitivity. Typically high concentrations are used (5–10% w/v) with volumes of 20–50 μl (Blume, 1988). Nonetheless high sensitivity calorimeters, such as described by Privalov et al. (1975) were recommended for the study of biomembranes (Mabrey & Sturtevant, 1978). These typically use a large cell volume (approximately 1 ml) and

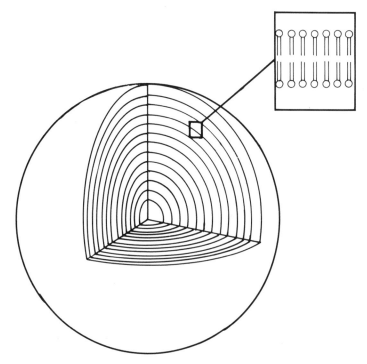

Fig. 12.2 — Representative structure of a typical liposome, demonstrating concentric layered structure in cut-away. Inset demonstrates bilayer structure of phospholipid molecules.

a concentration of 0.02–1% w/v (Blume, 1988) and detect either excess heat or the corresponding power used to keep the sample and reference at an identical temperature (Privalov *et al.*, 1975).

Measured parameters include the temperature at the peak (T_m), the onset temperature of the transition (T_o) and the extrapolated onset temperature. Additionally $\Delta T_{1/2}$, the width of the transition at half peak height, may be determined. One advantage of high-sensitivity calorimetry or DSC is that endotherms may be used to determine the co-operativity of a transition. The enthalpy of the process (ΔH) is proportional to the area under the curve and has units of for example, cal g. However a thermodynamic estimate, termed the van't Hoff enthalpy (ΔHvh), may also be determined. Mabrey & Sturtevant (1978) reviewed several methods for estimating it but the simplest is equation (12.1):

$$\Delta Hvh \simeq \frac{6.9 T_m^2}{\Delta T_{1/2}} \qquad (12.1)$$

Its units are for example, cal/mole. The ratio of the two enthalpies has units of g/mole and, if the molecular weight is known, can be converted to molecules giving an estimate of the co-operativity of the process. The co-operative unit is defined

(Mabrey & Sturtevant, 1978) as that quantity of a substance such that, if the sample were composed of independent units of that size undergoing a two-state transition, the transition curve would have the observed shape. The co-operative units are affected by the presence of other chemicals such as proteins or other lipids, ionic strength and the presence of antimicrobial agents. It is these changes that make thermal analysis such a useful tool in determining the sites of action of antimicrobial substances.

DSC provides a rapid technique which can be used to characterize a system without the modification or attachment of spectroscopic probes. It gives quantitative information without disturbing the transition of interest. However, DSC is unable to survey the behaviour of the different parts of the phospholipid molecules, information which is given by electron spin resonance and spin labelling. Careful use of suspending solvent is necessary. Distilled water may be used to examine transitions above the ice point but for low temperature transitions a solvent containing 50% ethylene glycol and 0.15 M NaCl is ideal (Steim *et al.*, 1969) although ethylene glycol may depress T_m by 3 to 5°C.

Hinz & Sturtevant (1972) were among the first to apply DSC to the study of transitions in dilute aqueous suspensions of liposomes using 0.04–0.66% w/w phospholipid dispersions. More recent studies have used concentrations up to 50% w/w (Fildes & Oliver, 1978) particularly when studying drug–bilayer interactions.

12.3 CHARACTERIZATION OF HYDRATED PHOSPHOLIPID BILAYERS

12.3.1 Single-component systems

The 1,2-diacyl-L-phosphatidylcholines are usually isolated from wet organic solvents in the hydrated (monohydrate) form. When subjected to DTA they show a large endothermic transition. If heated in open cells water loss from the monohydrate is observed. The endotherm corresponds to the transition from the solid crystalline form to a liquid crystalline form. As water is introduced into the sample, phospholipid and water interact and the transition temperature is lowered until at water content in excess of 20% w/w no further lowering is seen. The transitions seen under these conditions would be those observed in the case of phospholipids presented as a dilute liposomal dispersion and are considerably sharper and more well defined than the transitions seen with the dry phospholipids.

When phospholipids are dispersed in aqueous solutions and shaken vigorously they arrange themselves in suspension as lipid bilayers or liposomes. Pure lipids when dispersed in water form multilayer liposomes which consist of several concentric circles. On thermal analysis of these phospholipid bilayers, especially phosphatidylcholines, two transitions are apparent. The first, the pretransition, is due to a rotation of the polar head portion of the liquid molecules and is followed by the major transition which is a gel to liquid crystal transformation (Hinz & Sturtevant, 1972). The pretreatment is broader and smaller than the chain melting endotherm and is not found in all phospholipids but is present in phosphatidylcholine bilayers. The pretransition of aqueous dispersions of DPPC (dipalmitoylphosphatidylcholine) is associated with a structural transformation from one-dimensional to two-dimensional monoclinic lattice (Janiak *et al.*, 1976) involving the formation of a rippled phase (Berleur *et al.*, 1985). It is associated with motion of the phospholipid polar head groups (Ladbrooke & Chapman, 1969) or alternatively with co-operative movement of the rigid acyl side-chains in a transition between crystal forms below their melting temperature (Hinz & Sturtevant, 1972). A third possibility is that the pretransition is associated with a tilting of hydrocarbon chains before melting (Chapman *et al.*, 1974). The second transition comes from disordering of an all *trans*-configural gel state to a fluid state, i.e. hydrocarbon chain melting. This gel to liquid transition has been measured by DSC for a number of synthetic phospholipids (Chapman *et al.*, 1967, 1974; Hinz & Sturtevant, 1972; Vaughan & Keough, 1974; Mabrey & Sturtevant, 1976) and is found to depend on the nature of the polar head groups and the length and degree of unsaturation in the fatty acyl chains (Ladbroke & Chapman, 1969) (see Table 12.1).

Table 12.1 — Gel to liquid crystalline transition temperatures for hydrated 1,2-diacyl phosphatidylcholine dispersions (data abstracted from Ladbrooke & Chapman, 1969)

Acyl chain	Transition temperatures (°C)
C_{22} (behenoyl)	75
C_{18} (stearoyl)	58
C_{16} (palmitoyl)	41
C_{14} (myristoyl)	23
C_{12} (lauroyl)	$\simeq 0$
C_{18-2} (oleyl)	-22

In general, thermal analysis of phospholipids and membranes will produce an endothermic transition whose T_m varies with the hydrocarbon chain attached to the phosphate moiety, the head group and the presence of other solutes. Generally an increase in chain length increases the value of T_m whereas an increase in double bonds will decrease the T_m. The gel–liquid crystalline transition that this endotherm indicates is a partial melting representing a *trans* to *gauche* rotational isomerization arising from the hydrocarbon chains and involving lateral expansion and a decrease in bilayer thickness and additional changes in van der Waals interaction between the chains and polar interactions at the lipid bilayer–water interface. A DSC curve showing the main transition and the pretransition for a 50% w/w dispersion of DMPC in water is given in Fig. 12.3.

Major transitions in saturated diacylglycerophospo-ethanolamines occur some 20–25°C higher than for the corresponding phosphatidylcholines due to crowding of the larger phosphatidyl head groups (Chapman *et al.*, 1974). Pretransitions are absent. They will only form bilayers below a certain temperature (Melchoir, 1982). Impurities broaden the transition which in the highest purity lipids is isothermal.

The temperature difference between the pretransition and gel–liquid crystalline transition of homologues of the same head group decreases with increasing chain

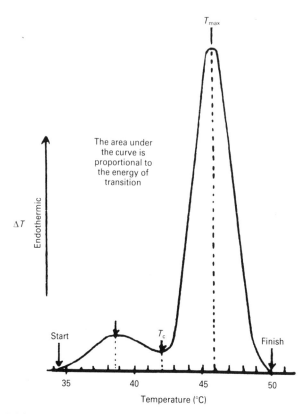

Fig. 12.3 — DSC heating curve of dipalmitoylphosphatidylcholine–water system (50:50 mixture) (reprinted with permission from Chapman *et al.*, 1974, © American Society of Biological Chemists).

length of attached fatty acid. DMPC displayed a pretransition at 13.5°C and the major transition at 23.7°C with corresponding heats of transition of 1.1 and 6.26 kcal/mol and distearolyphosphatidylcholine (DSPC) at 49.1 and 58.24°C with energies of transition of 1.4 and 10.84 kcal/mol (Hinz & Sturtevant, 1972). Liposomal particle size influences the calorimetric scans. Mutlilamellar aggregates of DMPC gave two peaks in their DSC curves but the enthalpy of transition of vesicles of 18 nm diameter was one-third the value of these aggregates (Gruenewald *et al.*, 1979).

12.3.2 Multiple component systems
Phospholipid mixtures, e.g. different phosphatidylcholines or a phosphatidylcholine with a phosphatidylglycerol, are often used in liposome formulations. Artificial membranes containing more than one component have thermal properties different to the individual ingredient. Phase diagrams reveal ideality of mixing. Generally when two phospholipids containing the same head group differ by two carbon atoms in their fatty acids nearly ideal mixing will take place. A difference of four carbon atoms shows deviations from ideality and once the chain length differs by six or more

carbons non-ideality occurs and monotectics may be formed indicative of phase separation. It is assumed tht phospholipids are fully hydrated and consequently phase diagrams are regarded as binary systems. Where the ingredients behave ideally the resultant system will be a continuous series of solid solutions. Close homologues of phosphatidylcholines in combination show solid solution behaviour. For the combination DSPC and DPPC there was no evidence of compound or eutectic formation on examination of DSC heating and cooling curves (Phillips *et al*, 1970). On increasing the difference in chain length to the DSPC–DMPC combination there was significant differences in the DSC heating and cooling curves (Phillips *et al.*, 1970). This was attributed to migration of phosphatidylcholine molecules within the liposome bilayers to give crystalline regions corresponding to the two compounds. Similar behaviour was observed with mixtures of saturated and unsaturated phosphatidylcholines (dioleylphosphatidylcholine with debehenoyl-phosphatidylcholine, DSPC, DPPC or DMPC) (Ladbrooke & Chapman, 1969).

Deviations from ideality produce monotectics, eutectics, and even peritectics and complex formation, all with concomitant broadening of the transition range. Mixing of DMPC and DPPC (Fig. 12.4) resulted in assymetric transitions appearing intermediate to those of the two components (Mabrey & Sturtevant, 1976). The phase diagram showed complete miscibility in both the liquid and solid phases. As DPPC concentration was increased the transitions shifted to higher temperatures and peak broadening was noticable at 1:1 ratios (Chapman *et al.*, 1974). The melting range was increased by 4°C over that of the sharpest melting pure component.

Similar, although less ideal, results were obtained for DMPC–DSPC and DSPC–DMPE (dimyristoylphosphatidylethanolamine) systems (Mabrey & Sturtevant, 1976). DSC of DMPC–DMPE (Fig. 12.5) showed broader assymetric peaks indicating that a lowering of the transition co-operatively occurred (Chapman *et al.*, 1974). Gel and liquid crystals therefore exist over a wide temperature range (Mabrey & Sturtevant, 1976). DLPC (dilauroylphosphatidylcholine)–DSPC mixtures showed monotectic behaviour, i.e. the DLPC melting transition remained constant over nearly all the composition range (Fig. 12.6). This indicated that DSPC is only slightly soluble in DLPC while DLPC has a significant solubility in DSPC. Rapidly cooled homogeneous bilayers of DMPC and DSPC displayed two endotherms (Mechoir, 1986) suggesting a fluid–fluid immiscibility between the two phospholipids. During rapid freezing the bilayer crystallized so rapidly that the lateral mobility of the lipids was too slow to allow co-crystallization into the equilibrium state. Storage resulted in loss of the two peaks indicating a return to equilibrium distribution.

Scanning at different speeds may produce anomalous results. DMPC–DPPC mixtures showed solid immiscibility at 10–35 moles% DPPC when scanned at 5°C/min which prodcued non-equilibrium conditions (Van Dijck *et al.*, 1977). Pre-incubation and slow scanning speeds removed this error.

Incorporation of fatty acids into the bilayers modifies the transition. Long-chain saturated or *trans*-unsaturated fatty acids increase and broaden the main transition of phosphatidylcholine membranes and bilayers (Oritz & Gomez-Fernandez, 1987), whereas fatty acids of 10 or fewer carbons, *cis*-unsaturated long-chain fatty acids and fatty acid derivatives probe lower and broaden the transition, all fatty acids removing the pretransition. DPPC–palmitic acid mixtures form a peritectic (Schullery *et al.*, 1981) rather than a simple mixture (Mabrey & Sturtevant, 1977). Stearic acid

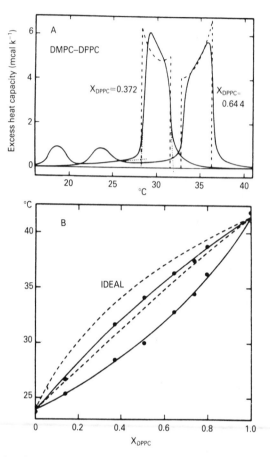

Fig. 12.4 — (A) (——) Observed calorimetric transition curves for two mixtures of DMPC and DPPC; (–––) transition curves calculated on the basis of the phase diagram in panel B. (B) (——) Phase diagram constructed from initiation and completion temperatures read from observed transition curves; (–––) ideal phase diagram. X_{DPPC} is mole fraction DPPC. Reprinted with permission from Mabrey, S. & Sturtevant, J. M. 1976).

removed the main transition in DMPC and DSPC to higher temperatures (Oritz & Gomez-Fernandez, 1987) and more than one phase developed in their phase diagrams. A new peak was apparent above the main transition and grew at the expense of the latter shifting to a higher temperature. A peritectic existed at 23.5°C indicating gel state immiscibility but lipid miscibility in the fluid phase. The liquidus and solidus lines intersected at 70 mole% indicative of a pure component comprising 1:2 phospholipid:fatty acid ratio. The fatty acid is excluded from the pure lipid below T_o. The stearic acid–DSPC system behaved similarly, the acid preferentially affecting the higher melting component giving rise to an additional peak at higher temperature. Thus stearic acid preferentially partitioned into the solid-like domains until a critical concentration was reached beyond which a single phase was formed. Oritz & Gomez-Fernandez (1987) concluded that the saturated fatty acids parti-

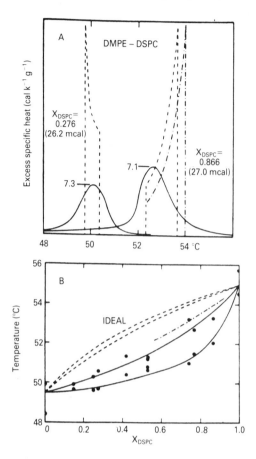

Fig. 12.5 — (A) (——) Observed calorimetric transition curves for two mixtures of DMPE and DSPC; (– – –) calculated transition curves; (– · – · –) transition curves calculated on the basis of the (– · – · –) curve in the phase diagram in panel B. (B) (——) Phase diagram constructed from initiation and completion temperatures read from observed transition curves; (– – –) ideal phase diagram. Reprinted with permission from Mabrey, S. & Sturtevant, J. M. 1976).

tion preferably into solid-like domains while *cis*-unsaturated fatty acids partition preferentially into the fluid-like domains.

Addition of palmitic acid raised the upper transitions of DPPC (Mabrey & Sturtevant, 1977). At 0.67 mole fraction palmitic acid there was a sharp transition (Fig. 12.7) and higher quantities further broadened what had become a very narrow transition to give a complex scan shape indicative of instability in the bilayer. The acid fitted into the hexagonal lattice of the DPPC chains in the bilayer to form an isothermally melting mixture at a 1:1 ratio.

DSC indicated that the combination of DPPC with dipalmitylphosphatidylglycerol at the 2:1 ratio were fully miscible in the bilayer structures (Findlay & Barton, 1978; Offringa *et al.*, 1987).

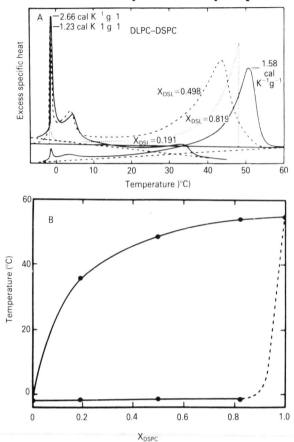

Fig. 12.6 — (A) (Solid curves, dashed baselines) transition curves for DLPC–DSPC mixtures with X_{DSPC} = 0.191 and 0.819; (dashed curve, solid baseline) transition with X_{DSPC} = 0.498; (dotted curve) calculated transition curve for X_{DSPC} = 0.498. (B) Phase diagram constructed from calorimetric transition curves. Reprinted with permission from Mabrey, S. & Sturtevant, J. M. 1976).

The inclusion of protein into phospholipid membranes may vastly alter their properties. The protein, bacteriorhodopsin, included into liposomes containing DPPC or MDPC has been examined as a model system (Alonso *et al.*, 1982). Above the gel to liquid crystalline transition, protein aggregates were formed at phospholipid:protein molar ratios lower than 50:1. Below this temperature extensive protein aggregation occurred. Incorporation of small amounts of the protein reduced the pretransition and broadened the main transition. Once the pretransition had been abolished the main endotherm continued to broaden and decrease in size (Alonso *et al.*, 1982). The main transition width was around 1.5 K for the pure lipid but was 6 K and 5 K in samples containing 20–30 moles DPPC or DMPC respectively per mole of bacteriorhodopsin. One molecule of bacteriorhodopsin removed 19 molecules of DMPC or 22 molecules of DPPC from the co-operative transition. Two melting processes occurred, firstly from high-protein lipid patches and secondly from the

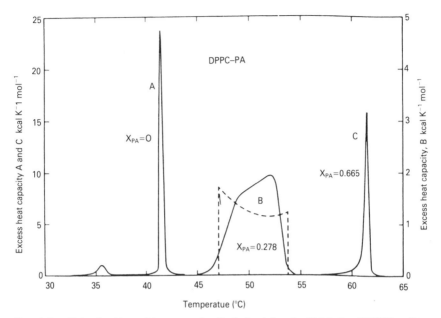

Fig. 12.7 — Gel to liquid transition curves for dipalmitoylphosphatidylcholine (DPPC) and two mixtures of DPPC with palmitic acid (PA). The dashed line is a calculated transition curve. (Reproduced with permission from Mabrey & Sturtevant, 1977).

remaining lipid. The latter broadened with increasing lipid concentrations, i.e. the patches became larger and the remaining areas of pure lipid decreased and the co-operativity of the transition decreased. The transition due to the high-protein lipid patches, possibly because the patches were tightly packed, was not able to undergo a co-operative melting process. Increased protein concentrations, during crystalliza-tion of the membrane phospholipids, took increasing levels of phospholipid to solvate itself (Alonso *et al.*, 1982), thereby reducing the quantity of lipid availability to crystallize and participate in the gel to liquid crystal transition on subsequent heating.

Sterols, notably cholesterol, are often incorporated into liposomal formulations to modify the permeability of the bilayers. Their effect can be detected in DSC curves for DPPC when increasing amounts of incorporated cholesterol (run as 50% w/w dispersions in water) show elimination of the pretransition peak on first addition of the sterol, followed by a broadening and decrease in area of the main phospholipid transition peak as the cholesterol level is incorporated. At a 1:1 molar ratio of phospholipid:cholesterol no transition was observed (Ladbrooke *et al.*, 1968). The 1:1 ratio is in fact the maximum level of cholesterol that can be incorporated before the separation of excess crystalline cholesterol occurs. The thermal data is explained by, in the presence of water, cholesterol disrupting the orderly array of hydrocarbon chains in the gel phase, resulting in a fluidity of these chains to a state intermediate between the gel and liquid crystalline phases of the pure phospholipid.

Examination of the water phase of these systems has been undertaken by

Offringa *et al.* (1987). Thermal analysis of DPPC/cholesterol/water systems subjected to a repeated freeze–thaw cycle showed a reproducible 'ageing' phenomenon of the bilayer. Melting characteristics of frozen water were used to provide information on the fraction of non-freezing and freezing (subzero-melting) hydration water and excess of bulk water. Offringa *et al.* (1987) concluded that incorporation of DPPG or cholesterol into the DPPC structure increased the subzero temperature interval compared with pure DPPC. About 20 mol water/mol of phospholipid was the maximum number of molecules of hydration water for DPPC and DPPC/cholesterol systems. Morris & McGrath (1981) have reported reduced integrity of DPPC/dicetyl phosphate liposomes on incorporation of cholesterol and application of a freeze–thaw cycle. Such information is essential in the development of freeze-drying cycles for the preparation of liposome formulations for long term storage.

12.4   EVALUATION OF PHOSPHOLIPID–DRUG INTERACTIONS

The interaction of small molecules such as drug substances with phospholipids can be investigated by DSC and can aid in designing liposomal formulations for drug delivery, and may also be used to evaluate how drugs might interact with the membranes of living cells.

12.4.1   Liposomal drug delivery systems

Hydrophilic water-soluble drugs are likely to concentrate in the aqueous compartments between the phospholipid bilayers and as such are unlikely to perturb the organization of molecules in the bilayers. Lipophilic molecules are likely to interact with the bilayers and disturb their organization. Bilayer organization might be assessed from DSC curves which are thus useful in assessing drug–phospholipid interactions in liposomes. Jain & Wu (1977) have indicated the nature of a variety of small molecule–phospholipid interactions by DSC, thus underlining this approach, although the compounds they utilized were mostly not drug substances. Over 90 lipid-soluble compounds were examined. Modified transitions depended on the changes in bilayer fluidity and were characterized by their distribution in DPPC. Small molecules were accomodated in-between the acyl side-chains and their disruptive effect influenced the mode of packing in the hydrocarbon chains and the order–disorder transition. The disruptive effect is passed on to the chain next to it and so on to several molecules thus forming a co-operative unit (Jain & Wu, 1977). Molecules form an additive block to this transmitting system so that the changes in the membrane will be damped out and remain relatively local.

A number of workers have evaluated phospholipid–steroid interactions. This is as a result of interest in the use of steroid–containing liposomes in the local treatment of rheumatoid arthritis (Shaw *et al.*, 1976). Fildes & Oliver (1978) utilized DSC in formulation of liposomes containing cortisol-21-palmitate for this therapeutic purpose. As steroid concentration in the liposome formulation was increased the width of the DSC phase transition peak at half its height ($\Delta T_{1/2}$) increased to a maximum, then showed a decrease. This was interpreted as saturation of the liposome bilayer with steroid at the maximum, with subsequent separation of a new steroid-rich phase on further increase in steroid level. All studies were carried out on samples adjusted to 50% w/w water. Thus this type of approach might be used in devising liposome formulations of lipophilic drugs to ensure drug in excess of the capacity of the phospholipid carrier is not utilized.

Arrowsmith *et al.* (1983) studied steroid–phospholipid interactions further, investigating the effect of steroid ester chain length, and found an increasing extent of interaction as ester chain length increased. Both $\Delta T_{1/2}$ and T_o temperatures versus percentage cortisone ester incorporated were utilized to assess saturation of bilayer with drug and gave comparable results of 11.25 mole% cortisone hexadecanoate in DPPC liposomes. Plots of enthalpy of phase transition versus steroid ester level were not discriminating enough to define saturation level for the added steroid ester.

Both Fildes & Oliver (1978) and Arrowsmith *et al.* (1983) evaluated phase transitions in DSC curves in an attempt to determine the nature of the phospholipid––steroid interaction, with Fildes & Oliver (1978) suggesting a simple model of the acyl chain 'dipping' into the bilayer and the steroid nucleus associated with the bilayer but not excluded from it. Arrowsmith *et al.* (1983) suggest a simple model where some bilayer–steroid nucleus interaction may occur.

12.4.2 Mode of action of drugs affecting membranes

Thermal analysis has played an important role in assessing the mode of action of drugs with antimicrobial activity, especially antibiotics and disinfectants. Many of these exert their mode of action by disrupting the bacterial membrane. Although transitions in bacteria are documented (Miles *et al.*, 1986) they are too complicated to use in mechanistic studies and consequently artificial membranes containing various phospholipids are used. Although phosphatidylcholines are not found in bacterial cells they are very often used as models of drug activity. Polymixin B decreased the endothermic transition of DPPC at 42°C and removed it at 1:1 and 2:1 lipid–drug ratios (Pache *et al.*, 1972). ESR (electron spin resonance), NMR and ORD (optical rotary dispersion) showed that electrostatic interaction took place between the amino-groups of the antibiotic and the phosphate groups of the DPPC and that the tails of the polymixin penetrated into the lipid layer. Gramicidin S did not fully remove the transition but shifted it to lower temperatures (Fig. 12.8) (Pache *et al.*, 1972). This interaction was only electrostatic in nature. The gramicidin S–DPPC complex separated and broke into the bilayer. Although bacterial membranes do not contain DPPC, other negatively charged phospholipids are present. The interaction is similar and therefore DPPC proved a resonable model.

The pretransition of DPPC is removed by gramicidin A even at a 1:200 drug:lipid molar ratio (Chapman *et al.*, 1974). At higher concentrations the peak maximum shifted to lower temperatures and the energy of the transition was lowered. This suggests that packing of the DPPC polar groups is affected, the loss of energy indicating that the molecule interdigitated among the lipid chains preventing chain crystallization from occurring. Whilst the bulk of the lipid below the transition temperature was rigid, the lipid immediately adjacent to the gramicidin may be fluid (Chapman *et al.*, 1974).

Chlorothricin, when added to DPPC (Fig. 12.9), behaved similarly to polymixin with DPPC (Pache & Chapman, 1972) as did gramicidin A when added to DMPC and DPPC liposomes (Chapman *et al.* 1977). The enthalpy of the lipid transition decreased with increasing amounts of gramicidin A. At a molecular ratio of 20

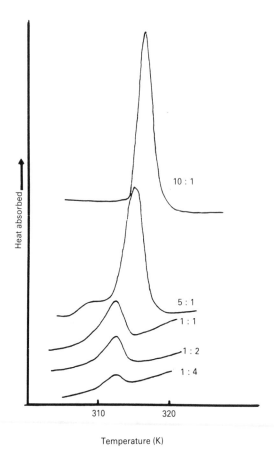

Fig. 12.8 — DSC curves for dipalmitoylphosphatidylcholine–gramicidin S mixtures in water, compared with pure dipalmitoylphosphatidylcholine. The ratios shown are dipalmitoylphos-phatidylcholine:gramicidin S. (Reproduced with permission from Pache *et al.*, 1972.)

lipids:1 polypeptide the width of the transition and its enthalpy were less sensitive to the futher addition of gramicidin A (Fig. 12.10). This corresponded to the onset of an aggregation process which produced localized polypeptide lipid clusters within the plane of the membrane (Chapman *et al.*, 1977). DLPC and DPPC mixtures displayed two transitions, equivalent to one region being rich in DLPC and the other DPPC. In the presence of gramicidin at a ratio of 1:50 gramicidin:lipid the lower endotherm was removed (Fig. 12.11). At higher concentrations an intermediate endotherm appeared. At polypeptide contents above 1:10 only one main endotherm was apparent. Slow cooling showed that gramicidin A at low concentrations moved preferentially into the lower melting regions of a bilayer containing DLPC and DPPC but at higher drug levels mixing of the two lipids took place (Chapman *et al.*, 1977). Interpretation was aided by Raman spectra, freeze-fracture electron micro-scopy, X-ray diffraction, and ESR studies. When the polypeptide is present at low concentrations and the temperature is lowered lipid chains, including those near to

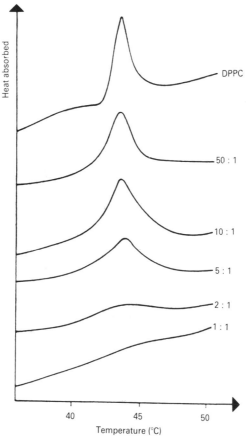

Fig. 12.9 — Endothermic transitions of dipalmitoylphosphatidylcholine (DPPC) and DPPC–chlorothricin mixtures. The ratios indicate the DPPC:chlorothricin ratio. (Reproduced with permission from Pache & Chapman, 1972.)

but not immediately next to the gramicidin, crystallize (Chapman *et al.*, 1977). This disturbs the lipid chains in the intermediate environment causing the observed reduction in total lipid enthalpy and increasing the transition width. Packing faults, adjacent to and radiating from the gramicidin, occur. These faults close some distance from the gramicidin A. Increasing its concentration leads to a higher density in the plane of the bilayer and increases random gramicidin–gramicidin contacts. When the temperature is lowered and the lipids crystallize the number of packing faults increase which leads to the exclusion of gramicidin from the crystallizing lipids forming clusters and aggregates. Exclusion of gramicidin from the crystallizing higher melting lipid concentrates the polypeptide into the lower melting region of the bilayer.

The study of Ikeda *et al.* (1984) is an excellent example of the role of thermal analysis in mechanistic studies of antimicrobial drugs. It compared the activity of poly(hexamethylenebiguanide hydrochloride) (PHMB) with its monomer diamino-hexylbiguanide hydrochloride (DAHB). The possible sites of activity of PHMB are

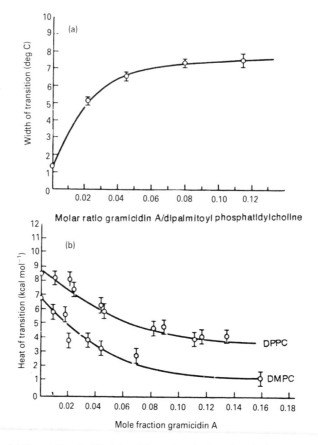

Fig. 12.10 — (a) The width at half-height of the main calorimetric endothermic as a function of the mole fraction of gramicidin A–dipalmitoylphosphatidylcholine. (b) The enthalpy (ΔH) of the main calorimetric endotherm as a function of the mole fraction of gramicidin A in dipalmitoylphosphatidyl choline (DPPC) and dimyristoylphosphatidylcholine (DMPC) dispersions in excess water. (Reproduced with permission from Chapman *et al.*, 1977.)

membrane bound proteins and phospholipids. The neutral lipid phosphatidyletha-nolamine constitutes 80% of total lipids in *Escherichia coli* and the acidic phospholipid, diphosphatidyl glycerol (DPG) and its dimer, cardiolipin are present to 10%. PHMB and DAHB were added to liposomes of DPPC, DPPE, PE or PG. The DPPC dispersion gave a sharp endotherm with T_m at 44°C. The presence of 20% PHMB did not significantly change T_m or peak shape. DPPE gave an endotherm with T_m at 56°C. In DPPE bilayers containing 20% the T_m was reduced by 1° to 55°C (Ikeda *et al.*, 1984). An egg PE dispersion was less complicated than that of DPPE due to the variety of acyl chains. Its T_m was 16°C and 20% w/w PHMB induced a depression in T_m by 2°C.

Egg PG dispersion gave a broad endotherm at $T_m = -5°C$, with PG alcyl chain length heterogenicity giving a broad phase transition. Twenty per cent PHMB depressed T_m to $-15°C$ and caused precipitation. DAHB caused depression to

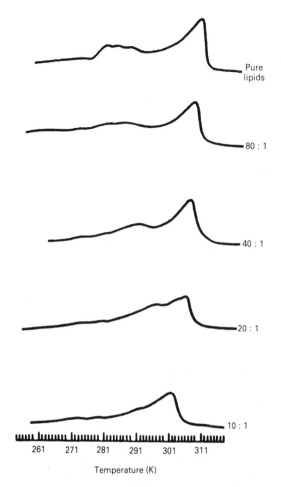

Fig. 12.11 — The calorimetric heating curves for equimolar mixtures of dipalmitoylphosphatidylcholine and dilauroylphosphatidylcholine with the indicated amounts of lipid to gramicidin A. The samples were in 50% ethylene glycol. The heating rate was 5 K min^{-1}. (Reproduced with permission from Chapman *et al.*, 1977.)

– 15°C but no precipitation. Mixtures of DPPC and egg PG gave endotherms with peak temperatures intermediate between the pure lipids, e.g. 50%:50% DPPC:egg PG gave a peak at $T_m = 27$°C and was used as a model system. 17% PHMB raised the T_m from 27 to 32°C and produced a second endotherm with T_m at – 15.5°C (Fig. 12.12). This corresponds to PHMB added to pure PG dispersion. Therefore PHMB produced isothermal phase separation of the mixture into a PHMB–PG complex domain and a DPPC-enriched domain. DPPC molecules are excluded from the PHMB–PG complex domain. DAHB (17%) only shifted T_m to a slightly lower temperature and no new peaks were formed, indicating phase separation was not induced. Thus PHMB has a large effect on negatively charged bilayers as compared with neutral lipid bilayers and interacted with negatively charged species in mixed bilayers of neutral and acid phospholipids.

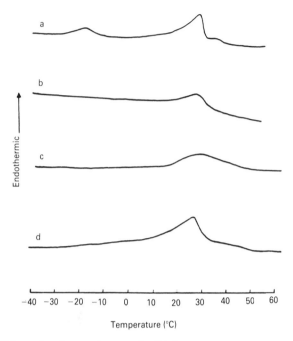

Fig. 12.12 — Effect of various cations on the DSC scans of a mixed lipid membrane of dipalmitoylphosphatidylcholine (DPPC) and egg phosphatidylglycerol (PG) (50:50 (w/w). a, 2 mg PHMB; b, 2 mg DAHB; c, 2 mg MgCl$_2$.6H$_2$O; d, none. The membrane consists of 5 mg DPPC and 5 mg egg PG dispersed in 30 μl of Tris-HCl buffer/ethylene glycol mixture (1:1, w/v) at pH 7.4. (Reproduced with permission from Ikeda *et al.*, 1984.)

Similar studies by Wang *et al.* (1984) examined interactions of neomycin with anionic phospholipid–lecithin liposomes. The DSC scan of phosphatidylinositol biphosphate (PIP$_2$)–DSPC liposomes was complicated giving a T_m greater than either pure component, probably due to hydrogen bonding. These forces probably stabilize the gel state more than the liquid crystalline state. Addition of neomycin induced downward shifts in the T_m which was considered to be due to the formation of a strong ionic complex between neomycin and PIP$_2$. This indicated that neomycin fluidized the bilayer and was related not to antimicrobial activity but to aminoglycoside toxicity.

A few other studies on the mode of action of non-antimicrobial drugs require mention. Berleur *et al.* (1985) examined DPPC liposomes in the absence and presence of 10% vinblastine using DSC. DPPC liposomes possessed a pretransition peak with T_o at 32.5 ± 0.25°C and T_m at 36°C with enthalpy of 5.8 ± 0.8 kJ/mole and a main transition with T_m at 42 ± 0.25°C and an enthalpy of 34.2 ± 0.8 kJ/mole with co-operativity of about 150. Vinblastine abolished the pretransition and the major peak T_m occurred at 41.25 ± 0.25°C with $\Delta T_{1/2}$ of 2.5°C and co-operativity of 37. A small peak was apparent ending at 45°C. The molar enthalpy of the main transition was about 34.7 ± 0.8 kJ/mole. The co-operativity in the presence of vinblastine was decreased about four times from 150 units. Slight modification in the gel state was detected by electron spin labelling but not DSC and was associated with disruption of the DPPC polar heads.

Berleur *et al.* (1984) similarly examined disruption of DPPC bilayers with the neurotropic drug isaxonine. A 1 mol:100 ml isaxonine; DPPL the pretransition was shifted from 32 to 34.5°C, its peak area trebled and the co-operativity was increased to 200. The main transition weakly shifted upward by 0.5°C. At the 30:100 ratio only one transaction occurred with T_o at $35 \pm 0.25°C$, T_m at $36.5 \pm 0.25°C$ and co-operativity reduced to 31. Domain formation was indicated by ESR studies. At low concentrations isaxonine reinforced the structural organization of the membrane as indicated by the pretransition amplitude and the main transition cooperativity. ESR labels indicated more accurately the mode of interaction of isaxonine with the different regions of DPPC. An increase in rigidity of the DPPC bilayer occurred near its polar head group in the gel. At 30% mol/100 ml disorganization was demonstrated by a shift down in transition temperature, i.e. broadening, and decrease in co-operativity.

REFERENCES

Alonso, A., Restall, C. J., Turner, M., Gomez-Fernandez, J. C., Goni, F. M. & Chapman, D. (1982) *Biochim. Biophys. Acta*, **689**, 283–289.

Arrowsmith, M., Hadgraft, J. & Kellaway, I. W. (1983) *Int. J. Pharm.*, **16**, 305–318.

Berleur, F., Roman, V., Jaskierowicz, D., Leterrier, F., Esanu, A., Braquet, P., Ter-Minassian-Saraga, L. & Madelmont, G. (1984) *Biochem. Pharmacol.*, **33**, 2407–2417.

Berleur, F., Roman, V., Jaskierowicz, D., Daveloose, D., Leterrier, F., Ter-Minassian-Saraga, L. & Madelmont, G. (1985) *Biochem. Pharmacol.*, **34**, 3081–3086.

Blume, A. (1988) Applications of calorimetry to lipid model membranes, Chapter 3 in *Physical properties of biological membranes and their functional implications* Hidalgo, C. (ed.) Plenum Press, New York, 71–121.

Chapman, D., Williams, R. M. & Ladbrooke, B. D. (1967) *Chem. Phys. Lipids*, **1**, 445–475.

Chapman, D, Urbina, J. & Keough, K. M. (1974). *J. Biol. Chem.*, **249**, 2512–2521.

Chapman, D., Cornell, B. A., Eliasz, A. W. & Perry, A. (1977) *J. Mol. Biol.*, **113**, 517–538.

Fildes, F. J. T. & Oliver, J. E. (1978) *J. Pharm. Pharmacol.*, **30**, 337–342.

Findlay, E. J. & Barton, B. G. (1978) *Biochem.*, **17**, 2400–2405.

Gruenwald, B., Stankowski, S. & Blume, A. (1979) *FEBS Letters*, **102**, 227–229.

Hinz, H-J. & Sturtevant, J. M. (1972) *J. Biol. Chem.*, **247**, 6071–6075.

Ikeda, T., Ledwith, A., Bamford, C. H. & Hann, R. A. (1984) *Biochem. Biophys. Acta*, **769**, 57–66.

Jain, M. K. & Wu, M. W. (1977) *J. Memb. Biol.*, **34**, 157–201.

Janiak, M. J., Small, D. M. & Shipley, G. G. (1976). *Biochem.*, **15**, 4575–4580.

Ladbrooke, B. D. & Chapman, D. (1969) *Chem. Phys. Lipids*, **3**, 304–356.

Ladbrooke, B. D., Williams, R. M. & Chapman, D. (1968) *Biochim. Biophys. Acta*, **150**, 333–340.

Mabrey, S. & Sturtevant, J. M. (1976) *Proc. Nat. Acad. Sci., U.S.A.*, **73**, 3862–3866.

Mabrey, S. & Sturtevant, J. M. (1977) *Biochim. Biophys. Acta*, **486**, 444–450.

Mabrey, S. & Sturtevant, J. M. (1978) *Methods Memb. Biol.*, **9**, 237–274.

Melchior, D. L. (1982) Lipid phase transitions and regulation of membrane fluidity in prokaryotes. Chapt. 3 in *Current topics in membranes and transport*, **17**, Razim, S. & Rottem, S. (ed.), Academic Press, 263–316.

Melchior, D. L. (1986) *Science*, **234**, 1577–1580.

Miles, C. A., Mackey, B. M. & Parsons, S. E. (1986) *J. Gen. Biol.*, **132**, 939–952.

Morris, G. J. & McGrath, J. J. (1981) *Cryobiology*, **18**, 390–398.

Offringa, J. C. A., Plekkenpol, R. & Crommelin, D. J. A. (1987) *J. Pharm. Sci.*, **76**, 821–824.

Oritz, A. & Gomez-Fernandez, J. C. (1987) *Chem. Phys. Lipids*, **45**, 75–91.

Pache, W. & Chapman, D. (1972) *Biochim. Biophys. Acta*, **255**, 348–357.

Pache, W., Chapman, D. & Hillaby, R. (1972) *Biochim. Biophys. Acta*, **255**, 358–364.

Phillips, M. C., Ladbrooke, B. D. & Chapman, D. (1970) *Biochim. Biophys. Acta*, **196**, 35–44.

Privalov, P. L., Plotnikov, V. V. & Filimonov, V. V. (1975) *J. Chem. Thermodyn.*, **7**, 41–47.

Schullery, S. E., Seder, T. A., Weinstein, D. A. & Bryant, D. A. (1981) *Biochem.*, **20**, 6818–6824.

Shaw, I. H., Knight, C. G. & Dingle, J. T. (1976) *Biochem. J.*, **158**, 473–476.

Steim, J. M., Tourtellotte, M. E., Reinert, J. C., McElhaney, R. N. & Rader, R. L. (1969) *Proc. Nat. Acad. Sci., U.S.A.*, **63**, 104–109.

Van Dijck, P. W. M., Kaper, A. J., Oonk, H. A. J. & De Gier, J. (1977) *Biochim. Biophys. Acta*, **470**, 58–69.

Vaughan, D. J. & Keough, K. M. (1974) *FEBS Letters*, **47**, 158–161.

Wang, B. M., Weiner, N. D., Ganesan, M. G. & Schacht, J. (1984) *Biochem. Pharmacol.*, **23**, 3787–3791.

13

The use of thermal analysis in freeze-drying

13.1 INTRODUCTION

Freeze-drying, otherwise termed lyophilization, is a process whereby water is removed from a frozen solution by sublimation to leave a dry porous solid that is easy to redissolve and is often used to produce a dry solid for injection for reconstitution or antibiotic powders intended to be resuspended or dissolved for oral use. Thermal analysis is used in the optimization and characterization of freeze-drying cycles and their products. The principles behind freeze-drying need to be fully understood in order to optimize the process. Prior knowledge of the position of the eutectic is essential since it governs the temperature at which a solution must be held for efficient freeze-drying. This temperature should be low enough to prevent simple melting yet high enough to allow sublimation to proceed at an economic rate (DeLuca & Lachman, 1965). Often a lyophilization cycle is developed without knowledge of the eutectic temperature and a temperature below −35°C is routinely used for freeze-drying of pharmaceutical products (DeLuca & Lachman, 1965).

Thermal analysis has various uses in freeze-drying. These are examination of the phase equilibria by DSC or DTA of a drug(s) and possible excipients with water to determine the eutectic temperature of the formulation to be freeze-dried, and the determination of drug fusion (since freeze-drying commences with a rapid freezing process a glassy solid may be formed on the completion of drying, or a change in the polymorphic form of the drug may occur). Additionally TG has application to the quantification of residual solvent in the final product and electrical resistance studies to the determination of phase equilibria.

13.1.1 Theory of freeze-drying

An understanding of the phase diagram of water is vital to the understanding of the freeze-drying process and how the presence of a solute may alter the processing temperatures. The fundamentals of the phase diagram (Fig. 13.1) were outlined by Travers (1988). The boiling point of water is lowered by a reduction of the external

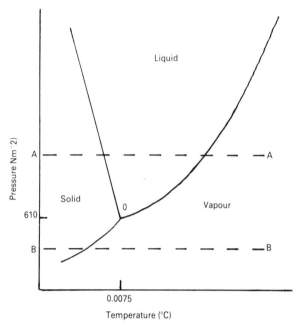

Fig. 13.1 — The phase diagram for water. The explanation of the symbols is given in the text.

pressure above the water, the melting point of ice slightly increases as the external pressure above it reduces and the vapour pressure exerted by the ice reduces as the temperature is lowered. The enclosed areas each correspond to a single phase but two phases coexist along the boundary lines. The point 0 is unique and represents that composition of temperature and pressure when all three phases co-exist. It is known as the triple point and occurs at a pressure of 610 Nm^{-2} and a temperature of 0.0075°C.

Consideration of two pressures in Fig. 13.1 indicates the principles of freeze-drying. Increasing the temperature at constant pressure corresponding to point 'A', a pressure above the triple point, produces at the first phase-boundary line water–liquid and then, once the second line is reached, water–vapour. A similar temperature increase from point 'B', a pressure below that corresponding to triple point, results in crossing one phase boundary only and the conversion of the solid ice directly into the vapour phase, a process known as sublimation. It is the aim of freeze-drying to accomplish this direct transition from the solid state into the vapour phase. This heat of vaporization is considerably greater than the heat of vaporization at atmospheric temperature and therefore some heat is required to effect drying. Solutes lower the temperature of the triple point to below 0.0075°C and freeze-drying is accomplished at a temperature below the triple point, usually in the range −30 to −10°C.

13.1.2 The freeze-drying cycle
The cycle consists of several stages. Initially the samples are frozen, often under rapid rotation and as rapidly as possible to produce a shell of very small ice crystals

coating the inside of the container. A vacuum is applied to reduce the pressure to below the triple point. With the careful application of heat large volumes of water vapour are produced and removed giving a porous solid. This forms a part of the cycle known as primary drying. The solid contains a low quantity of moisture (about 0.5%) which must be removed during secondary drying. This consists of raising the temperature up to 50–60°C and, with the use of a suitable desiccant, produces a dry, stable product. Thermo-labile products should not be affected by secondary drying due to the presence of the low residual moisture. Following freezing and the application of a vacuum the shelf temperature of a freeze-drier is increased somewhat to provide the energy for sublimation. The amount of heat has to be carefully controlled to avoid melting. To avoid foaming or puffing during freeze-drying the temperature used does not exceed that of the eutectic temperature for a crystalline solute or the collapse temperature of an amorphous solute.

A further problem may exist that, on cooling, the solute or eutectic may not crystallize out and a glassy deposit will remain in the ice-crystal matrix. The result of freeze-drying such a system will be the production of an amorphous product during a very long drying process. Raising the heat to allow recrystallization of the sample prior to reduction and sublimation may result in a crystalline product. The stability of the product may depend on the form of the drug in the completed product.

13.2 PHASE EQUILIBRIA DETERMINATION

The use of thermal analysis in phase equilibria is discussed in Chapter 2. A phase diagram representing the typical interactions between a salt and water is given in Fig. 13.2. The phase boundary AC represents the depression in the melting point of water caused by the addition of the salt. The line BC represents the solubility of the drug in water. Cooling liquid of composition 'X' results in ice formation at the temperature predicted by the intersection with line AC. Continual cooling forms more ice, giving an increase in the salt concentration in the residual solution until the line depicting the eutectic temperature is reached when the remaining liquid, corresponding to the eutectic composition, solidifies. The net result is the formation of a solid in which the eutectic is dispersed in excess ice crystals. A similar occurrence will take place when liquid of composition 'Y' is cooled. Initially once the line 'BC' is reached excess salt precipitates from the solution, until as the temperature is lowered through the eutectic temperature the eutectic finally crystallizes giving a dispersion of it in salt crystals of the salt. Cooling a liquid corresponding to the eutectic composition results in one rapid crystallization only when the eutectic freezes as a solid mass. Although theoretical relationships may be used to predict the eutectic temperature of a system it is advisable to determine their values experimentally since it is unlikely that a system intended to be freeze-dried will behave ideally. The eutectic temperature is not significantly effected by a change in pressure (DeLuca & Lachman, 1965).

DTA, DSC and the change of electrical resistance with temperature are suitable methods for determining the eutectic temperatures of systems intended for freeze-drying. Using a specially designed low conductance bridge DeLuca & Lachman (1965) defined the eutectic temperatures of several solute–water systems and found electrical resistivity to be a more sensitive method than DTA. Plots of the log resistivity against temperature produced very sharp inflections corresponding to the

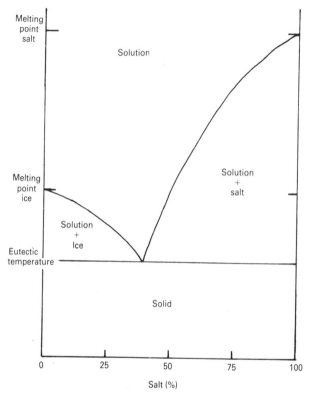

Fig. 13.2 — Typical phase equilibria for a salt and water showing a simple eutectic.

eutectic temperature. Plots, such as Fig. 13.3, gave values of -21.6, -12.9 and $-11.1°C$ for the eutectic temperatures for aqueous solutions of sodium chloride, potassium bromide and potassium chloride respectively.

Whilst the phase equilibrium of inorganic salts with water is simple, organic materials may provide complications. The measured resistance of a drug solution may be a function of the heating or cooling rates employed (Fig. 13.4). Rapid thawing involved raising the temperature from -35 to $-2°C$ in 20 min whereas slow warming required $1\frac{1}{2}$–2 hours. Rapid warming masked the eutectic point possibly due to either poor instrument sensitivity or the rapid temperature changes masking the endothermic heat of reaction at the eutectic point. Electrical resistivity measurements give higher estimates of the eutectic temperature than DSC or DTA (DeLuca & Lachman, 1965). The discrepancies reflect the greater inaccuracy of freezing curve analysis and indicate that the more reliable data of resistivity should be used in developing lyophilization cycles.

This early study of DeLuca & Lachman (1965) highlighted some very important points. Other factors were considered to be important in the cycle such as the degree of supercooling and the thermal conductances of the frozen mass. A eutectic temperature of $-10°C$ and a supercooling of $10°C$ would necessitate the use of

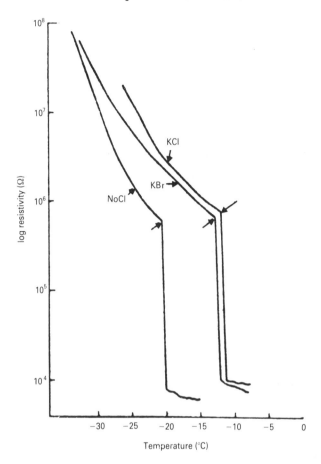

Fig. 13 3 — Determination of the eutectic temperature by plotting log resistivity as a function of temperature. Values at indicated points on curves are; NaCl, −21.6°C, KBr, −12.9°C and KCl, −11.1°C. (Reproduced with permission from DeLuca & Lachman, 1965.)

temperatures less than −20°C to avoid frothing and prevent problems at higher temperatures due to water removal from a liquid rather than a solid. Alternatively the product may be cooled to the lower temperature and then raise the temperature to −15°C to allow crystallization (DeLuca & Lachman, 1965).

Ito (1970a) used microscopy to follow the changes occurring on freeze-drying and electrical resistances to predict lyophilization schedules. On cooling, ice crystals were apparent once the liquidus temperature was reached. The changes visualized were classed as (1) no change, (2) a slight but not sharp increase in opacity, (3) the formation of clear spots which were dark opaque in transmitted light and which possessed only low growth rates or (4) similar spots with very high growth rates. Examples of class 4 are frozen samples of sodium chloride, potassium chloride and acetic acid (Ito, 1970a). γ-Aminobutyric acid (GABA) was an example of class 3 behaviour when low growth rates were apparent at −40 to −20°C. The temperatures at which the spots disappeared on heating corresponded to the eutectic

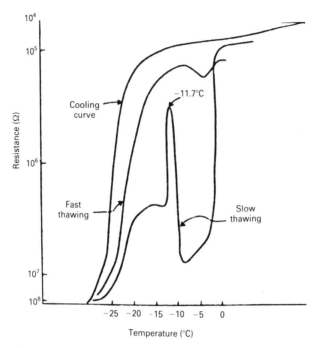

Fig. 13.4 — Cooling and warming curves for a 0.3 molal methylphenidate hydrochloride solution (0–10^8 Ω decade) (reproduced with permission from DeLuca & Lachman, 1965).

temperatures. These were also detected (Ito, 1970a) by electrical resistance changes on heating and cooling. The eutectic temperatures of solutions of strong electrolytes (e.g. NaCl) appeared as sudden decreases in resistance in the warming curves but for weak electrolytes and non-electrolytes the changes occurred slowly. The resistance curve of sucrose solutions (Fig. 13.5) is representative of the behaviour shown by mannitol, sorbitol, glucose and lactose solutions which displayed large increases in resistance near 0°C on rewarming. This may have been due to recrystallization of ice but no further explanation was apparent from microscopy studies. The electrical resistance of 10% GABA (Ito, 1970a) displayed a characteristic resistance increase on warming at about −30°C and then a sharp drop at around −20°C. The maximum, absent at fast heating rates, was attributed to the formation of a new phase or eutectic crystals and was representative of recrystallization. Ito (1970a) used microscopy to evaluate the collapsing temperature of freeze-dried systems. When the temperature exceeded this temperature puffing or collapsing of the sample occurred. Its value was characteristic and reproducible. The values approximated to the values of resistance changes and the disappearance of spots for the materials which could be classed as groups (3) and (4) but the maximum resistance temperatures of groups (1) or (2) did not correlate with the collapsing temperatures. Sucrose, glucose and lactose solutions represent group (1) behaviour and gelatine and PVP were characteristic of group (2) (Ito, 1970a). These latter systems are characterized by an ability to

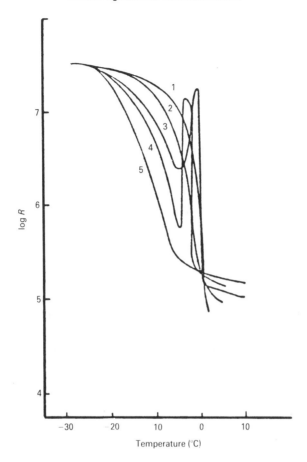

Fig. 13.5 — Variation of electrical resistance (R) of sucrose solution during rewarming. Key; 1: 1%, 2: 5%, 3: 10%, 4: 25%, 5: 50% sucrose. (Reproduced from Ito, 1970a, with permission of the publishers.)

supercool and do not recrystallize until subsequent reheating. For many the temperature of recrystallization during heating was close to that of the collapsing temperature. The temperatures were dependent on the visco-elastic properties of the solutions (Ito, 1970a). These properties prevented crystallization on cooling giving a supercooled liquid dispersed within an ice crystal meshwork. On lyophilization once a softening of the supercooled liquid occurs (as the ice sublimes) the matrix collapses. It is inevitable that all systems will display at least a modicum of supercooling when they are cooled and Ito (1970b) considered that between 5 and 15°C below the eutectic was the optimum range below the eutectic to promote crystallization. Electrical resistance measurements do not give an accurate estimate of crystallization rate.

Patel & Hurwitz (1972) considered that both DTA and DSC were suitable methods in the evaluation of suitable solvents to freeze-dry drugs. They examined the ability of sodium chloride to modify the eutectic temperatures of several buffer systems to render them suitable for freeze-drying. Fig. 13.6, representative of their

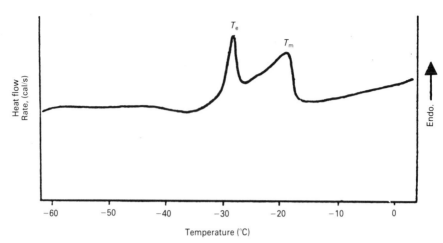

Fig. 13.6 — DSC scan of system B (drug–sodium citrate–lactose 10:5:85) with 9.8% sodium chloride and 56.71% w/w water (reproduced with permission from Patel & Hurwitz, 1972).

findings, is a DSC scan of a drug:sodium citrate:lactose (10:5:85) system with 9.8% w/w sodium chloride and 56.71% water. Both the eutectic and melting point transitions are apparent. As the water content in the sample increased the latter endotherm disappeared until only the eutectic endotherm was present. The phase

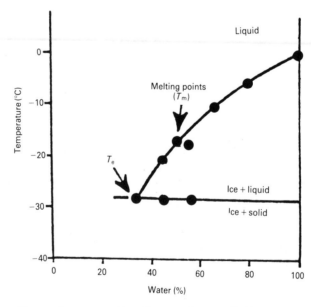

Fig. 13.7 — Temperature–composition diagram of system B (drug–sodium citrate–lactose 10:5:85) containing 9.8% sodium chloride (reproduced with permission from Patel & Hurwitz, 1972.)

diagram of the system is indicated in Fig. 13.7. In the absence of sodium chloride the endotherm corresponding to the melting of the eutectic was inapparent. This was also characteristic of the phase diagram for a drug : sodium citrate : lactose : sodium acetate system $(10:5:75:10)$ containing 5.36% sodium chloride (Fig. 13.8). No

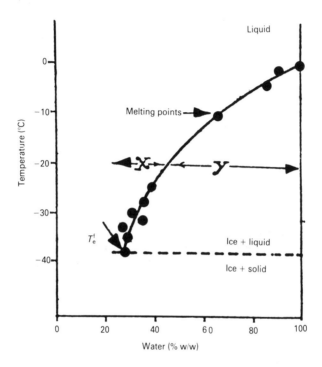

Fig. 13.8 — Temperature–composition diagram of system A (drug–sodium citrate–lactose–sodium acetate $10:5:75:10$) containing 5.36% sodium chloride (reproduced with permission from Patel & Hurwitz, 1972.)

endotherm corresponding to the eutectic was seen and that corresponding to the melting ice disappeared as the supposed eutectic point was reached. This composition contained no freezable water and represented the eutectic temperature (Patel & Hurwitz, 1972). Although the use of all these adjuncts may seem excessive, many components are often required to bulk up the drug or for isotonicity, and the formulator must understand the influence of these added ingredients on the percentage melted at any given temperature. Fig. 13.8 also indicates how the percentage melted may be calculated, which is represented by equation (13.1).

$$\text{Percentage at given temperature} = 100x/(x+y) \qquad (13.1)$$

Patel & Hurwitz (1972) added sodium chloride to three buffer systems in an attempt to increase their eutectic temperatures. System A (drug : sodium citrate : lactose : sodium acetate system $(10:5:75:10)$), system B (drug : sodium citrate : lactose $(10:5:85)$) and.system C (drug : sodium citrate $(75:25)$) had eutectic temperatures of -30, -22 and $-29°C$ respectively. Patel & Hurwitz (1972) calculated that although the eutectic temperatures were similar for systems A and C they differed considerably in terms of the molten fraction at given temperatures. Thus at $-20°C$, 6.7% and 9.5% fractions were melted for systems A and C respectively. Addition of sodium chloride to system A decreased the eutectic temperature. However the eutectic temperature of sodium chloride was masked by the buffer ingredients and no improvement in lyophilization properties was predicted (Patel & Hurwitz, 1972). A gradual increase of sodium chloride to system B initially decreased the apparent eutectic temperature to $-36°C$ but at higher concentrations the eutectic temperature increased. Addition of sodium chloride to system C resulted in an increasse in the temperature of and appearance of a true eutectic transition. Thus for this system the addition of sodium chloride would be expected to permit lyophilization at higher temperatures resulting in a shorter cycle. A high eutectic temperature and a low percentage melted at temperatures near the eutectic were desirable for successful freeze-drying vehicles (Patel & Hurwitz, 1972).

The debate of whether thermal analysis or resistance/resistivity measurements should be used in the characterization of systems intended for freeze-drying was developed by Jennings (1980a, b) utilizing the concept of D_2 values to describe phase transitions. Resistivity measurements were preferable to resistance measurements (Jennings, 1980a) because the latter is dependent on the dimensions of the cell and the resistivity of the ice product. Resistivity is an intrinsic bulk electrical property of the substance and is independent of the amount of the material. The D_2 ratio (Jennings, 1980a), represents the ratio of the resistivity of the product matrix to that of the resistivity of ice. High D_2 values indicate incomplete freezing, and a value of 1 represents complete freezing. Values lower than 1 mean that fewer ice surfaces for conduction are present than for ice at the same temperature (Jennings, 1980a). Fig. 13.9 shows the variation of the D_2 value plotted as a function of heat treatment. High values of D_2 during cooling indicate that the freezing process was incomplete (Jennings, 1980a). Line B implies that the matrix is completely formed but there is still conduction in interstitial regions. At point C the value of D_2 is unity and the matrix is completely frozen. The line D (warming) indicates the occurrence of a phase change. Extrapolation of this line back to $D_2=1$ gives an approximation of the melting point of the matrix. The diagram of the resistivity of distilled water indicated that the ice matrix, on cooling, did not completely form until about $-15°C$ (Jennings, 1980a). As the temperature approached $-35°C$ the resistivity of the ice began to decrease due to cracking within the ice, forming conduction boundaries. On warming the resistivity decreased still further. Recooling the same sample to $-43°C$ moved the sharp increase in resistivity to $-30°C$. The explanation appeared to be that a large number of conducting interfaces formed during the previous thawing process and fusion of these interphases, because of stress in the matrix, resulted in the sudden increase in resistivity. The resistivity changes on warming were the same on either treatment and were reproducible.

When the D_2 values are determined during the cooling treatment the specimen being examined must have identical treatment as the reference. For a potassium chloride matrix (Fig. 13.10) the matrix was not completely formed by $-20°C$

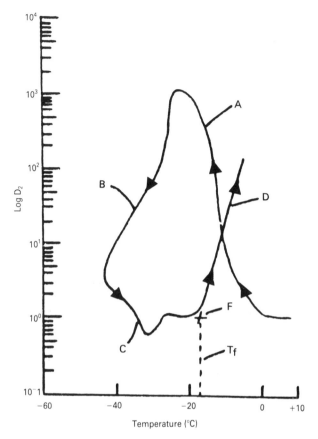

Fig. 13.9 — Illustration of log D_2 as a function of temperature (reproduced with permission from Jennings, 1980a).

(Jennings, 1980a). On warming the solution D_2 decreased below 1 indicating that there is less conducting surface in the matrix than in a pure ice matrix. Near $-13°C$ there was a large increase in the value. Extrapolation of the eventual straight line back to $D_2=1$ indicated a eutectic temperature of $-11.6°C$. A similar treatment for a mannitol matrix indicated that matrix formation was not complete until $-28°C$ (Jennings, 1980a). A D_2 value of 1 was not reached until $-43°C$. On warming, the mannitol matrix had a higher resistivity than the pure ice possibly because there are fewer ice boundaries in the mannitol matrix. Extrapolation gave a D_2 value of 1 at $-6.6°C$. Similar treatment on a retreated matrix gave a value of $-5.5°C$ and freezing was complete by $-15°C$.

Although DSC and DTA may determine heat change during a transition, neither method is particularly useful in evaluating the correct freeze-drying temperature since they are unable to determine when a product is fully frozen or the melting temperature in the absence of a eutectic (Jennings, 1980b). This neglects their importance in determining the glass transition temperature. Jennings (1980b)

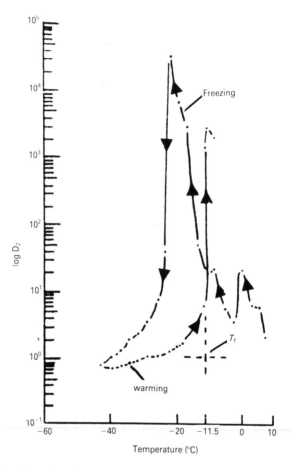

Fig. 13.10 — Log D_2 as a function of temperature for an ice–KCl matrix, determined with a
vertical cell, as a function of temperature during the initial freezing and warming cycle
(reproduced with permission from Jennings, 1980a).

described the design and use of a combined resitivity and DTA apparatus. The
criticism that thermal analysis does not detect interstitial water was raised by
Williams & Polli (1984) who considered that the formation of such water does not
involve the melting of a true eutectic and that the energy change is so small that it
does not register on a DSC or DTA scan.

Nail & Gatlin (1985) advocated resistance measurements rather than DSC or
DTA because the former technique determined more transitions, was suited for
derivative analysis and was found to be particularly suitable for organic solutes
which, unlike inorganic materials, did not show a sharp drop in the resistance curve.
Examples of the conversion in the resistance curve by derivative analysis are shown
in Figs. 13.11 and 13.12. The resistance curve for cytabarine showed little change
below −25°C, and between −25 and −10°C its slope was nearly constant, as
evidenced by a broad truncated peak in the derivative curve. The inflections in the

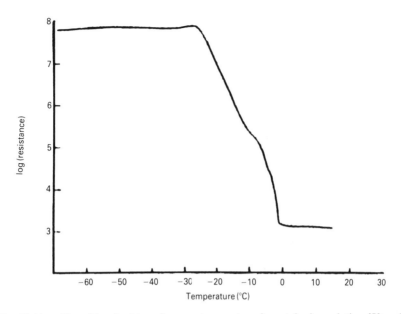

Fig. 13.11 — Plot of log (resistance) versus temperature for cytabarine solution (50 mg/ml) (reproduced with permission from Nail & Gatlin, 1985).

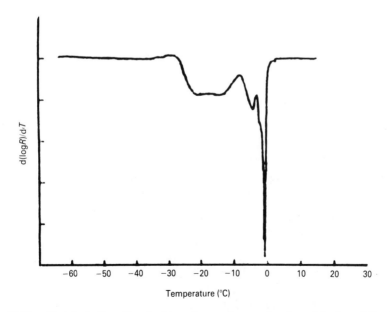

Fig. 13.12 — First derivative of log (resistance) versus temperature for cytabarine solution (50 mg/ml (reproduced with permission from Nail & Gatlin, 1985).

resitance curve at −10 and −4°C were very evident on the derivative plot although their meaning was unclear. Similar improvements on differentiation of the resistance data were achieved for methylprednisolone sodium succinate solutions. DSC of the cytabarine solution (Fig. 13.13) showed a small exotherm on the leading edge of the

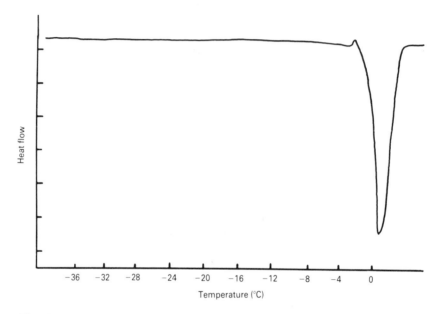

Fig. 13.13 — DSC scan for cytabarine solution (reproduced with permission from Nail & Gatlin, 1985).

melting endotherm but no transitions corresponding to the resistance changes were noted. DSC of the prednisolone derivative gave some transitions corresponding to resistance changes but a resistance change at −45 to −20°C did not have a DSC equivalent.

A simple approach to phase equilibria between a drug and water may not be realistic (Bogardus, 1982). Nafcillin sodium–water was a complicated system which included the presence of liquid crystals. DSC of the system could only be used for concentrations of 70% drug or below due to high viscosity and reproducibility problems. Typical data (Fig. 13.14) indicated that differences existed in the heating, cooling and reheating curves. For the 15% drug content a small endotherm at −5°C on the shoulder of the broad endotherm corresponding to the major ice portion melting was present on first heating. On cooling an exotherm existed due to recrystallization. On rewarming the small endotherm was still present but the major endotherm was narrower. A lack of melting point depression was attributed to micelle formation (Bogardus, 1982). In the 35% drug scans a slight melting point depression of the major endotherm was noted and although the small endotherm was present on the first run it was absent in the rewarming curve. The transition was due

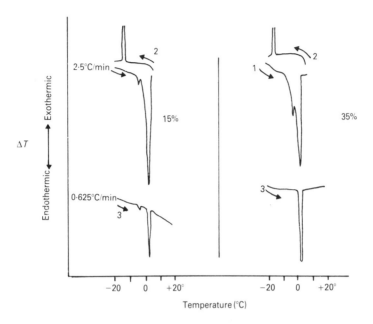

Fig. 13.14 — DSC scans for 15 and 35% nafcillin sodium–water mixtures. Scans 1 and 2 were
run at 2.5°C/min while scan 3 was obtained at 0.625°C/min. Prior to scan 1 the samples were
cooled from ambient temperature to −20°C at 2.5°C/minute. (Reproduced with permission
from Bogardus, 1982.)

to a metastable eutectic state (Bogardus, 1982). Addition of further nafcillin
produced even more complicated DSC scans. Microscopy was required to relate the
transitions to specific events and three crystalline phases and a lamellar meshophase
were apparent. On heating the 60% drug system an α-phase converted to a β-phase
at 9°C, which converted to the γ-phase at 22°C and which itself converted into the
lamellar liquid crystal phase at 27°C. Complete dissolution to the isotropic liquid
occurred at 30°C. The implications to freeze-drying (Bogardus, 1982) was that
following freezing warming was required prior to lyophilization to produce a
crystalline product. The phase changes may produce differences in the freeze-dried
product depending on the initial drug concentration. Temperature increases up to as
high as 20°C were required at certain drug levels to produce the correct crystal form.

13.2.1 Moisture content
The importance of thermal analysis in the detection and quantification of water in
freeze-dried products was highlighted by May *et al.* (1982) with reference to viral
vaccines. Limit for the residual moisture content of freeze-dried biological products
are usually determined by the loss on drying by gravimetric methods. TG was
compared with gravimetric and Karl–Fischer methods of analysis (May *et al.*, 1982).
Two TG methods were used. A TG(profile) method was employed when the samples
were heated from 23 to 250°C at 5°C/min. A 60°C hold method was used when
samples were heated from 23 to 60°C and then subjected to isothermal hold. This

temperature avoided sample charring. During the T_g(profile) method effluent samples were continuously analysed by mass spectrometry monitoring at ion intensities of the mass peaks at $m/e = 18$ and 44 which corresponded respectively to water and carbon dioxide. This differentiated between decomposition and water release. The two TG methods gave comparable results which confirmed data provided by the Karl–Fischer method. The temperature profile of a measles virus vaccine, (Fig. 13.15) showed a clearly defined weight loss between 23 and 110°C and

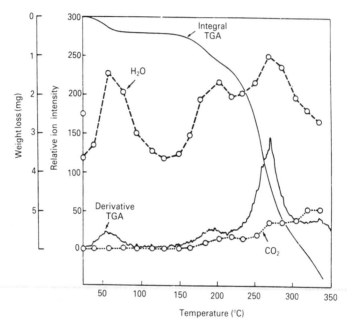

Fig. 13.15 — Temperature profile scan, a plot of weight loss versus temperature (°C), labelled 'Integral TGA' and mass spectral relative ion intensities (I) for water ($m/e = 18$) and carbon dioxide ($m/e = 44$) versus temperature for measles virus vaccine, live, attenuated freeze-dried in buffered sorbitol–gelatin. The derivative TG curve marks each change in slope in the integral TG curve (from May *et al.*, 1982).

its TG–mass spectrum indicated that it corresponded to water loss. Samples above 150°C showed weight loss attributable to sample degradation since both water and carbon dioxide were detected by mass spectroscopy. Once established that the weight loss at lower temperatures was solely due to evolved water and moisture content could be determined from TG data. For a similar measles vaccine, freeze-dried in a protein hydrolysate and sorbitol stabilizer, a clearly defined weight loss transition was not apparent up to 100°C. TG/MS indicated that at slightly higher temperatures both water and carbon dioxide were evolved due to decomposition but below this temperature moisture loss only occurred. The advantages (May *et al.*, 1982) of the TG method include the use of small sample size (less than 1–2 mg) and the rapid production of results (less than 30 min). Vial to vial variability in moisture content was large and due to location differences in the freeze-drier during their production.

May *et al.* (1986) extended their work to examine the residual moisture levels in non-viral-based freeze-dried vaccines. Results obtained from a combination of TG and MS were again comparable with those obtained by the Karl–Fischer method. MS showed that the weight lost in the analysis of typhoid vaccine up to 150°C was solely due to moisture loss but above this temperature decomposition was presumed due to the detection of carbon dioxide. Similarly three mass peaks due to water were observed for meningococcal polysaccharide vaccine when TG did not display a well defined weight loss. Mass peaks due to carbon dioxide were apparent as the higher transitions and consequently only weight loss up to 175°C was equivalent to residual moisture. Similarly sample decomposition for honey bee venom allergenic extract occurred above 150°C. MS was essential to determine which weight loss in ill-defined transitions are solely due to water (May *et al.*, 1986). The combined method was considered superior to Karl–Fischer methodology since not all biological products are soluble in the reagents normally employed and to gravimetric analysis where the pooled contents of up to 200 vials are often required.

13.2.2 Optimization

The product of a freeze-drying process should be in an acceptable form. For those products which do not form a eutectic, variations of temperature throughout a freeze-drier may result in problems such as melt-back and collapse. Since transitions may occur in the frozen state it is important to optimize conditions prior to the application of the vacuum. Additionally one particular polymorphic form of the drug may be the preferred form on the basis of stability. Rapid cooling of a system may produce glass-like structures within the pellet. Gatlin & DeLuca (1980) illustrated these problems with reference to three antiobiotics, cefazolin sodium, cephalothin sodium and nafcillin sodium. The former was available as a commercial freeze-dried product and cepahlothin was considered an unsuitable candidate for freeze-drying. Samples of the drugs in solution were frozen to −70°C at a rate of 10°C/min. prior to DSC analysis at rates between 0.625 and 10°C/min. The DSC scan of a typical cefazolin sodium solution (Fig. 13.16) depicts an endothermic shift (point B) at −20°C probably corresponding to the glass transition, an irreversible recrystalliza- tion endotherm beginning at point C at −11°C and ice melting commencing at −4°C (point F) (Gatlin & DeLuca, 1980). Fig. 13.16 illustrates the phase diagram of the thermal events for compositions up to 40% drug which presented the highest composition that could be readily prepared. The same composition as Fig. 13.16 but pretreated following cooling by warming to −6°C and recooling to −25°C presented a different DSC scan displaying only the transition due to ice melting. This indicated a conversion of the amorphous drug to a crystalline form (Gatlin & DeLuca, 1980). The product was amorphous without the pretreatment but following the treatment was crystalline. Similar trends occurred for nafcillin sodium solutions when DSC showed an initial endothermic event at −10°C, a reversible endotherm at −7°C and an irreversible exotherm at −4°C leading directly into the melting endotherm of ice. If the freeze-dried product was prepared following cooling below −10°C without thermal treatment an amorphous product was obtained but if the mass was warmed to between −10 and −4°C the product was crystalline (Gatlin & DeLuca, 1980).

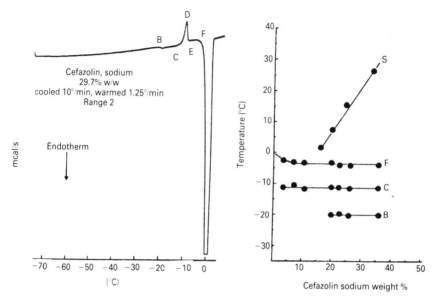

Fig. 13.16 — DSC scan for the warming of a frozen cefazolin solution and phase diagram of thermal events as a function of concentration (reproduced with permission from Gatlin & DeLuca, 1980).

Solutions of cephalothin sodium behaved differently. DSC of a representative sample showed a small endothermic event at −22°C and ice melting between −15 and −7°C (Gatlin & DeLuca, 1980). Thermal treatment did not effect these characteristics and a crystalline spray-dried product could not be produced by thermal treatment as predicted by thermal analysis. Gatlin & DeLuca (1980) emphasized however that thermal analysis does not predict that aseptic seeding may lead to a crystalline product.

Gatlin & DeLuca (1980) examined similarly treated mannitol solutions (mannitol is a common bulking agent for freeze-dried products). The frozen solution displayed a T_g at −65°C, another endothermic shift at −32°C, an exotherm at −29°C and an ice melting endotherm commencing at −4°C. This indicated that a glassy form of mannitol could only be prepared below the T_g. It is therefore possible to freeze-dry at any temperatutre below −4°C and produce a crystalline product; this should represent the effective upper freeze-drying temperature, the lower limit being −29°C.

A combination of electrical conductivity and DTA proved suitable in the optimization of sodium ethacrynate lyophilization (Yarwood et al., 1986). A combined cell containing 3 ml volumes (distilled water as reference) was used to simultaneously perform each analysis and the effect of drug concentration, sample volume and cooling rates were considered in the light of the known rapid degradation of ethacrynic acid in its amorphous form. A sample rapidly cooled to −140°C, warmed at 1.5°C/min and containing 4% w/v drug showed a T_g between −48 and

−14°C in both analytical traces but changes in the resistance curve occurred at −8°C which were not detected by DTA. Fig. 13.17 shows the DTA profiles of 0.5% w/v

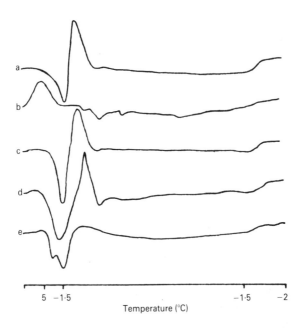

Fig. 13.17 — DTA scans of 0.5% w/v aqueous solutions of sodium ethacrynate under selected cooling and heating conditions.

Sample	Cooled to (°C)	Cooling rate (°C/ min)	Heating rate (°C/ min)
a	−140	20	1.5
b	− 50	0.5	0.2
c	− 45	1.5	1.5
d	−140	1.5	1.5
e	−140	20	0.2

(Reprinted from Yarwood *et al.* (1986) p. 2162, by courtesy of Marcel Dekker, Inc.)

solutions. Generally a shallow endotherm starting at −1.8°C was apparent often followed by an exotherm and the liquidus endotherm. Additional endotherms were present in the slow cooled–slow heated sample with no liquids endotherm but a large exotherm occurred at −1.5 to 5°C (Yarwood *et al.*, 1986). The resistance data displayed a two-step decrease except for the slow cooled–slow heated sample when a smooth resistance decrease with time was noted. The presence of glass transitions indicate that the form of the drug on processing was temperature-dependent (Yarwood *et al.*, 1986). Slowly frozen samples were more stable than the fast frozen products because the latter were amorphous. Larger fill volumes similarly produced a more stable product because of the difficulty in rapidly cooling the whole of the sample and thereby preventing the formation of a totally amorphous product. Additionally stability decreased with a lowering of drug concentration because the presence of 4% drug propagated nucleation and crystal growth, thereby favouring the crystalline product.

Phillips *et al.* (1986) produced a suitable freeze-drying cycle for drug E3816 which tended to form glasses when acetone was included in the vehicle. DTA showed that crystallization commenced at −62°C on warming the frozen solution but was slow due to the high viscosity of the solution. A eutectic endotherm was noted at −21°C and was accompanied by a fall in resistance. The successful cycle designed from this information involved cooling to −40°C to rapidly freeze, warming to −15°C to allow crystallization and then cooling back to the lower temperature to effect freeze-drying. Acetone remained as an interstitial liquid, was removed under vacuum, and induced crystallinity because the drug was soluble in aqueous acetone but not in the dry product. This aided crystallization because on cooling ice crystals formed, therby 'drying' the acetone, and once the solubility of the drug was exceeded it precipitated.

REFERENCES

Bogardus, J. B. (1982) *J. Pharm. Sci.,* **71**, 105–109.
DeLuca, P. & Lachman, L. (1965) *J. Pharm. Sci.,* **54**, 617–624.
Gatlin, L. & DeLuca, P. P. (1980) *J. Parent. Drug Assoc.,* **34**, 398–408.
Ito, K. (1970a) *Chem. Pharm. Bull.,* **18**, 1509–1518.
Ito, K. (1970b) *Chem. Pharm. Bull.,* **18**, 1519–1525.
Jennings, T. A. (1980a) *J. Parent. Drug Assoc.,* **34**, 109–126.
Jennings, T. A. (1980b) *Med. Device Diagn. Ind.,* **2**(11), 49–57.
May, J. C., Grim, E., Wheeler, R. M. & West, J. (1982) *J. Biol. Stand.,* **10**, 249–259.
May, J. C., Wheeler, R. M. & Grim, E. (1986) *J. Therm. Anal.,* **31**, 643–651.
Nail, S. L. & Gatlin, L. A. (1985) *J. Parent. Sci. Technol.,* **39**, 16–27.
Patel, R. M. & Hurwitz, A. (1972) *J. Pharm. Sci.,* **61**, 1806–1810.
Phillips, A. J., Yarwood, R. J. & Collett, J. H. (1986) *Anal. Proc.,* **23**, 394–395.
Travers, D. N. (1988) Drying, Chapter 38 in *Pharmaceutics, the science of dosage form design*, Aulton, M. E. (ed.) Churchill Livingstone, London, 629–646.
Wiliams, N. A. Polli, G. P. (1984) *J. Parent. Sci. Technol.,* **38**, 48–59.
Yarwood, R. J., Phillips, A. J. & Collett, J. H. (1986) *Drug Dev. Ind. Pharm.,* **12**, 2157–2170.

14

The use of thermal analysis in the evaluation of miscellaneous pharmaceutical processes

Thermal analysis has proved a useful tool in the evaluation of many pharmaceutical processes intended to provide the drug or excipient in a form acceptable to use. Freeze-drying has been discussed in the previous chapter. This chapter outlines the applications of thermal analysis to some miscellaneous processes such as spray-drying, spherical aggregation, co-grinding and solvent deposition.

14.1 SPRAY-DRYING

The major use of thermal analysis in evaluating spray-dried products is in the identification of the polymorphic or crystal form of the drug in the product since spray-drying may result in a polymorphic change. Spray-drying has many uses including the preparation of amorphous or high energy forms of a drug, the preparation of free-flowing excipients or granules prior to their tabletting, drying pharmaceutical powders, the preparation of microencapsulated agglomerates for parenteral administration and in the manufacture of sustained action tablets. Whenever the crystal form of the product has to be determined thermal analysis is used in conjuction with other established processes, e.g. X-ray diffraction and IR spectroscopy.

The use of spray-drying in the preparation of polymorphic forms of drug substances is exemplified by a study on the preparation and characterization of phenylbutazone polymorphs (Matsuda *et al.*, 1984). Careful control of the spray-drying conditions allowed modification of the polymorph obtained. Rather than spray-drying from aqueous suspensions or solutions, Matsuda *et al.* (1984) used solutions in methylene chloride. Crystallization was controlled by varying the temperature of the droplets sprayed from the atomizing nozzle of the spray-drier. A 5% solution was sprayed at inlet temperatures varying from 30 to 120°C. The corresponding outlet temperatures varied from 27 to 64°C. A combination of X-ray diffraction, IR spectroscopy, HSM, TG and DTA was used to characterize the product. TG indicated that no residual solvent was present and therefore no solvate was formed. X-ray diffraction indicated that decreasing the nozzle temperature resulted in a change of drug form and that probably two or three crystal forms were present. The sample prepared at 120°C was the δ-form, at 80 and 100°C was a mixture of two forms, and a third form was apparent at 70°C; at below 60°C the X-ray diffraction patterns were dissimilar to those from samples prepared at higher temperatures. Mixtures containing two forms contained the β- and δ-forms and when three forms occurred β-, δ- and ε-forms were present. Samples prepared at 100 and 120°C gave a DTA endotherm at 103°C (Fig. 14.1) equivalent to the δ-form. Samples

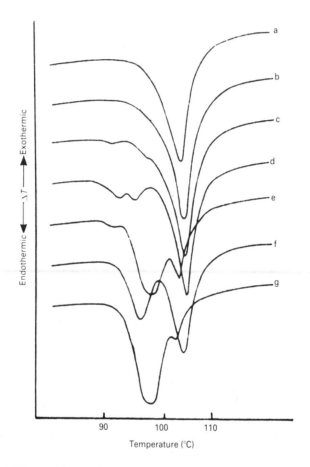

Fig. 14.1 — DTA scans of spray-dried samples of phenylbutazone prepared at drying temperatures of 120°C (a), 100°C (b), 80°C (c), 70°C (d), 60°C (e), 40°C (f) and 30°C (g). (Reproduced with permission from Matsuda *et al.*, 1984.)

obtained at 70 and 80°C displayed melting endotherms at 91–92°C which were followed by a recrystallization exotherm at 93°C equivalent to conversion to the δ-form, which melted at 103°C. A drying temperature of 70°C produced an extra endotherm at 95°C. At lower drying temperatures this transition became more apparent at the expense of the endotherm at 92°C (Fig. 14.1).

Therapeutically a mixture of polymorphs is undesirable because of their different dissolution rates and bioavailability and possible polymorphic interconversion. The later is a major problem in spray-dried products, especially when amorphous or glassy forms are produced. Corrigan and his coworkers have used the incorporation of polyvinylpyrrolidone (PVP) into the solution to be spray-dried as a retardant to crystallization. Typical of their studies were data on hydroflumethiazide (Corrigan & Holohan, 1984), bendrofluazide and hydrochlorothiazide (Corrigan *et al.*, 1984). Fig. 14.2 is illustrative of the changes induced by the spray-drying of hydroflumethia-

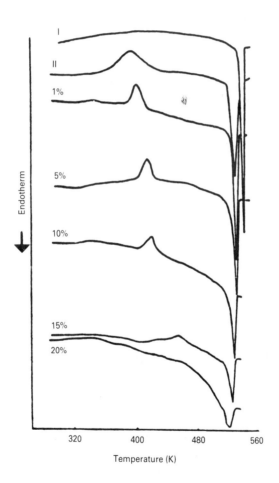

Fig. 14.2 — DSC scan of hydroflumethiazide spray-dried with PVP. I crystalline drug, II spray-dried drug. Percentage values are the percentages of PVP present in the final spray-dried material. (Reproduced with permission from Corrigan & Holohan, 1984.)

zide from ethanolic solutions. Spray-dried products displayed an exotherm corresponding to recrystallization of the amorphous sample. Its peak temperature increased with increasing levels of PVP. These values were considered to define the temperature at which spontaneous crystallization of the amorphous phase occurred and its increases suggested that PVP stabilized the transformation (Corrigan & Holohan, 1984). The area of this exotherm was representative of the stability of the system on storage at room temperature. It was lost after 10 days in the system containing no PVP and after six months in the system containing 1% PVP, but was not lost in systems with higher levels of PVP after one year, indicating physical stability. Application of the method of Theeuwes *et al.* (1974) (see section 8.3.3) gave an apparent solubility of 69% drug in PVP. Similar data (Corrigan *et al.*, 1984) indicated that, if analysed immediately after preparation, spray-dried cyclopenthiazide, polythiazide, bendrofluazide, cyclothiazide or hydrochlorothiazide passed through a glassy or amorphous state whereas spray-dried chlorothiazide formed a crystalline product. This was evidenced by a crystallization exotherm in the samples on DSC analysis before the melting endotherm. The sample of hydrochlorothiazide lost this exotherm within 24 hours showing that the sample was unstable but the remaining four drugs were stable in their amorphous or glassy form after 12 months. Bendrofluazide and polythiazide displayed small endothermic peaks corresponding to their glass transition (Corrigan *et al.*, 1984). Corrigan *et al.* (1985) similarly examined spray-dried indomethacin and naproxen with and without PVP. Indomethacin spray-dried from its ethanolic solution solidified to form a glass. DSC of the freshly prepared sample showed a recrystallization exotherm followed by melting endotherms of indomethacin form I and form II. Crystallization during storage was noted by a reduction in the endotherm corresponding to form II. Co-spray-drying with PVP maintained the amorphousness of the drug. DSC traces confirmed that naproxen spray-dried to a crystalline material. PVP reduced this crystallinity and samples containing 40% PVP were amorphous.

13.1.1 Microencapsulation

Spray-drying is sometimes used to microencapsulate drugs. Takenaka *et al.* (1981) used spray-drying to microencapsulate sulphamethoxazole with cellulose acetate phthalate (CAP), colloidal silica, montmorillonite or talc. IR spectroscopy suggested that the form I sulphamethoxazole was converted to form II by spray-drying with CAP. DSC of form II gave sharp endotherms at 166 and 171°C indicative of the transition of form II to form I and the melting of form I respectively. The spray-dried CAP–sulphamethoxazole samples displayed an exotherm at 140–150°C characteristic of drug crystallization from an amorphous form (Takenaka *et al.*, 1981). Hydrogen bonding between the drug and the polymer and a steric hinderance induced by the polymer probably restricted the intermolecular bonding of the sulphonamide inhibiting recrystallization. X-ray diffraction confirmed these findings. The effects of spray-drying slurries in either water or ammoniacal solutions were examined (Takenaka *et al.*, 1981). The talc product gave a DSC scan typical of the form II (Fig. 14.3) and the silica product resulted in a scan showing a broad endotherm at 150–160°C characteristic of amorphous drug. The montmorillonite clay scans were dependent on the pH of the solution used to prepare the spray-dried product (Takenaka *et al.*, 1981). Form I characteristics were noted in the product prepared

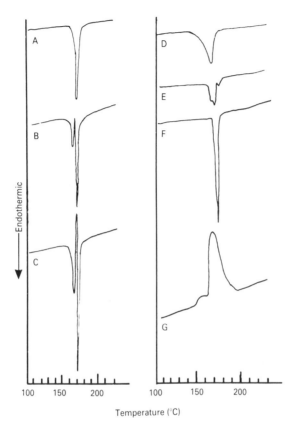

Fig. 14.3 — DSC scans for original sulphamethoxazole and spray-dried products. Key: A, form I; B, form II; and C–G, spray-dried products prepared from formulations containing 30 g of talc (C), 30 g of colloidal silica (D) and 30 g of montmorillonite clay (E–G). Media were 5% NH_4OH (C–E), distilled water (F) and pH 1.2 solution (G). (Reproduced with permission from Takenaka *et al.*, 1981.)

from distilled water. This was not suprising since sulphamethoxazole was relatively insoluble in water and form I was the starting material. Acidic media resulted in an exothermic peak at 164°C indicating that an interaction may occur between the drug and the surface of the clay. The ammoniacal slurry of the clay gave a similar DSC scan to that of talc microcapsules, the broadness being indicative of amorphousness (Takenaka *et al.*, 1981).

14.2 SPHERICAL AGGREGATION

Kawashima *et al.* (1982) developed a novel method of preparing spherical aggregates of salicylic acid crystals using a three-solvent system. Thus needle crystals of salicylic acid were formed from a mixture of three partially miscible liquids, such as water, ethanol and chloroform with agitation. Because of their spherical nature it was possible to directly compress the crystals directly into tablets. Spectroscopy and X-ray diffraction, but not thermal analysis, were used to confirm the absence of polymorphic modification and solvate formation. The work was expanded (Kawashima *et al.*, 1984) to prepare spherical agglomerates of aminophylline. This drug is a complex of theophylline and ethylenediamine and it is essential that each molecular species is present in the spheres. The thermal bahaviour of the crystals before and after treatment was examined by DSC. X-ray diffraction indicated that three forms of the aminophylline were present in the aggregates, the β-form being synonymous with aminophylline and the α and γ-forms were other crystallite forms of the complex. DSC traces of the spheres showed an endotherm at around 120°C corresponding to the release of ethylenediamine (Kawashima *et al.*, 1984). The water of crystallization was released in different ways depending on the crystalline form present. TG confirmed both these weight losses. It was suggested that the water content was 1 and 2.5 molecules of water of crystallization per molecule of complex for the β- and γ-forms respectively. These findings are summarized in Fig. 14.4 which

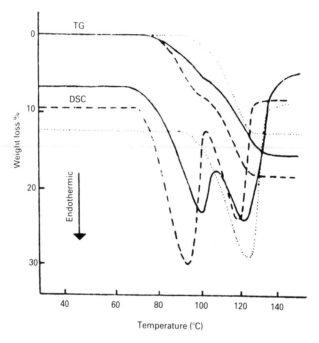

Fig. 14.4 — DSC–TG scans of agglomerates and aminophylline. Key: (......) α-form of agglomerates, (——) β-form of agglomerates, aminophylline; (–––) γ-form of agglomerates. (Reproduced with permission from Kawashima *et al.*, 1984.)

indicates that the α-form had a low water content. The molar water content was 0.5 mol or below.

14.3 CO-GRINDING

Several studies have examined ball-milling or grinding mixes of drugs and polymers
with a view to altering the polymorphic form of the drug. Giordano *et al.* (1985)
ground two unnamed drugs (A and B) with cross-linked PVP manually with a mortar
and pestle following blending in a Turbula apparatus. Fig. 14.5 indicates the

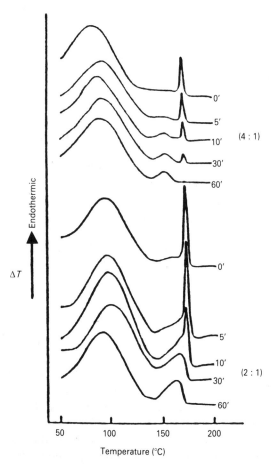

Fig. 14.5 — DSC traces of two polymer (Polyplasdone XL)–drug A mixtures recorded after
different grinding times: scan speed 10°C/min; start temperature 50°C; end temperature 210°C.
(Reproduced with permission from Giordano *et al.*, 1985.)

influence of grinding time on thermal analysis of two polymer : drug A ratios.
Immediately following blending a broad endotherm due to the polymer and the
endotherm due to the drug were apparent. On trituration the drug melting endotherm slowly decreased at the expense of the development of a smaller, broader
endotherm near 155°C indicating a polymorphic transformation. Similar findings

were noted for drug B although the newer endotherm produced was at least partially initiated by the thermal analysis itself.

Ball-milling (Ikekawa & Hayakawa, 1982) effected a change in amylobarbitone when accomplished in the presence of various diluents. This work followed an earlier study (Kaneniwa et al., 1978) which demonstrated that various particle size reduction methods such as ball-milling, automatic trituration with a mortar and pestle, mortar-grinding and stamp-milling produced both a decrease in particle size and an alteration of the crystal form of amylobarbitone. DTA scans of the pure drug, milled in the absence of diluents, were unchanged. Following milling in the presence of diluents such as activated charcoal, precipitated silica or ethylcellulose the DTA scan showed two or three endotherms with a concomitant decrease in intensity and a move to lower peak temperatures, indicative of the presence of at least two polymorphic forms and an increase in the amorphousness (Kaneniwa et al., 1978). IR spectroscopy and X-ray diffraction confirmed these findings. Ikekawa & Hayakawa (1982) ball-milled amylobarbitone at the 40% level with methylcellulose, at the 10 and 30% levels with microcrystalline cellulose and at the 30% levels with dextran using milling times of up to more than 200 hours. X-ray diffraction of the various carriers indicted a conversion of the initial form I material to the form III polymorph. The fraction of the latter form increased rapidly on milling and was nearly constant within 10 minutes. X-ray diffraction results were used to compute the fraction of crystalline material present with the ball-milled sample. The fraction was expressed as a function of results obtained from amylbarbitone that had been milled for 60 hours in the absence of diluents. This had an assumed crystallinity of 82%. This was compared with the fraction of crystalline material obtained from DSC measurements based on the heat of fusion, and calculated from equation (14.1)

$$\Delta H_{obsd}=29.3x_{cr}+5(1-x_{cr}) \tag{14.1}$$

where ΔH_{obsd} was the obtained heat of fusion and x_{cr} was the crystallinity. The values of crystallinity obtained by the two methods were comparable except in the case of the methylcellulose samples where the estimate based on thermal analysis was smaller. This was attributed to an interaction between amylobarbitone and this carrier. The fusion peak of amobarbital was broardened and changed into a double endotherm on ball-milling (Fig. 14.6). Five estimates of the thermal behaviour were used. T_1 was the start of drug fusion, T_2 and T_3 peak melting temperatues, T_3^* the temperature at the beginning of the shoulder and T_4 represented the end of fusion (Ikekawa & Hayakawa, 1982). T_1 was lowered by ball milling in the presence of diluents and was lower when the degree of crystallinity was less (Ikekawa & Hayakawa, 1982). The lowering of T_1 was greatest with methylcellulose and even occurred in physical mixtures. Similarly the values of T_3 and T_4 were lowered with methylcellulose but were relatively unaffected in the microcrystalline cellulose and dextran mixtures. The interaction of methylcellulose was also evidenced by pretreatment (Ikekawa & Hayakawa, 1982). Scanning up to 200°C, followed by cooling to room temperature and then reheating produced nearly identical scans for the dextran and microcrystalline cellulose mixes but in the case of the methylcellulose samples no singlet due to the drug appeared (Ikekawa & Hayakawa, 1982). Even

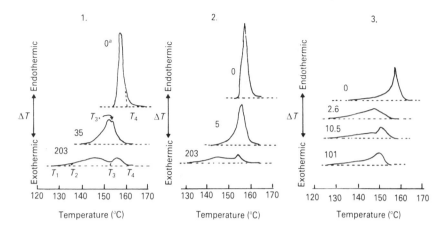

Fig. 14.6 — Influence of ball-milling with diluents on DTA scans of amylobarbitone. The content of amylobarbitone in the mixture with microcrystalline cellulose is 30%. Diluent: 1, dextran; 2, microcrystalline cellulose; 3, methylcellulose. The figure on each of the curves is the ball milling time (hours). (Reproduced with permission from Ikekawa & Hayakawa, 1982.)

storage at room temperature for a few days between the runs only generated small drug peaks. Therefore the polymer retarded the recrystallization of amylobarbitone.

14.4 SOLVENT DEPOSITION

Solvent deposition is the precipitation of a solute, i.e. the drug, onto a solid surface or matrix by evaporation of its solvent. It is possible that the polymorphic form of the drug precipitated onto the solid will not be the same ploymorphic form as the starting material. Consequently thermal analysis has found favour as a method of analysis of the systems.

Monkhouse and Lach (1972) examined the properties of several drugs solvent-deposited onto silicon dioxide. The drugs were analysed on the basis of the symmetry of the DSC curves either side of the melting transition. The slopes of the traces immediately before and after transition were calculated as a ratio. The ratio for pure drugs and their physical mixes with the carrier approximated to unity. Asymmetry was reflected as a high slope ratio. For seven drugs (hydrochlorothiazide, aspirin, sulphaethidole, chloramphenicol, oxolinic acid, griseofulvin and reserpine) symmetrical curves were noted for the physical mixes and asymmetrical curves for the solvent-deposited material (Monkhouse and Lach, 1972). The former was due to the narrow range of melting of the drug when present solely as large crystals. However, the drug in the deposited product was partly present as an amorphous form bound onto the silicon dioxide by forces such as Van der Waals forces and hydrogen bonds. The disparity in the energy required to disrupt these bonds was reflected in a wide melting range and consequently, following initial melting, the larger crystallities would commence to melt (Monkhouse and Lach, 1972). The overall effect was to depress the final melting point, i.e. the temperature at the peak apex, by 2–6°C.

During evaporation, the surface acted as a nucleating agent for these drug substances and, due to the high viscosity, very small crystallites would be favoured thereby leading to high dissolution rates. Their existence was reflected by the melting point depressions. For two further drugs (indomethacin and probucol) double melting endotherms were apparent suggesting the existence of two polymorphs. Increasing the content of silicon dioxide within the sample favoured the formation of the lower melting point polymorph. Such an occurrence may be due to a preferred molecular orientation of the drug in polymer solution being altered in the gel network produced by the silicon dioxide resulting, during evaporation, in the precipitation of the less stable polymorph at the surface of the silicon dioxide (Monkhouse & Lach, 1972).

Solvent deposition of spironolactone onto glass and lactose was examined using DTA as the analytical technique by Salole & Al-Sarraj (1985). Their data are represented in Fig. 14.7. Form III, deposited from acetone onto glass, displayed a

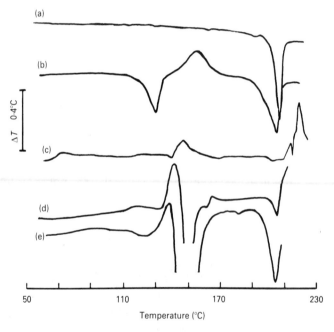

Fig. 14.7 — DTA curves of spironolactone deposited on glass from (a) acetone and (b) methanol, and on lactose from (d) acetone and (e) methanol; (c) DTA curve of lactose using an 'active reference'. (Reprinted from Salole & Al-Sarraj (1985) p. 2064, by courtesy of Marcel Dekker, Inc.)

single melting endotherm but the sample similarly deposited from methanol displayed an endotherm–exotherm complex transition prior to the final melting endotherm. This, following HSM analysis in silicone oil, was attributed to a desolvation–recrystallization reaction of the form D material which is a solvate (Salole & Al-Sarraj, 1985). Dissolution of this latter system lead to a supersaturation prior to precipitation of form I. Because of large lactose endotherms at 147 and 219°C the technique of using lactose as an active reference to cancel out these effects was promoted (Salole & Al-Sarraj, 1985). The lactose sample so run is also seen in Fig. 14.7. Even allowing for this correction the acetone sample showed a complex transition at 160°C concomitant with the formation of the poorly soluble form A at the lactose surface. This was due to moisture on the lactose sample surface. The formation of form D, not unexpected since methanol dehydrates lactose, was detected in the sample not by thermal analysis, since the complex reaction in the scan would be masked by the lactose, but by IR spectroscopy. Neither system produced supersaturation during dissolution (Salole & Al-Sarraj, 1985). However following particle size separation a 53 μm and below fraction showed only a low dissolution rate but a 150–210 μm fraction displayed a rapid dissolution rate and achieved supersaturation. The difference was due to form A spironolactone being present in the finer-sized fraction but the highly soluble form D was present in the larger fraction. Again DTA was unable to detect the difference due to the lactose dehydration. The work of Salole & Al-Sarraj (1985) serves to illustrate that it is unwise to rely solely on one technique to detect polymorphic forms and that a thermally active excipient may mask the true identity of an added ingredient.

REFERENCES

Corrigan, O. I. & Holohan, E. M. (1984) *J. Pharm. Pharmacol.*, **36**, 217–221.
Corrigan, O. I., Holohan, E. M. & Sabra, K. (1984) *Int. J. Pharm.*, **18**, 195–200.
Corrigan, O. I., Holohan, E. M. & Reilly, M. R. (1985) *Drug Dev. Ind. Pharm.*, **11**, 677–695.
Giordano, F., Bettinetti, G. P., La Manna, A. & Giuseppetti, G. (1985) *Calorim. Anal. Therm.*, **16**, 220–223.
Ikekawa, A. & Hayakawa, S. (1982) *Bull. Chem. Soc. Jpn.*, **55**, 1261–1266.
Kaneniwa, N., Ikekawa, A. & Sumi, M. (1978) *Chem. Pharm. Bull.*, **26**, 2734–2743.
Kawashima, Y., Aoki, S., Takenaka, H. & Miyake, Y. (1984) *J. Pharm. Sci.*, **73**, 1407–1410.
Kawashima, Y., Okumura, M. & Takenaka, H. (1982) *Science*, **216**, 1127–1128.
Matsuda, Y., Kawaguchi, S., Kobayashi, H. & Nishijo, J. (1984) *J. Pharm. Sci.*, **73**, 173–179.
Monkhouse, D. C. & Lach, J. L. (1972) *J. Pharm. Sci.*, **61**, 1435–1441.
Salole, E. G. & Al-Sarraj, F. A. (1975) *Drug Dev. Ind. Pharm.*, **11**, 2061–2070.
Takenaka, H., Kawashima, Y. & Lin, S. Y. (1981) *J. Pharm. Sci.*, **70**, 1256–1260.
Theeuwes, F., Hussain, A. & Higuchi, T. (1974) *J. Pharm. Sci.*, **63**, 427–429.

Index

Index